Thin Films

Advances in Research and Development

VOLUME 21

Serial Editors

MAURICE H. FRANCOMBE
Department of Physics
The University of Pittsburgh
Pittsburgh, Pennsylvania

JOHN L. VOSSEN
John Vossen Associates
Technical and Scientific Consulting
Bridgewater, New Jersey

Contributors to this Volume

K. M. S. BANDARA

K. K. CHOI

S. D. GUNAPALA

R. P. GAMANI KARUNASIRI

JENNY NELSON

J. S. PARK

A. G. U. PERERA

K. L. WANG

Thin Films

Advances in Research and Development

HOMOJUNCTION AND QUANTUM-WELL INFRARED DETECTORS

Volume Editors

Maurice H. Francombe

Department of Physics
Georgia State University
Atlanta, Georgia

John L. Vossen

John Vossen Associates
Technical and Scientific Consulting
Bridgewater, New Jersey

VOLUME 21

ACADEMIC PRESS

San Diego   New York   London   Sydney   Tokyo   Toronto

Copyright © 1995 by ACADEMIC PRESS, INC.

Academic Press, Inc.
A Division of Harcourt Brace & Company
525 B Street, Suite 1900, San Diego, California 92101-4495

United Kingdom Edition published by
Academic Press Limited
24-28 Oval Road, London NW1 7DX

International Standard Serial Number: 1079-4050

International Standard Book Number: 0-12-533021-9

Printed and bound in the United Kingdom
Transferred to Digital Printing, 2011

Contents

Physics and Novel Device Applications of Semiconductor Homojunctions

A. G. U. Perera

Progress of SiGe / Si Quantum Wells for Infrared Detection

R. P. Gamani Karunasiri, J. S. Park, and K. L. Wang

Recent Developments in Quantum-Well Infrared Photodetectors

S. D. Gunapala and K. M. S. Bandara

Multiquantum-Well Structures for Hot-Electron Phototransistors

K. K. Choi

Quantum-Well Structures for Photovoltaic Energy Conversion

Jenny Nelson

Contributors

Numbers in parentheses indicate the pages on which the authors' contributions begin.

K. M. S. BANDARA (113), Department of Physics, University of Peradeniya, Peradeniya, Sri Lanka

K. K. CHOI (239), U. S. Army Research Laboratory, Physical Science Directorate, Fort Monmouth, New Jersey 07703

S. D. GUNAPALA (113), Center for Space Microelectronics Technology, Jet Propulsion Laboratory, California Institute of Technology, Pasadena, California 91109

R. P. GAMANI KARUNASIRI (77), Department of Electrical Engineering, National University of Singapore, Singapore 0511

JENNY NELSON (311), Blackett Laboratory, Imperial College of Science, Technology and Medicine, London SW7 2BZ, United Kingdom

J. S. PARK (77), Jet Propulsion Laboratory, California Institute of Technology, Pasadena, California 91109

A. G. U. PERERA (1), Department of Physics and Astronomy, Georgia State University, Atlanta, Georgia 30303

K. L. WANG (77), Department of Electrical Engineering, University of California, Los Angeles, Los Angeles, California 90024

Preface

In recent years there has been a tremendous surge of interest in special semiconductor interface structures for use in high-performance devices for electro-optic, switching, solid-state laser, high-speed, infrared detection, solar cell, and other applications. A significant effort, for example, has been directed toward internal photoemission junction detectors, in which long-wavelength infrared sensitivity has been enhanced using heavy impurity doping to broaden the spectral response of a thin absorber/emitter layer. However, by far the major interest in infrared devices, since the earlier theoretical studies in 1983, has been concentrated on the physics and novel device applications of epitaxial III–V compound and group IV quantum-well structures. In these devices, a tailorable IR response is achieved by varying the energy of photoexcited intersubband transitions for quantum wells in either the conduction or the valence band, simply by adjusting the doping and thickness of the quantum well. Earlier developments in this field were reviewed by Coon and Bandara, in volume 15 of this serial.

Volume 21 of *Thin Films* addresses three main areas of this rapidly expanding field, i.e., photoemissive homojunction detectors in which the spectral response can be extended into the far infrared, novel device configurations based on multi-quantum-well detectors for imaging at wavelengths out to the VLWIR range, and incorporation of quantum wells into solar cell junction structures as a means of extending their spectral bandwidth. The first chapter by A. G. U. (Unil) Perera discusses the structure and physics primarily of heavily doped silicon homojunctions and Ge- and GaAs-based alloy junctions. In the case of silicon, it is shown that the infrared photoemissive behavior can be subdivided into three groups, depending on the impurity concentration of the absorber layer. Recent experimental results on Si homojunctions show spectral response with a long-wavelength threshold varying (with doping) from about 30 μm to about 200 μm. It is also demonstrated that, under bias, spontaneous pulsing occurs in these junctions, with characteristics emulating the behavior of biological neurons, and that the frequency of such pulses varies uniformly with incident IR intensity. These structures hold great promise for integration in large monolithic, multispectral IR imaging arrays.

The second chapter, by R. P. G. (Gamani) Karunasiri, J. S. Park, and K. L. Wang, reviews the current status of SiGe/Si quantum wells for infrared detection. Until recently, most of the research in this field was carried out using III–V-based material systems, such as GaAs/AlGaAs, and this work is reviewed extensively in the third and fourth chapters of this volume. In the SiGe/Si system the doped SiGe alloy quantum wells are separated by Si barriers. Optically excited intersubband transitions for structures with wells in the valence and conduction bands have been observed, with the emphasis to date being on valence–band transitions. The origin of different transitions is identified using polarization-dependence measurements. Depending on the device design parameters, the photoresponse is shown to cover atmospheric windows of both 3–5 μm and 8–12 μm. In addition to the potential versatility of Si-based technology, these materials also exhibit intersubband transitions at normal incidence.

In the third chapter, S. D. Gunapala and K. M. S. V. Bandara discuss key developments (primarily at AT&T Bell Laboratories) in the rapidly growing body of research that has appeared recently on quantum-well infrared photodetectors (QWIPs) based on III–V compound systems. The review deals mainly with multi-well QWIPs formed in the most extensively studied GaAs/AlGaAs (well/barrier) system, but it also presents exciting new results on, for example, InGaAs/InAlAs structures tailored for the MWIR window (3–5 μm) and InGaAs/GaAs structures suitable for response at wavelengths beyond 14 μm. The topics discussed include comprehensive and well-documented treatments of the physics of intersubband transitions and photodetector device parameters, bound-to-bound and bound-to-continuum QWIPs, asymmetrical QWIPs, miniband superlattice QWIPs for low-power and broadband application, indirect band-gap QWIPs, light-coupling methods, and the design and performance of imaging arrays. The chapter concludes with examples of imaging, at wavelengths out to about 15 μm, using 128 × 128 detector arrays, and it forecasts that sensitive, low-cost, large (1024 × 1024) LWIR and VLWIR focal plane arrays can be expected in the near future.

It is well known that detectivities of discrete QWIP detectors are limited to levels significantly less than those of intrinsic HgCdTe IR detectors, due to high values of dark current. In the fourth chapter, K. K. Choi reviews his development of a family of novel three-terminal, multi-quantum-well

QWIP devices, designed to improve high-temperature IR detectivity at long wavelengths. The basic concept involves a quantum barrier placed next to a QWIP in order to serve as an electron energy analyzer. The structure can be used to evaluate the energy spectra of the hot conducting electrons excited by thermal and optical stimulation. This approach not only yields a determination of the barrier heights and bias conditions needed to suppress thermal dark currents, but also leads to a new infrared hot-electron transistor (IHET) device which serves a wide variety of functions. For example, in addition to producing increased detectivity, it can increase detector impedance to lower readout noise, amplify photovoltage and photopower, control the IR cutoff wavelength, and reduce the generation–recombination noise and $1/f$ noise of a QWIP detector.

The fifth and final chapter, by Jenny Nelson, describes recent studies, mainly by the Imperial College group in London, aimed at using multi-quantum-well structures to achieve higher performance in solar cell devices based on materials systems such as AlGaAs/GaAs, GaAs/InGaAs, InP/InGaAs, InGaP/GaAs, and InP/InAsP. Previous efforts to obtained more efficient use of the solar spectrum have involved more than one photoconverter in tandem, allowing different solar spectral ranges to be preferentially absorbed by two or more junctions of different band gap which are electrically connected. In the alternative structure described here, quantum wells are inserted in the space charge region of a wide gap $p-i-n$ cell to ensure absorption of longer wavelengths (corresponding to intersubband carrier transitions), without the need for lossy electrical connections. This chapter begins with a coverage of the principles of photovoltaic conversion in diodes, then reviews the effects of quantum-well insertion on spectral response and photocurrent, and finally discusses the dark current and voltage behavior in various materials systems. The different approaches to calculation of limiting efficiency are considered, and other novel ways of using quantum-well structures for photoconversion are described. *Maurice H. Francombe* and *John L. Vossen.*

In-Press Note. It is with profound sadness that we report the sudden death of our coeditor John Vossen on May 21, 1995. John will be greatly missed by his many friends and colleagues in the scientific community. On behalf of Academic Press, we convey our deepest sympathy to his wife Joan and to the Vossen family. *Maurice H. Francombe* and *Zvi Ruder.*

Physics and Novel Device Applications of Semiconductor Homojunctions

A. G. U. PERERA

Department of Physics and Astronomy, Georgia State University, Atlanta, Georgia

1

I. Introduction and Background

A junction formed by two different electrical types of the same (band-gap) material can be classified as a homojunction. Similarly, a heterojunction is formed by two chemically different materials. These types of junction structures are well known and extensively discussed in the literature. A common example for a homojunction is the heavily used, well-understood silicon $p-n$ junction. Recently developed $GaAs/Al_xGa_{1-x}As$ structures are a good example of a heterojunction. Here our emphasis is on crystalline semiconductor homojunctions, especially Si. Since almost all the circuit components, such as resistors, capacitors, transistors, diodes, charged coupled devices (CCDs), charge integrated devices (CIDs), shift registers, and detectors, could be fabricated using standard Si technology, putting all those components in one single chip to fabricate an integrated circuit (IC) is a major advantage of using Si.

Semiconductor homojunction structures, especially the p^+-n-n^+ ($p-i-n$) structures on which we concentrate here, have been studied for a very long time and have been used in a variety of applications. The advent of the molecular beam epitaxy (MBE), chemical vapor deposition (CVD), and other thin film techniques has advanced both homo- and heterojunction design and fabrication to new levels. However, our studies in recent years have demonstrated that even simple and mundane p^+-n-n^+ junction structures can exhibit a variety of new electrical and optical phenomena, leading to novel and intriguing device applications consistent with twenty-first century research interests. Here our focus is on homojunctions for both electronic and optoelectronic applications, which mainly involve intraband processes rather than interband processes. These devices will include different types of infrared detectors and spontaneous pulse generators, that act like biological neurons. The fact that the same semiconductor material (with different dopants or concentrations) is used in the homojunction makes the fabrication of these samples much simpler than heterostructures. The materials-technology needs for the implementation of these devices in practical applications were met at least a decade ago, making the incorporation of these devices into high-performance integrated circuits just a routine exercise.

First we will address homojunction infrared detectors based on internal photoemission mechanisms. Various IR detector approaches, using interfacial workfunctions in homojunctions and other IR detector approaches based on homojunctions, such as a delta doped potential well approach and a room-temperature FIR detector approach, based on a p-type high−low Si junction and a charge storage approach, will be discussed. The interfacial workfunction-type structures will be subdivided into three groups based on their impurity concentrations. Recent experimental results on Si homojunctions showing spectral response with a long-wavelength threshold (λ_t) varying from around 30 to 200 μm confirm the wavelength tunability of these detectors. The main significance of this detector concept is in establishing a technology base for the evolution of large-area, uniform detector arrays with a multispectral capability for greatly improved NEΔT sensitivity using the well-established Si growth and processing technology.

Next we will discuss spontaneous pulsing in silicon p^+-n-n^+ homojunctions. Another mode of infrared detection will be discussed in connection with these pulses (spiketrains). This is the only mode of IR detection which does not need any preamplifiers. Compared to measuring very small (pico- or microampere) currents spread out over a long duration, counting pulses will be much easier. These spiketrains convey both analog and digital information (mixed character), which can carry more information than either situation alone. The interpulse time intervals convey analog information, whereas the presence or the absence of the almost-uniform-height pulses conveys digital information. Noise immunity is another advantage of this pulsing approach.

Another application of these pulses is in emulating biological neurons, opening up pathways to design brainlike parallel asynchronous multispectral focal-plane image/sensor processors for various missions. These low-power, high-resolution sensor/processor innovations may have a profound impact on techniques currently being explored for defense, geological and meterological survey systems, strategic and tactical IR systems, ground, airborne, and space-based FLIR systems, planetary probes, and medical diagnostic systems even though these are futuristic approaches.

II. Homojunction Internal Photoemission IR Detectors

In general, infrared detectors can be categorized as thermal and photon detectors. The latter can be further divided into photoconductors, photovoltaic detectors, and internal photoemission detectors. In photoconductors, the electrical conductivity of the material is changed by the incident photons. The photovoltaic effect needs an internal potential barrier and a built-in electric field. According to various photon detection mechanisms, photon detectors can also be classified as intrinsic, extrinsic, intersubband, and internal photoemission detectors. A detailed review of extrinsic IR photoconductors can be found in articles by Bratt (8) and later by Sclar (9). A well-known example of intersubband photoconductors is multiquantum well infrared photodetectors (MQWIPs), developed recently (10, 11). Although extrinsic photovoltaic detection is possible, most of the practical photovoltaic detectors utilize the intrinsic photoeffect. A typical photovoltaic structure is a simple p–n junction. Other structures include p–i–n junctions, Schottky barriers, and MIS structures. Recently, the concept of a photovoltaic intersubband detector was demonstrated for the long-wavelength IR range (12). In addition, blocked-impurity-band (BIB) detectors, developed first at Rockwell Science Center (13), can also be categorized as an extrinsic photovoltaic detector. There are several internal photoemission detectors under development (14), among which metal–semiconductor Schottky barrier detectors (15) and $Ge_x Si_{1-x}$/Si heterojunction detectors (16, 17) are two examples. Although a large body of research on the above topics exists, our aim is to concentrate mainly on homojunction internal photoemission detectors (HIP) where the term HIP follows the terminology of Lin and Maserjian (16), who used it for heterojunction internal photoemission. Typical spectral ranges obtained for Si and Ge homojunction type IR detectors are shown in Fig. 1 with the relevant reference also listed in the figure.

The basic structure of the HIP detectors (18) consists of a heavily doped layer for IR absorption and an intrinsic (or lightly doped) layer across which most of the external bias is dropped. According to the doping concentration level in the heavily doped layer (N_d), these HIP detectors can be divided into three types (labeled I, II, and III) due to their different photoresponse mechanisms and response wavelength ranges. More details of these three types of HIP detectors will be given after discussing the concentration dependence of the workfunction, which accounts for most of the wavelength tunability. The concentration-dependent interfacial workfunctions in type I and II structures will be used as the barriers involved in

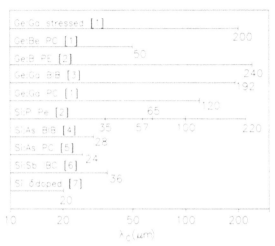

FIG. 1. Typical spectral sensitivity ranges for Si and Ge IR homojunction detectors. Shown are the shifts toward long wavelength of detector thresholds in doped detector structures relative to standard extrinsic PC structures. Also shown are the much larger shifts achieved for much more highly doped absorber layers used in photoemissive (PE) IR detectors. The impurity concentrations were 1×10^{16} cm^{-3} for Ge:Ga stressed (1), 1×10^{16} cm^{-3} for Ge:Be PC (1), $\sim 5 \times 10^{17}$ cm^{-3} for Ge:B PC (2), 3×10^{16} cm^{-3} for Ge:Ga BIB (3), 2×10^{14} cm^{-3} for Ge:Ga PC (1), $> 2 \times 10^{18}$ cm^{-3} for Si:P PE (2), with various values for λ_t observed for different detectors shown, 8×10^{17} cm^{-3} for Si:As BIB (4), 2×10^{15} cm^{-3} for Si:As PC (5), 1–5×10^{17} cm^{-3} for Si:Sb IBC (6), and 2×10^{20} cm^{-3} for Si δ doped (7).

the IR detection process, whereas in type III the barrier is due to the space charge effect (19). Since the barriers associated with type I and type II are due to the interfacial workfunctions, these HIP type I and type II detectors will be called HIWIP (homojunction interfacial workfunction internal photoemission) type I and HIWIP type II detectors.

The HIWIP detector is based on the electronic phenomena at interfaces between the two regions of a semiconductor (n^+-n, or p^+-n, although we will concentrate on n^+-n interfaces) in terms of analogies with traditional vacuum tube photocathodes and thermionic cathodes. The relevant energy scale (the interfacial workfunction) is sufficiently small that low temperatures and low photon energies are required to make the analogies valid (20). Such analogies can motivate consideration of semiconductor device concepts based on vacuum tube concepts. The device concepts are based on the classic photoelectric effect of Einstein (21, 22) who, among other things, showed that an energy quanta labeled photon could excite an electron from a bound state. In addition to regarding n^+-n interfaces as photocathodes, p^+-n interfaces can be considered to be photoanodes (hole emitters). When the impurity concentration is above the metal−insulator transition the cathode can be regarded as a metal cathode; otherwise it can be regarded as a semiconductor cathode. The distinction between these two cases hinges on the absence or presence of an impurity bandgap separating occupied and unoccupied impurity band states. The forward biased mode of operation in the HIWIP detector is radically different from the conventional use of reverse biased detection of near infrared, visible, or ultraviolet radiation using interband transitions. This method requires low-temperature operation in order to block the dark current, and the long wavelength threshold (λ_t) plays a significant role in determining the operating temperature. In order to see either a photocurrent or a thermionic current, the device has to operate above the flatband voltage (V_0), which is 1.1 V in the case of Si. A sharp flatband cutoff in low temperature current data for a $p-i-n$ structure as seen in Fig. 2, gives a clear indication of this effect (18). The small deviations of the experimental data (especially for the threshold photocurrent) from the empirical formula $i = G(V - V_0)$ proposed (18) was attributed to the lowering of the workfunction associated with the interface sharpness (20) and hence

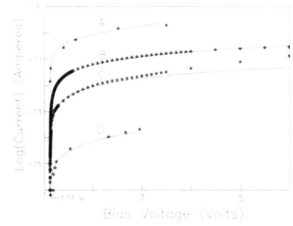

FIG. 2. (A) Modulated mercury arc infrared photocurrent for a p^+-n-n^+ structure at 4.2 K. (B) 300-K background photocurrent with the device at 4.2 K. (C) Filtered ($\lambda > 28$ μm) 300-K background photocurrent with the device at 4.2 K. (D) Thermionic dark current (device at 16 K). The dark current at 4.2 K is less than 10^{-16} amperes. Dots correspond to experimental data points. Lines correspond to an empirical fit given by $i = G(V - V_0)$, where i is the current, V is the applied bias voltage, and V_0 is the flatband voltage. After Ref. (18).

the threshold at high voltages (*18*). The applied bias voltage is given by *V*; V_0 is the flatband, while *G* is linearly dependent (*2*) on the incident intensity. More recent developments (*23*) indicating a dependence of this interface sharpness on the current support this idea.

A. WORKFUNCTION VARIATION WITH CONCENTRATION

A significant advantage of HIP (HIWIP) detectors is their ability to cover a wide infrared spectral range with Si detectors. Due to the compatibility with other electronic components and due to the highly mature Si technology, this is indeed an asset to the optoelectronic community. In the standard mode of operation, a photoconductor cutoff wavelength (at which the response drops to 50% of the peak response) λ_c is determined by the single impurity binding energy values in the semiconductor (see Fig. 3). The wavelength tunability is obtained by varying the doping (impurity)

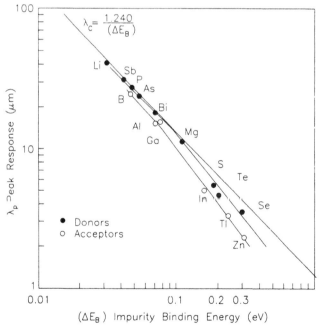

FIG. 3. The peak spectral wavelength vs. the impurity binding energy for donors and acceptors in silicon. After Ref. (*9*).

concentration in the heavily doped absorption (emitter) layer. This binding energy would approach zero as the concentration is increased above the metal–insulator or Mott transition (24). For example, Si:Ga, Si:Al, and Si:Bi detectors with λ_c's varying from 17.8 to 18.7 μm, operating around 28 K, could be used to extend the wavelength to 50 μm or 100 μm by increasing carrier concentration, while operating at higher temperatures. The same is true for Si:In which operates at 78 K. Similarly Ge:B, Ge:Sb detectors with λ_c's between 130–138 μm operating at 4.2 K could be used to extend the wavelength to 250 μm and beyond (9). The binding energies for GaAs:C, GaAs:Be and GaAs:Mg are 26, 28, and 28 meV respectively. Based on these results, III–V materials such as GaAs in this mode of operation should also yield extremely long cutoff wavelengths associated with low dark currents, and high detectivities at low temperatures. The cutoff wavelength (λ_c) is defined as

$$\lambda_c = \frac{1.24}{\Delta},$$

where Δ is the energy gap (binding energy in a single impurity case) in meV. In other words, Δ is the energy of a photon with a wavelength λ_c. In the case of a detector, the wavelength can be extended by barrier lowering effects and tunneling effects. For an internal photoemission-type detector, the quantum efficiency will increase with the energy, giving rise to a monotonically decreasing spectral response curve with increasing wavelength. In this case, where no clear peak is present, the cutoff wavelength (λ_c) is the wavelength where the response reaches zero. However, the interfacial workfunction detectors, even though they also qualify as internal photoemission detectors, give rise to a clear peak due to the high absorption efficiency at lower wavelengths. Hence, following the standard photoconductor literature, we will use the term long-wavelength threshold (λ_t) to denote the zero response wavelength and λ_c to denote the wavelength at 50% of the peak response.

   The workfunction dependence on the concentration can be estimated by the tight binding approach (25) or by a modified high density theory (25), depending on whether the concentration is above or below the Mott transition. Qualitatively, as the doping increases, a semiconductor tends to behave as a metal, in effect reducing the activation energy (workfunction) until reaching the metal-insulator transition point where the workfunction approaches zero due to free carriers. Here we briefly summarize three methods used (26) or proposed (27) for the workfunction calculations.

1. Tight Binding Theory. Using a scaled hydrogenic ground state wavefunction for the donors, the impurity to conduction band energy gap could be estimated (25) from the tight binding (TB) scheme through the energy transfer integral (28),

$$J(\mathbf{R}) = J(\mathbf{R}_i - \mathbf{R}_j) = \int \frac{q^2}{4\pi\epsilon_s\epsilon_0|\mathbf{r} - \mathbf{R}_i|}\phi_0(\mathbf{r} - \mathbf{R}_i)\phi_0(\mathbf{r} - \mathbf{R}_j)\, d^3r, \quad (1)$$

where $\mathbf{R}_i$ and $\mathbf{R}_j$ are the positions of adjacent impurities, $\phi_0(\mathbf{r})$ is the donor ground-state wavefunction, and $\mathbf{R}$ is the donor nearest-neighbor distance. The scaled hydrogen-like donor ground-state wavefunction is given by $\phi_0(\mathbf{r} - \mathbf{R}_i) = (\xi^3/\pi)^{1/2}\exp(-\xi|\mathbf{r} - \mathbf{R}_i|)$, where $\xi = (1/a_H)$ $(E_D/R_H)^{1/2}$, E_D is the donor ionization energy for the low-donor concentration case, $R_H = 13.6m_d/\epsilon_s^2$ (eV) is the effective Rydberg energy, and $a_H = 0.53\epsilon_s/m_d$ (Å) is the effective Bohr radius. Now the energy transfer integral becomes

$$J(R) = \frac{q^2\xi}{4\pi\epsilon_s\epsilon_0}(1 + \xi R)\exp(-\xi R). \quad (2)$$

Assuming a random spatial distribution of donors (Poisson's), the average energy transfer integral is given by

$$\langle J(R)\rangle = \int J(R)4\pi N_d R^2 \exp\left(-\frac{4}{3}N_d R^3\right) dR. \quad (3)$$

The TB model, with the Poisson distribution, gives the total impurity bandwidth $B = 2|\langle J(R)\rangle|$. The workfunction is $\Delta = E_D - B$, where E_D is 45 meV in the scaled hydrogenic model.

The results of numerical calculation on the donor level broadening could provide the dependence of the workfunction on the donor concentration, which is summarized in Fig. 4. From this model, one would expect an interfacial workfunction of about 25 meV, corresponding to an IR detection threshold of about 50 μm for a concentration of 10^{18} cm^{-3}. This is in agreement with the experimental data for an Si:P sample with a doping concentration of $\sim 10^{18}$ cm^{-3} (for low doping concentration values), as seen in Fig. 4. The experimental values, measured for several samples at temperatures near 4.2 K, indicated threshold wavelengths (λ_t) in the range 37 to 60 μm (18). However, the TB method starts to deviate from the experimental values as the concentration increases beyond the Mott transition values (28, 29).

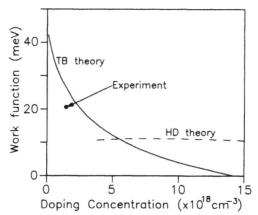

FIG. 4. Activation energy for Si:P with respect to doping concentration in the tight binding approximation. As expected, the results are in agreement with the experimental value for P doped Si at lower concentrations. Also shown is the curve obtained from high density theory, which is shown for a wider range in Fig. 5. After Ref. (25).

2. *High-Density Theory.* Above the Mott transition, with increasing doping concentration, the impurity band starts to shrink and finally becomes absorbed into the conduction band. The lowering of the conduction band edge, ΔE_c, can be described using the high-density (HD) theory (30), where electron–electron interactions (many-body effects) cause a rigid downward shift of the conduction band (31). This gives rise to an exchange energy, which is denoted by ΔE_c^{ex}. The electron–impurity interaction causes an additional shift, ΔE_c^i, and also distorts the density of states function. The HD theory approximately describes (30) the behavior of heavily doped Si and Ge in the high-density regimes, i.e., above Mott's critical concentration, and even gives reasonable results at doping concentrations as low as 10^{18} cm^{-3}.

Recently Jain and Roulston (30) have derived a simple expression for the shift of the majority band edge, ΔE_{maj}, that can be used for all n- and p-type semiconductors in the high-density regime. By introducing two correction factors to take into account the deviations from the ideal band structure, ΔE_{maj} can be expressed as

$$\Delta E_{maj} = \Delta E_{maj}^{ex} + \Delta E_{maj}^i$$

$$= 1.83 \Lambda a_H \left(\frac{4\pi N}{3 N_m} \right)^{1/3} R_H + \frac{1.57 a_H}{N_m} \left(\frac{4\pi a_H N}{3} \right)^{1/2} R_H, \quad (4)$$

where N is the impurity concentration. The correction factor Λ takes into account the effect of anisotropy of the bands in n-type semiconductors and the effect of interactions between the light and heavy hole-valence bands in p-type semiconductors. Here N_m is the number of conduction band minima in the case of n-type Si and n-type Ge, and $N_m = 2$ for all p-type semiconductors.

The modified expression for the Fermi energy to take the multiplicity of the majority band (32) into account is given by

$$E_F = \frac{\hbar^2 k_F^2}{2m^*} = \frac{\hbar^2}{2m_d}\left(\frac{3\pi^2 N}{N_m}\right)^{2/3}. \tag{5}$$

The workfunction (Δ) is now given by $\Delta = \Delta E_{maj} - E_F$. Using the above equations, the doping concentration dependence of ΔE_{maj}, E_F, and Δ can be calculated. The critical concentrations (N_0) which correspond to $\Delta = 0$ were obtained from the curves shown in Fig. 5 with the parameters used shown in Table 1 for both n- and p-type Si, Ge, and GaAs. The Mott critical concentrations (33), N_c, are also listed in Table 1. Although the actual critical concentrations depend on the impurity selected for doping, the above calculations will give an estimate for the workfunction, which is important for the design of these homojunction detectors.

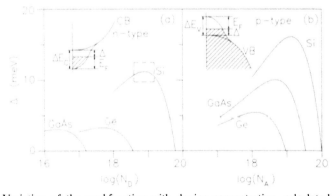

FIG. 5. Variation of the workfunction with doping concentration, calculated using HD theory for (a) n-type and (b) p-type semiconductors (Si, Ge, and GaAs), with the insets showing the shift of majority band edge and the position of Fermi level. Shaded areas represent the energy states filled with electrons. The boxed region inside n-type Si in (a) is also shown in Fig. 4. The parameter values used in this figure are shown in Table 1. After Ref. (25).

TABLE I

THE VALUES OF CRITICAL CONCENTRATIONS AND PARAMETERS (30) USED IN THE CALCULATIONS
FOR DIFFERENT SEMICONDUCTORS

	n-Si	p-Si	n-Ge	p-Ge	n-GaAs	p-GaAs
ϵ_s	11.4	11.4	15.4	15.4	13.0	13.0
m_d	0.33	0.59	0.22	0.36	0.0665	0.47
N_m	6	2	4	2	1	2
Λ	1	0.75	0.84	0.75	1	0.75
R_H (meV)	34.5	61.7	12.6	20.6	5.4	38.1
a_H (Å)	18.3	10.2	37.1	22.7	103.6	14.6
N_c (10^{17} cm^{-3})	26	150	3.1	13	.14	50
N_0 (10^{18} cm^{-3})	60	100	3.5	8.1	.15	33

Note. Here m_d is the density of states mass, N_m is the number of subbands, Λ is the correction factor for the band structure, N_c is Mott's critical concentration and N_0 is the concentration at which the workfunction goes to zero.

Although the HD theory is applicable for concentrations at or above the M–I transition point, it considers only the type (n or p) but not what the impurity is. Obviously this is a drastic simplification of the situation.

3. *Density of States Approach.* If a more detailed and rigorous theory is needed, one might have to consider a density of states calculation, which gives the Fermi level, and hence the workfunction. Even though this is a more accurate method, it will require much computing time.

Considering the dopants as centers for screened coulomb (Thomas–Fermi) potential randomly distributed throughout the sample, with uniform macroscopic density, the potential is given by $(e^2/\kappa r)e^{-\alpha r}$, where κ is the dielectric constant and α is the inverse screening length at zero temperature, given by

$$\alpha^2 = 2\left(\frac{e^2}{\kappa}\right)\left(\frac{2m^*}{\hbar^2}\right)\left(\frac{3N}{\pi}\right)^{1/3},$$

where N is the impurity concentration. The method of calculation for the density of states is based on the multiple scattering technique of Klauder (34), which treats the solution as a Dyson series summed to all orders in the interaction potential. The treatment for zero temperature can be done self-consistently by using a dressed propagator, which takes into account the effect of all other scattering centers when the solution for any given center is computed.

In the framework of the effective mass approximation, one may have the following Hamiltonian for a sample of a doped semiconductor:

$$H = -\sum_i \hbar^2 \frac{\nabla_i^2}{2m^*} + \frac{1}{2} \sum_{i,j;\, i \neq j} \frac{e^2}{\kappa |\mathbf{r}_i - \mathbf{r}_j|}$$
$$-\sum_{i,j} \frac{e^2}{\kappa |\mathbf{r}_i - \mathbf{R}_j|} + \frac{1}{2} \sum_{i,j;\, i \neq j} \frac{e^2}{\kappa |\mathbf{R}_i - \mathbf{R}_j|}.$$

Here m^* is the electron effective mass, κ is the dielectric constant, and $\mathbf{r}$ and $\mathbf{R}$ are the electron and impurity coordinates, respectively. The one-electron Green's function is given by

$$G(k, E) = [E - \epsilon_k - \Sigma(\mathbf{k}, E)]^{-1},$$

where ϵ_k is the electron eigen energy of the noninteracting system, and $\Sigma(\mathbf{k}, E)$ the electron self-energy. For a single conduction band with parabolic dispersion law one has

$$\epsilon_k = \frac{\hbar^2 k^2}{2m^*},$$

but the formalism allows consideration of any complex band structure. The electron self-energy has its exchange (xc) and electron–impurity (ei) parts, and it is combined as

$$\Sigma(\mathbf{k}, E) = \Sigma_{xc}(\mathbf{k}, E) + \Sigma_{ei}(\mathbf{k}, E).$$

After determining the Green's function, one may calculate the spectral density function as

$$A(\mathbf{k}, E) = -\frac{1}{\pi} \operatorname{Im} G(\mathbf{k}, E),$$

which allows one to calculate the density of states through

$$D(E) = \frac{1}{2\pi^2} \int_0^\infty dk\, k^2 A(k, E).$$

Following Klauder (34), Serre (35), and Lowney (36), one may get coupled nonlinear integral equations corresponding to the case of screened Coulomb potential, incorporating the effect of compensation. Although it may require some heavy computing, these coupled equations could be solved numerically by using a Gaussian elimination scheme to obtain the

Green's function. The density of states calculated using the Green's function can be used to obtain the Fermi energy.

B. Type I HIWIP Detectors $N_d < N_c$ $(E_F < E_c^{n^+})$

Type I is defined as having doping concentration N_d (N_a) in the n^+ (p^+)-layer below the Mott critical value (N_c), forming an impurity band where the Fermi level will be located at low temperatures. Although, both n^+ and p^+ layers can be effective with either electrons or holes respectively, we will concentrate on n^+ regions with electrons as photocarriers. The incident FIR light is absorbed due to the impurity photoionization, with a workfunction given by $\Delta = E_c^{n^+} - E_F$, where $E_c^{n^+}$ is the conduction band edge in the n^+-layer. An electric field is formed in the n-layer by an external bias to collect photoexcited electrons generated in the n^+-layer. Obviously, type I HIWIP (HIP) detectors are analogous to semiconductor photoemissive detectors (38) in their operation, which can be described by a three-step process (see Fig. 6): (1) Electrons are photoexcited from filled impurity band states below the Fermi level into empty states above the conduction band edge; (2) photoexcited electrons first rapidly thermalize into the bottom of the conduction band by phonon relaxation and then diffuse to the emitting interface, with the transport probability determined by the electron diffusion length; (3) those electrons reaching the emitting interface tunnel through an interfacial barrier (ΔE_c), which is due to the offset of the conduction band edge caused by the bandgap narrowing effect, and are collected by the electric field in the n-region. The collection

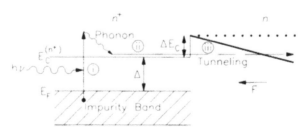

FIG. 6. Energy band diagram for Type I HIP (HIWIP) detector, $N_d < N_c$ ($E_F < E_c^{n^+}$). Here, N_c is the Mott's critical concentration and N_0 is the critical concentration corresponding to $\Delta = 0$. The conduction band edge of the n-layer is represented by a dotted line for $V_b = V_0$ (flatband) and by a solid line for $V_b > V_0$. The three steps involved are (1) photoexcitation. (2) phonon relaxation, and (3) tunneling through the barrier. After Ref. (37).

efficiency will depend on the tunneling probability, which in turn depends on the n-region electric field. The λ_c can be tailorable with the doping concentration to some extent because, with the increase of doping concentration, the impurity band broadens and its peak density of state moves rapidly toward the conduction band (30).

1. Spectral Response Analysis of Type I. Although it is straightforward, here we give a brief description of a model to predict the spectral response. The photoionization cross-section can be given as (39)

$$\sigma = \frac{1}{\sqrt{\epsilon_s}} \frac{\theta_0}{(m_d/m_0)\Delta} \frac{\theta_1^{3/2}}{(h\nu/\Delta)^5} \text{ cm}^2, \tag{6}$$

where Δ is the impurity ionization energy (workfunction), ϵ_s is the relative dielectric constant, m_d is the electron density of state mass, $\theta_0 = 32\hbar q^2/(\epsilon_0 m_0 c) = 7.46 \times 10^{-16}$ cm^{2}eV, and $\theta_1 = [(h\nu/\Delta) - 1]$. The photon energy dependence of σ has a peak at $h\nu = 1.43\Delta$, with the peak value given by

$$\sigma_p = \frac{3.52 \times 10^{-17}}{\sqrt{\epsilon_s}(m_d/m_0)\Delta} \text{ cm}^2. \tag{7}$$

If the doping concentration is N_d, the absorption coefficient is given by

$$\alpha = N_d\sigma. \tag{8}$$

Now, to calculate the ideal collection efficiency of electrons flowing into the i-region, the continuity equation can be written as

$$-D_n\frac{d^2n}{dx^2} + \frac{n}{\tau_n} = G(x), \tag{9}$$

where D_n and τ_n are the diffusion coefficient and lifetime of exited electrons. $G(x)$ is the generation rate of electrons induced by incident photons, given by $G(x) = I_0\alpha(1 - r)e^{-\alpha x}$, with I_0 the light flux intensity incident on the front surface (n^+-side), α is the absorption coefficient in the n^+-layer, and r is the front surface reflectivity. Two assumptions made are (1) that the backside surface reflectivity can be ignored and (2) that the optical absorption in the i-region has no contribution to the photocurrent.

The basic solution of Eq. (9) is

$$n = C_1 e^{-x/L_n} + C_2 e^{x/L_n} + \frac{\tau_n}{1 - \alpha^2 L_n^2} G(x) \tag{10}$$

with the boundary condition being $n|_{x=w_n} = 0$, $(dn/dx)|_{x=0} = 0$, where L_n is the electron diffusion length, $L_n = \sqrt{D_n \tau_n}$, and w_n is the n^+-layer thickness. The electron current flowing into the n^+-i interface region is given by $q D_n (dn/dx)|_{x=w_n}$. The ideal quantum efficiency (η_{id}) for the collection of electrons into the i-region is the ratio of this current to the maximum possible photogenerated current, $q I_0$. Thus, it can be shown that

$$\eta_{id} = \frac{\alpha L_n (1 - r)}{1 - \alpha^2 L_n^2} \left\{ \alpha L_n \exp(-\alpha w_n) + \frac{\exp(-\alpha w_n)\sinh(w_n/L_n) - \alpha L_n}{\cosh(w_n/L_n)} \right\}. \tag{11}$$

When $w_n \gg L_n$ and $\alpha L_n \ll 1$, which is similar to our experimental situation, Eq. (11) becomes

$$\eta_{id} \approx \frac{\alpha L_n}{1 - \alpha L_n}(1 - r)\exp(-\alpha w_n) \approx \alpha L_n(1 - r)\exp(-\alpha w_n). \tag{12}$$

In this case, η_{id} is proportional to the diffusion length and decreases exponentially with the increase of the n^+ layer thickness. Since not all photoexcited carriers will cross the n^+-i interface, the barrier tunneling probability has to be calculated. Assuming a triangle barrier, this probability can be written as (40)

$$p_t = \exp\left(-\frac{4\sqrt{2m_e^*}(\Delta E_c)^{3/2}}{3q\hbar F}\right) \tag{13}$$

where ΔE_c is the offset of the conduction band edge at the n^+-i interface, F is the electric field given by the ratio of the net applied voltage to the i-layer thickness, and m_e^* is the tunneling electron effective mass. Finally, the effective quantum efficiency is $\eta = \eta_{id} p_t$.

The photovoltage responsivity is given by

$$R_V = \frac{q\eta}{h\nu} r_l, \tag{14}$$

where r_l is the load resistance. The curves in Fig. 7 show the modeling output and the experimental data for one of our Ge structures. From spreading resistance measurements, the i-region impurity concentration

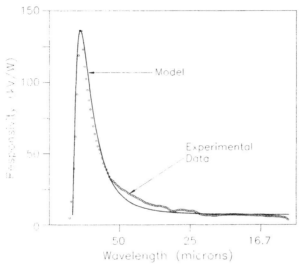

FIG. 7. Photoresponse spectrum for a Ge sample. The points show the experimental data. The line is calculated using the model. The threshold wavelength is around 200 μm (50 cm^{-1}). After Ref. (25).

and the thickness for this Ge sample was found to be about 10^{14} cm^{-3} and 120 μm, while the n^+- or p^+-layer concentration was around 10^{17} cm^{-3} and the thickness was about 1 μm. The flatband voltage was used as 0.46 V for this sample, which was obtained from I–V measurements. The spectra were measured (2) at 1.5 K with an applied voltage of 0.7 V and a load resistance of r_l = 10 MΩ. The diffusion length was estimated to be about 1000 Å for an assumed compensation concentration of 1×10^{14} cm^{-3}. It can be seen (Fig. 7) that this model agrees well with the experimental results given the following fitting parameters: the impurity ionization energy, ΔE = 8.4 meV; the offset of the conduction band energy, ΔE_c = 0.19 meV.

2. *Comparison with Blocked-Impurity-Band Detectors.* It is interesting to note the similarity and difference between the type I HIWIP detector and the blocked-impurity-band (BIB) detector. They have similar structures. However, they have different photoexcited electron collection mechanisms due to the difference of doping in the heavily doped layer. Due to compensation effects, especially in heavily doped semiconductors, an electric field can be induced in the n^+-layer by the external bias, giving rise to

BIB detectors (*13, 41, 4, 3*), for which the photoexcited electron collection mechanism is different from the processes (2) and (3) as shown in Fig. 6. The detection mechanism for a Si:As back-illuminated (BI) BIB detector may be summarized briefly by reference to the band diagram in Fig. 8 (*4*). The IR-active *n*-type, highly doped ($\sim 10^{17}$ cm^{-3}) layer contains a small concentration of residual acceptor impurities, which are ionized. With no field applied, charge neutrality requires the formation of an equal concentration of ionized donors. Whereas negative charges are fixed at the residual acceptor sites, the positive charges at the ionized donor sites (D^+) can migrate rapidly through the crystal when the neighboring donors are sufficiently close. Application of a positive bias to the top contact creates a field that drives the pre-existing D^+ charges toward the substrate, while the undoped blocking layer prevents the injection of new D^+ charges. This creates a depletion width that depends on the applied bias, the residual acceptor concentration, and the blocking layer width. In operation, an incident IR photon, absorbed by a neutral donor, produces a mobile D^+ charge and a conduction band electron. The electron is swept out (Fig. 8) through the blocking layer and is collected by the top contact. The D^+ charge drifts toward the substrate, and is considered collected when it enters the undeplated part of the IR active layer. Due to the high internal

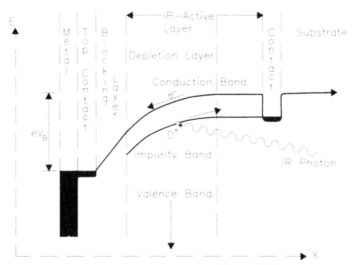

FIG. 8.   Simplified energy-band diagram of a positively biased BIBIB detector. Absorption of an IR photon by a neutral donor creates an electron–D^+ pair; both charges are swept out of the IR-active layer. After Ref. (*4*).

quantum efficiency (QE) associated with the field, the external QE will depend mainly on the absorption efficiency.

In contrast to the type I HIWIP detector, the BIB detector resembles a reverse-biased photovoltaic detector in its operation, with the collection efficiency, which depends on the compensated concentration, approaching 100% in the electric field region. With high compensated acceptor concentrations, the field induced will be limited to the interface, which resembles the type I HIWIP detector. At lower wavelengths, the BIB structure has an advantage over the type I, due to the higher collection efficiency achievable by the field in the IR active region. As the concentration increases, the ability to obtain low compensated acceptor concentrations becomes more difficult, tilting the advantage over to the HIWIP at longer wavelengths. In order to optimize the BIB detector, the IR active layer should be highly doped with very low compensation, and the blocking layer should be pure and highly compensated. Also the BIB detector is superior to standard photoconductors with respect to radiation damage due to the smaller thickness and fast recombination. However, type II should be superior to the BIB, since it has an even higher concentration.

C. Type II HIWIP Detector: $N_c < N_d < N_0$ ($E_c^{n^+} < E_F < E_c^i$)

When the doping concentration is above the Mott transition, the impurity band is linked with the conduction band edge, and the n^+-layer becomes metallic. Even in this case, the Fermi level can still be below the conduction band edge of the n-layer ($E_F < E_c^i$) due to the bandgap narrowing effect, giving rise to a workfunction $\Delta = E_c^i - E_F$ as shown in Fig. 9, unless N_d exceeds a critical concentration N_0 at which $\Delta = 0$. Type II HIWIP detectors are analogous to Schottky barrier IR detectors (15) in their operation. One of their unique features is that, in principle, there is no restriction on λ_c, which is tailorable in FIR range, because the workfunction can become arbitrarily small with increasing doping concentration. Unlike type I HIWIP detectors, the photon absorption in type II detectors is due to free carrier absorption. In spite of the fact that the free carrier absorption coefficient in the metallic n^+-layer is lower than in a metal, due to the lower electron concentration, the type II detector has a higher internal quantum efficiency than the Schottky barrier detector, due to the reduction of the Fermi energy. In addition, the hot electron scattering length in these detectors could be larger than in metals, due to the lower electron energy. The photoemission of photoexcited electrons from the n^+-layer into the n-layer is determined by the emission to the

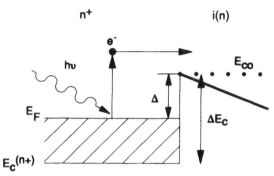

FIG. 9. Energy band diagram for Type II HIP (HIWIP) detectors, $N_c < N_d < N_0$ ($E_c^{n^+} < E_F < E_c^i$). Here, N_c is the Mott's critical concentration and N_0 is the critical concentration corresponding to $\Delta = 0$. The conduction band edge of the n-layer is represented by a dotted line for $V_b = V_0$ (flatband) and by a solid line for $V_b > V_0$. After Ref. (37).

interfacial barrier, hot electron transport and barrier collection process. The emission probability depends on the photon energy and the Fermi energy. The transport probability is governed by various elastic and inelastic hot-electron scattering mechanisms occurring in the n^+-layer. The collection efficiency is due to the image–force effect at the n^+–n interface, which gives rise to a voltage dependence on quantum efficiency. The type II detector is an FIR detector, and usually operates at temperatures lower than 77 K.

1. A Photoemission Model for Type II. Here we give a summary of the responsivity model proposed by Perera, Yuan, and Francombe (37). A homojunction n^+–n interface as shown in Fig. 9 is used in the calculation. First, the generation rate of photoexcited electrons is obtained (1) assuming that the contact layers make no contribution to the photocurrent, (2) neglecting the wavelength dependence of the front (R_F) and back (R_B) reflection coefficients, and (3) taking the multiple internal reflections into account, which is given by

$$G(x) = G_1 \exp(-\alpha_d x) + G_2 \exp(\alpha_d x), \tag{15}$$

where

$$G_1 = \frac{\alpha_d I_0 (1 - R_F)}{1 - R_F R_B \exp[-2(\alpha_d W_d + \alpha_i W_i + \alpha_b W_b)]},$$

$$G_2 = G_1 R_B \exp[-2(\alpha_d W_d + \alpha_i W_i + \alpha_b W_b)],$$

W_d, W_i, and W_b are the thicknesses, N_d, N_i, and N_b, the doping concentrations and α_d, α_i, and α_b, the absorption coefficients of the heavily doped emitter layer, the intrinsic (or lightly doped) layer, and the bottom contact layer, respectively. Absorption in the n-layer was ignored for obvious reasons. The photon absorption efficiency is given by

$$
\begin{aligned}
\eta_a &= \int_0^{W_d} G(x)\,dx/I_0 \\
&= \frac{(1 - R_F)(1 + R_B \exp[-(\alpha_d W_d + 2\alpha_b W_b)])[1 - \exp(-\alpha_d W_d)]}{1 - R_F R_B \exp[-2(\alpha_d W_d + \alpha_b W_b)]},
\end{aligned}
\tag{16}
$$

with I_0 being the incident flux. The absorption coefficient, which is experimentally verified in the range of 2.5–4.0 μm (42, 43, 44) and has negligible temperature dependence (45), is given by (46)

$$
\alpha = \frac{q^3\lambda^2 N}{4\pi^2\epsilon_0 c^3 n m^{*2}\mu},
\tag{17}
$$

where λ = wavelength, N = density of free carriers, which is also the doping concentration, n = refractive index, m^* = effective mass, and μ = mobility.

The responsivity is given by

$$
R = q\eta\lambda/hc,
\tag{18}
$$

where the total quantum efficiency (η) is the product of photon absorption efficiency (η_a), internal quantum efficiency (η_i), and barrier collection efficiency (η_c); q is the electron charge and λ is the wavelength.

The collection efficiency in first approximation, is given by (47, 48)

$$
\eta_c = \exp(-x_m/L_s),
\tag{19}
$$

in which L_s is the electron scattering length in the image-force well (located in the n-layer) and x_m is the distance from the interface to the barrier maximum.

a. Internal Quantum Efficiency The internal quantum efficiency (η_i) was obtained following a simplified Vickers Mooney model (49, 50) as

$$
\eta_i = \frac{\eta_0}{1 - \gamma + \gamma\eta_0/\eta_M},
\tag{20}
$$

where $\eta_M = N_M/N_T$ is the ratio of capturable electrons to photoexcited electrons and $\gamma = L_e/(L_e + L_p)$ is the probability that an excited electron will collide with a phonon before it collides with a cold electron. The inelastic scattering with cold electrons, which is assumed to "cool" the excited electrons to below the barrier energy, is characterized by the scattering length L_e; elastic scattering with phonons and impurities is characterized by the scattering length L_p. The fraction of electrons captured prior to any bulk scattering events ($e-e$ or $e-p$) is given by (50)

$$\eta_0 = \frac{L^*}{W_d} U(W_d/L^*)\eta_{id}, \tag{21}$$

with $L^* = 1/L_e + 1/L_p$ and η_{id} being the ideal internal quantum efficiency described using an escape cone model (51) and $U(W_d/L^*) \sim [1 - \exp(W_d/L^*)]^{1/2}$. The curves in Fig. 10 show the long-wavelength threshold, quantum efficiency, and responsivity for different device parameters calculated from the model.

2. *A Dark Current Model for Type II.* The dark current of a detector, i.e., the carrier (electron in this case) transport across the barrier without any incident photons, is one of the major performance limiting factors, which could be due to thermionic emission (20, 52), thermally assisted field emission (also known as thermionic field emission), or tunneling [field emission (53)]. This nomenclature depends on the energy level of the electron, as shown in Fig. 11.

The thermionic emission current density is given by the well-known Richardson–Dushmann equation

$$J_{TE} = A^{**}T^2 \exp\left(-\frac{\Delta - \Delta\phi}{kT}\right), \tag{22}$$

where $A^{**} = f_p A^*$ is the effective Richardson constant, $A^* = 4\pi q m^* k_B^2/h^3 = 120(m^*/m_0)A$ cm^{-2}K^{-2} is the Richardson constant, f_p is the barrier escape probability, k_B is the Boltzmann constant, and m^* is the effective electron mass. The barrier lowering that results from image force effects ($\Delta\phi$) is given by (52)

$$\Delta\phi = \left(\frac{qF}{4\pi\epsilon_0\epsilon_s}\right)^{1/2} \tag{23}$$

where $F = (V_b - V_0)/W_i$ is the electric field in the i-region, V_b the applied bias voltage, V_0 the flatband voltage (18), and ϵ_s the relative dielectric constant. For n-type Si, $m^* \approx 2.1m_0$ (52). f_p is determined by

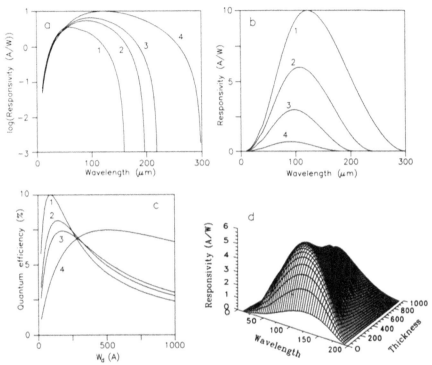

FIG. 10. (a) Spectral response calculated for different emitter-layer doping concentrations and corresponding optimum thicknesses. **1** $N_d = 1 \times 10^{19}$ cm^{-3}, $w_d = 540$ Å. **2** $N_d = 2 \times 10^{19}$ cm^{-3}, $w_d = 180$ Å. **3** $N_d = 2.32 \times 10^{19}$ cm^{-3}, $w_d = 150$ Å. **4** $N_d = 3 \times 10^{19}$ cm^{-3}, $w_d = 100$ Å. (b) Spectral response calculated for different field strengths with $N_d = 3 \times 10^{19}$ cm^{-3} and $w_d = 100$ Å. **1** $E = 1000$ V/cm, **2** $E = 500$ V/cm, **3** $E = 200$ V/cm, **4** $E = 100$ V/cm. (c) Quantum efficiency vs. emitter layer thickness calculated for different doping concentrations and corresponding peak wavelengths. **1** $N_d = 3 \times 10^{19}$ cm^{-3}, $\lambda_p = 124$ μm. **2** $N_d = 2.32 \times 10^{19}$ cm^{-3}, $\lambda_p = 100$ μm. **3** $N_d = 2 \times 10^{19}$ cm^{-3}, $\lambda_p = 92$ μm. **4** $N_d = 1 \times 10^{19}$ cm^{-3}, $\lambda_p = 60$ μm. (d) 3-D plot of responsivity as a function of wavelength and emitter layer thickness calculated for $N_d = 2 \times 10^{19}$ cm^{-3}.

the electron–phonon scattering in the image force potential well. As a first approximation, f_p is given by (52)

$$f_p = \exp\left(-\frac{x_m}{L_s}\right), \tag{24}$$

where L_s is the scattering length. Due to f_p, the thermionic emission current is voltage dependent, especially in the low electric field. The

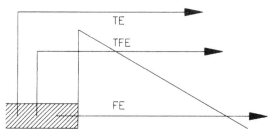

FIG. 11. Energy-band diagram showing the three dark current components due to the energy level of the carrier.

quantum mechanical reflection effect is negligible in the low electric field range ($F < 10^4$ V/cm) (54).

The field emission and the thermionic field emission current density flowing through the barrier can be written as

$$J_t = \frac{2q\hbar}{m^*} \int \frac{k_x}{(2\pi)^3} f(E) T_t(E_x) \, d^3\vec{k}, \tag{25}$$

where $f(E) = \{1 + \exp[(E - E_F)/k_B T]\}^{-1}$ is the Fermi–Dirac distribution function, E is the total energy of the electron, k_x and E_x are the wave vector and associated energy in the tunneling direction, and T_t is the tunneling probability. Converting the integration in the $\vec{k}$-space to that for energy, one obtains

$$J_t = \frac{qm^*}{2\pi^2\hbar^3} \int f(E) T_t(E, E_\perp) \, dE \, dE_\perp , \tag{26}$$

where $E_\perp$ is the energy associated with the momentum perpendicular to the direction of tunneling (or transverse momentum), given by $E_\perp = E - E_x$. The effect of $E_\perp$ is to reduce the tunneling probability. Transverse momentum must be conserved in direct tunneling (52).

The tunneling probability can be calculated using the WKB approximation, given by

$$T_t(E, E_\perp) = \exp\left[-2 \int_{x_1}^{x_2} |k_x| \, dx\right], \tag{27}$$

with

$$|k_x| = \sqrt{\frac{2m_t}{\hbar^2}(PE - E_\perp + E)} ,$$

where $PE = \Delta E_c - qFx - q/16\pi\epsilon_0\epsilon_s x$ is the potential energy and m_t is the tunneling effective mass component. For Si, $m_t = 0.26\,m_0$. In Eq. (27), x_1 and x_2 are the classical turning points, given by

$$x_{1,2} = \frac{\Delta E_c - E_\perp + E}{2qF}\left[1 \mp \sqrt{1 - \frac{q^3F}{4\pi\epsilon_0\epsilon_s(\Delta E_c - E_\perp + E)^2}}\right]. \quad (28)$$

The electron–phonon scattering in the image force potential well will further reduce the tunneling probability of the electron. Thus, the effective tunneling probability should be

$$T_t^* = T_t\exp\left(-\frac{x_1}{L_s}\right), \quad (29)$$

where x_1 is given by Eq. (28).

In Eq. (26), the integration limits for $E_\perp$ are given by $0 < E_\perp < \min(E - E_c^n, E)$, where $E_c^n = \Delta E_c - qV_b$ is the conduction band edge of the collector layer. The integration limits for E are given by $\max(E_F, E_c^n) < E < \Delta E_c - \Delta\phi$ for the thermionic field emission current J_{TFE}, and $\max(0, E_c^n) < E < E_F$ for the field emission current J_{FE}.

The total dark current is the sum of thermionic emission, thermionic field emission, and field emission currents,

$$I_d = A_D(J_{TE} + J_t) = A_D(J_{TE} + J_{TFE} + J_{FE}), \quad (30)$$

where A_D is the detector area. For photoemission detectors, the main noise mechanism is due to the shot noise (55, 56). The total noise current is contributed by the combination of the background photon noise and the thermal noise, given by

$$i_n = \sqrt{2qG(I_B + I_d)}\ \text{A}/\sqrt{\text{Hz}}, \quad (31)$$

where $I_B = q\eta Q_B A_D$ is the background photocurrent, Q_B is the background photon flux density, and η is the quantum efficiency, which has been described. The total noise current (dark current) variation with the bias voltage for different temperatures and the bias dependence of the different current components at 4 K (i.e., thermionic, thermionic field emission, and field emission) for a detector with a $\lambda_t = 144\ \mu$m at $V_b = 25$ mV, is given in Fig. 12. The temperature dependence of the dark current at three different bias voltages and the current components are shown in Fig. 13.

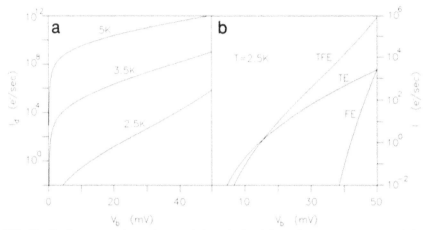

FIG. 12. Dark current-voltage characteristics calculated for a detector structure consisting of an Si n^+-i interface with a 500-Å emitter layer doped to 1×10^{19}. (a) Results at different temperatures, and (b) comparison of the three current components at $T = 2.5$ K. Structure parameters are the same as for curve 1 in Fig. 10(a).

3. Background Limited Performance. At low temperatures, the dark current will be dominated by the tunneling component, which depends on the barrier height and the electric field. Hence, as the temperature is lowered, the dark current will decrease until it reaches the current due to the

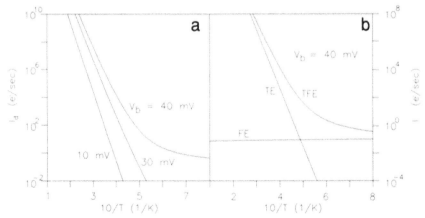

FIG. 13. Temperature dependence of dark current calculated for the same structure considered in Fig. 12. (a) Results at different biases, and (b) comparison of various current components at $V_b = 40$ mV.

background photon flux (noise), at which point the detector has reached its background limited performance (BLIP). The detector noise equivalent power (NEP), which is an important figure of merit to represent the performance, is given by

$$(NEP)_\lambda = \frac{i_n}{R_\lambda} \; W/\sqrt{Hz}, \tag{32}$$

where $R_\lambda = q\eta\lambda/hc$ is the responsivity (A/W). The background limited NEP is given by

$$(NEP)_{BLIP} = \frac{hc}{\lambda} \sqrt{\frac{2Q_B A_D}{\eta}} \; W/\sqrt{Hz}. \tag{33}$$

The background limited temperature, T_{BLIP}, is determined by the equation $I_B(\lambda, V_b) = I_d(V_b, T_{BLIP})$. Calculated NEP curves are given in Fig. 14.

4. *Single and Multilayer Structures.* Published results (2, 57) on a p^+-n interface confirm FIR detection by holes, and the comparison with a p^+-n-n^+ confirms that both carriers can contribute significantly. The bias (V) dependence of the long wavelength cutoff λ_C (50% of peak) was attributed to the barrier (Δ) lowering effect given by (20) $\Delta^* = \Delta - edF$, where F is the field across the sample. A fit to the data gives an estimate for d, inversely proportional to the sharpness of the doping profile (20) (see Fig. 15).

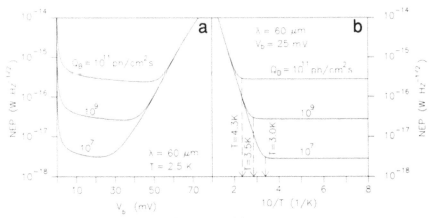

FIG. 14. NEP dependence on (a) bias and (b) temperature, calculated for the same structure considered in Fig. 12 at $\lambda = 60 \; \mu$m and under different backgrounds.

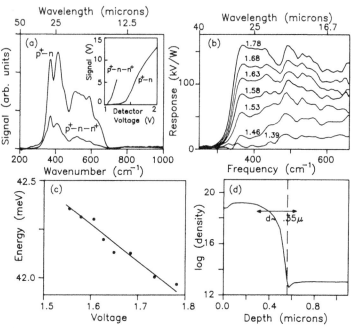

FIG. 15. (a) Raw data for silicon p^+-n-n^+ (SDC-150) and p^+-n structures at 4.2 K obtained with 1.2-V and 2.0-V bias across the sample, respectively. For both structures λ_t is 35 ± 3 μm. The inset shows the signal detected for both structures vs. the detector bias voltage. The high bias needed was due to a non-ohmic contact produced by Al contact on damaged i-region. (b) The bias dependence of the spectral response for the silicon $p-i$ structure at 1.5 K. (c) Variation of the cutoff energy with the detector bias at 1.5 K. The slope gives a value of ~ 0.35-μm for the transition region thickness d. (d) The concentration and the thicknesses of the p^+ and n regions of the sample obtained by the spreading resistance measurements showing a possible transition region. Fig. (a) after Ref. (2) and (b). (c) and (d) after Ref. (26).

In principle, based on published results (2, 18, 57), it is possible for a photoemissive-type silicon homojunction detector with P or As doping concentrations significantly in excess of 1×10^{18} cm^{-3}, to exhibit fairly high spectral responsivity all the way from the long wavelength infrared (LWIR) range to well beyond 200 μm. Unfortunately, to achieve this wide spectral response through impurity-band broadening in a single detector involves unnecessary performance degradation, because of additional background noise and reduced collection efficiency (limited by diffusion length). However, thinner absorber (electron emitter) regions can compensate for the collection efficiency problems, in turn reducing the IR absorption. The solution for this will be to introduce more layers for higher absorption.

The detection mechanism in a multilayer structure is shown schematically in Fig. 16. The IR photoionized electron in the first active layer drifts through the i-region and enters into the second active region to induce an impact ionization, thus resulting in a photoconductive gain. From the impact-ionization cross-section curves adapted to this situation, one can conclude that the maximum secondary ionization will occur when the incident energy is about twice the ionization energy. Since the applied bias needs to be about twice each barrier height for the optimal photoconductive gain, the Schottky barrier lowering effect may be appreciable, resulting in a higher dark current. This effect has been empirically incorporated (20) in a modified Richardson–Dushman equation with the effective bias dependent barrier Δ^*. Since $F = V_1/D$, where V_1 is the bias across and D the thickness of the i-region, the ionization barrier change can be expressed as $\delta\Delta = (d/D)qV_1$, where d is a measure of the interface sharpness. Since $qV_1 \simeq 2\Delta$ in our case, we need to have $d/D \ll 1$ to maintain the required barrier and effectively block the dark current; i.e., the n^+-region should be much thicker than the transition region. We have observed $d \leq 0.01$ μm in pulsing diode experiments, where barrier lowering and restoring is believed to be the main mechanism of the pulsing phenomena (58). In MBE growth the thickness can be controlled to about ~ 10 Å, and can define a much sharper interface. When $d \simeq 10$ Å, a D of ~ 1000 Å will be enough to block the dark current. These dimensions are also similar to those currently used in the design of MBE-grown 50-layer multiquantum well IR detector structures.

Photoexcitation in a doped layer can be followed by relaxation before the excited carrier reaches an interfacial barrier. Hence the doped layer thicknesses should be less than the carrier mean free path to maximize the

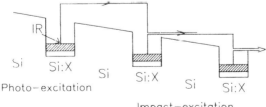

FIG. 16. Proposed multilayered homojunction detector structure to be grown by MBE, showing both photoionization and impact ionization. Although both n^+- and p^+-types are possible, only n^+-type is shown here. After Ref. (26).

collection efficiency and should be much larger than the Bohr radius of an electron. Otherwise the active region will not be effectively defined, since the quantum size effect pushes the impurity levels up to overlap with the conduction band. In addition, a thin active region may result in a low absorption of photons. In this sense, the active region of about one scattering length should be appropriate. The electron Bohr radius is ~ 20 Å in Si, ~ 100 Å in GaAs, and ~ 40 Å in Ge. When the concentration is at or above the Mott transition value, the mean free path ranges from tens of Å to a few hundred Å (40). These are again within the range of MBE growth.

The temporary local electric field created by the space charge due to impact ionization is given by

$$F_{local} = \frac{\sigma_{local}}{\epsilon_0} = \frac{qN_+}{\epsilon_0 A}. \tag{34}$$

To sustain the ionization process, this temporal field should be less than the overall electric field due to bias given as $F = V_1/D \simeq 2\Delta/qD$. Since the impact ionization gain is ~ 2 in the ideal condition described above, we can assume $2^{(N_L-1)}$ positive charges will be created in the N_L-th active layer. Then the local field in the N_L-th layer is $F_{local} = 2^{(N_L-1)}q/\epsilon_0 A$, and we get the following condition on the number of layers: $2^{(N_L-2)} \leq \epsilon_0 \Delta A/q^2 D$. With $\Delta \sim 30$ meV, $D \sim 1000$ Å, and $A \sim (100 \ \mu m)^2$, a value for the minimum number of layers will be given by $N_L \leq 20$.

5. *Phonoelectric Effect.* Another possible gain mechanism proposed for this multilayer homojunction system is through phonons. For the very low excitation energies for the devices discussed here, phonon energies may exceed excitation energies. In this case, phonons generated during carrier drift through the superlattice can contribute to the photocurrent (see Fig. 17). This electron–phonon cascade concept is interesting as a new concept and as a feature that is special to far infrared detection. The requirement that excitation energies be less than maximum phonon energies indicates that the electron–phonon cascade is potentially useful for infrared detection $\geq 20 \ \mu m$ in silicon and $\geq 34 \ \mu m$ in germaninum and gallium arsenide. In GaAs the longitudinal optical (LO) phonon intersubband lifetime has been predicted and measured (59, 60). The typical value is ~ 1 ps when the transition energy is higher than the LO phonon energy (~ 37 meV), and ≤ 100 ps when it is lower. This transition rate is much

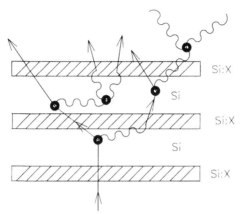

FIG. 17. Schematics of electron phonon shower production. Straight lines indicate the electrons. The donuts indicate the impurity atoms in the lightly doped n-region and the wavy lines the phonons. After Ref. (26).

higher than the radiative transition rate (61) and opens the possibility of a phonoelectric gain.

D. TYPE III HIP DETECTOR: $N_d > N_0$ $(E_F > E_c^i)$

When the Fermi level is above the conduction band edge of the n-layer, the n^+ (p^+)-layer becomes degenerate, and a barrier associated with a space-charge region is formed at the n^+-n interface due to the electron (hole) diffusion, as shown in Fig. 18. The barrier height depends on the doping concentration and the applied voltage, giving rise to an electrically tunable λ_c. This type of device on Si was first demonstrated by Tohyama *et al.* (63, 62), using a structure composed of a degenerate n^{++} hot carrier emitter, a depleted barrier layer (lightly doped p, n, or i), and a lightly doped n-type hot carrier collector for ~ 12-μm operation. As the bias voltage is increased, the barrier height is reduced, the spectral response shifts toward longer wavelength, and the signal increases at a given wavelength. The photoemission mechanism of type III HIP detectors is similar to that of type II HIP detectors, the major difference being different response wavelength ranges and different operating temperature ranges. The type III HIP detectors are expected to operate near 77 K and have responses in the MWIR and LWIR ranges (14).

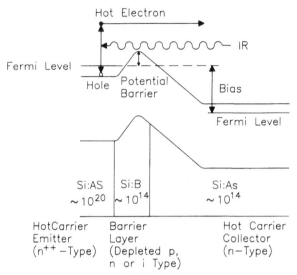

FIG. 18. Energy band diagram for the LWIR homojunction sensor (type III) active area, illustrating the IR detection mechanism. Concentrations are shown for a p^+-type barrier layer. After Ref. (62).

E. DELTA DOPED POTENTIAL WELL IR DETECTORS

Liu *et al.* (7) have demonstrated another homojunction IR detector approach. One way to look at this structure (see Fig. 19) is as a p^+-n-n^+ structure, where the intrinsic region is delta doped with boron creating alternating highly doped and undoped regions. The p and n regions are contact regions, and also will increase the device impedance due to the $p-n$ (or p^+-n-n^+) junction. These epitaxially grown test structures consist of a 0.9-μm undoped buffer layer, followed by a 0.6-μm, 5×10^{18}-cm^{-3} boron doped bottom contact layer (p layer), an 85-nm undoped layer, and 20 units consisting of 3 nm δ doped with boron and a 30-nm undoped region, a 30-nm undoped Si layer, and the top contact layer doped to 2×10^{19} cm^{-3} with arsenic to a depth of 0.3 μm. The δ doping concentration was varied to obtain the wavelength tunability as shown in the interfacial workfunction detectors (18, 2, 57, 64). The peak response wavelengths were estimated by the expected depths of the potential wells, which are mainly due to the many-particle effects of the heavy doping (31, 65, 66). Cutoff wavelengths (9) (wavelength at which the response reduces to 50% of the peak response) of 10, 12, and 20 μm have been observed

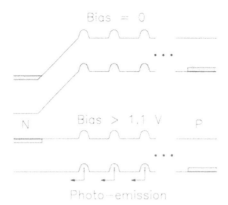

FIG. 19. Band diagram for a delta doped structure under zero and forward bias. After Ref. (7).

for samples with doping concentrations of 2×10^{20}, 1×10^{20}, and 6×10^{18} cm^{-3}, respectively. In addition to photoexcitation in the δ doped potential wells, one would expect the p and n interfaces to respond similarly to the HIP type I, II, or III detectors, depending on the concentration.

F. UNCOOLED AVALANCHE INJECTION IR DETECTOR

Most of the semiconductor LWIR detectors need some cooling to achieve low dark currents and high sensitivity. Room temperature operation of these detectors even with a lower responsivity and a higher dark current compared to cooled detectors will have some use in various situations. Recently some of the compound semiconductor structures have been optimized to use as LWIR detectors. However, as mentioned before, Si will have an immense advantage over these compound semiconductor structures due to the ease in integrating with other Si components.

Kikuchi (67) has recently reported an Si homojunction structure which could detect 10.6-μm radiation at room temperature. The structure consists of a 3×10^{17}-cm^{-3} p^+ layer for IR absorption, and a 200-μm p-type substrate with a hole concentration of 2×10^{14} cm^{-3}, forming a homojunction as seen in Fig. 20. Incident IR radiation from a CO_2 laser will excite the holes, creating the photocurrent. The detection mechanism here is similar to the HIWIP detectors discussed before. This detector will also show wavelength tunability with the concentration of the p^+ layer and bias

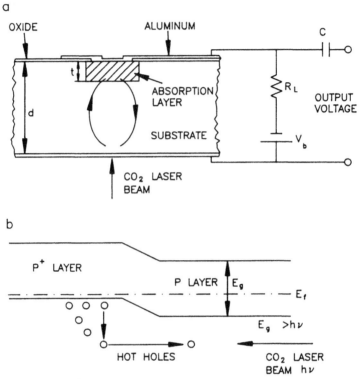

FIG. 20. (a) The Si $p^{+}-p$ junction used for room temperature operation of the 10.6 μm detector with the operating circuit. (b) The zero bias energy band diagram for the high-low junction shown above. After Ref. (67).

voltage. The observed specific detectivity (D^*) of 5×10^5-cm Hz$^{1/2}$/W (500 V/W responsivity and 5.5×10^{-9} W NEP) shows that much more has to be done for this detector to be competitive in high-responsivity applications. Although it is predicted that the performance could be improved by designing the structure for optimum reflection coefficient and the reduction of the dark current, the latter may require highly creative device design.

G. OTHER MATERIAL SYSTEMS

Although we have concentrated only on Si homojunctions so far, there exists also a large amount of literature on other material systems. Even though Ge was thought to be the material of the future at the beginning of

the semiconductor age, Si became the material of choice. However, Ge
devices are still active in various areas.

1. Ge Homojunctions for FIR Detection. One would expect most of the
effects observed with Si to be present in Ge due to their similarities,
except for the direct gap for Ge, which might have advantages as far as
speed is concerned. The idea of wavelength tunable IR detectors classified
as type I, II or III was extended to cover other materials (*2*). Perera *et al.*
have demonstrated the HIWIP (Type I or Type II) concept in Ge using
commercially available p^+-n-n^+ samples (*2*). Raw spectral data obtained
from one sample showing a λ_t of 240 ± 20 μm is shown in Fig. 21. Also
visible are Fabry–Perot oscillations superimposed as peaks and valleys
(*68*). As expected, the λ_t moves to longer wavelengths as the bias is
increased due to barrier lowering effects. The lowest bias voltage of 0.7 V
at which a photoresponse still exists roughly agree with the flatband
voltage of Ge, which is 0.66 V.

Another Ge p^+-n-n^+ sample, as shown in Fig. 22, displayed a λ_t of
175 ± 10 μm. The peak sensitivity of 1.4×10^5 V/W was obtained at
around 100 μm. The NEP at this frequency was about 10^{-11} W/$\sqrt{\text{Hz}}$ and
$D^* = 3 \times 10^{10}$ Hz$^{1/2}$/W cm^2 with unrestricted bandwidths. The peak

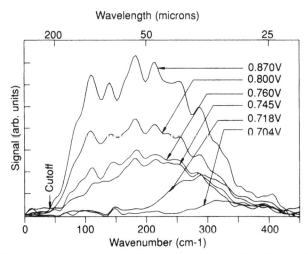

FIG. 21. Raw spectra obtained with a Ge sample at 4.2 K. The threshold at high bias
corresponds to about 240 ± 20 μm and shifts to shorter wavelengths as the bias is decreased.
Also visible are Fabry–Perot oscillations superimposed as peaks and valleys on the raw data.
After Ref. (*2*).

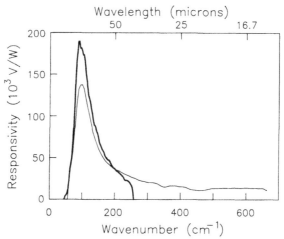

FIG. 22. The spectral response of a Ge sample at 1.5 K with a detector voltage of 0.7 V showing a λ_c = 175 ± 10 μm with a response bandwidth at half height points 70 μm. The dark line corresponds to the response observed when a longpass filter with a cut-on of 40 μm was placed on the light path to reduce the incident bandwidth. The sensitivity is obtained by dividing the raw data by the bolometer spectrum. After Ref. (2).

sensitivity and the threshold was increased to 1.9×10^5 V/W and 200 ± 10 μm, respectively, by limiting the incident IR bandwidth to greater than 40 μm, indicating a limited number of carriers for photoexcitation (2). This observation brings up the issue of the effective photoactive region thickness, which in this case may be only a fraction of the total photoactive thickness. Hence the optimization of these detectors would require multiple, thinner absorber regions (26).

Although the Ge BIB detector development is not as successful (69) as the Si BIB detector development due to the lack of high-purity Ge epitaxial growth techniques (69) and structural defects and impurities (70), some progress has been made toward fabricating linear arrays (70). Ga doped Ge detectors have shown threshold spectral response up to 220 μm (70, 71, 72).

2. InGaAs Homojunctions for FIR Detection. For an InGaAs $p-i-n$ structure as shown in Fig. 23, the threshold wavelength λ_t comes to about 90 μm. In contrast to Si or Ge (2), which does not exhibit a short wavelength threshold inside the measurement range, this detector gives a short wavelength threshold of about 42 μm, which may be due to bulk

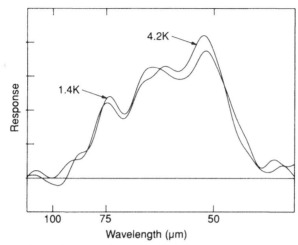

FIG. 23. Spectra for an InGaAs p^+-n-n^+ sample showing λ_t of 90 ± 4 μm. There is not much change for 4.2-K to 1.4-K temperature variation. The shorter wavelength cutoff may well be due to bulk absorption. After Ref. (*25*).

absorption. This confirms that this workfunction mechanism is also viable on III–V material structures.

3. Possibilities with GaAs. GaAs is characterized by very shallow donor levels, ~ 5.5 meV for high-purity material and even lower values when doped (*73*) (e.g., 1.89 meV for $N_d = 4.7 \times 10^{15}$), making this a candidate for the FIR detectors. However, the ~ 80 Å ground state Bohr radius gives rise to large dark currents (*69*). One would expect wavelength tunability up to 400–500 μm from the *n*-type GaAs HIWIP detectors (see Fig. 5a). However, no GaAs detectors are yet reported either in this HIWIP or BIB mode.

III. Other Homojunction IR Detectors

So far we have concentrated on homojunction detectors, which can be described as HIP detectors. Here we briefly describe some other homojunction detectors either developed recently or not covered well in previous review articles.

A. CHARGE STORAGE MODE IR DETECTORS

Another mode of operation of these homojunctions has produced an alternative approach to IR detection in an integrating mode, which utilizes charge storage in impurity levels as opposed to storage in potential wells as in CIDs or CCDs. Banavar *et al.* initially demonstrated (*74*) this charge storage effect in silicon p^+-n-n^+ structures with phosphorus impurities at 4.2 K. Then they went on to demonstrate the tunneling effects under an applied field from these impurity levels (*75*) labeled as field ionization.

At very low temperatures, the localized impurity levels can act as traps for charge carriers (*76*). In the absence of photoionization, the storage time will depend on the temperature and also on the quantum mechanical tunneling (*77*). This mode of IR detection is based on a combination of field and photoionization techniques (*78*). For a hydrogenic system, the binding energy is $m^*/m_0[Z/\kappa]^2 \times 13.6(eV)$. Phosphorus in Si:P is similar to a proton bound to a Si atom giving rise to shallow levels, which can be observed by IR spectroscopy (*79*).

For most of the time during operation, the p^+-n-n^+ structure will be in the reverse biased condition as shown in Fig. 24. However, the phosphorus impurities in the n-region are neutralized by first applying a forward pulse. Then the structure is exposed to IR photons for some time while under reverse bias. During this photoexcitation, most of the carriers will be swept out by the field, reducing the amount of charge stored in the n-region. Applying a high field (> 1 V/μm for Si:P) can remove the remaining stored charge in the n-region via field ionization. Measuring this field-ejected charge and calculating the total charge capacity in the n-region will indicate the amount of photoexcited carriers.

The electric field during the IR exposure (or the integration time) can also be used to modulate the infrared response due to field-assisted tunneling (*81*). This can easily be understood by looking at the field dependence of the potential near an n-region phosphorous impurity as seen in the inset in Fig. 25. A -3-V applied bias will not pull down the potential enough for the $2P_0$ level to be free. As the bias is increased to -6 V, the $2P_0$ level becomes free, showing a peak in the spectral response curve in Fig. 25. This method provides handles on (1) ground state binding energy through ground state field ionization; (2) excited state energy levels through photoexcitation and field ionization of excited states; (3) the impurity potential and binding energy through the field dependence of the photoionization threshold.

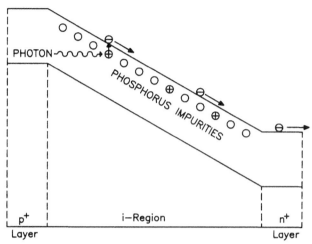

FIG. 24.   Band diagram for a reverse biased p^+-n-n^+ structure showing impurity photoionization and excited carrier swept out by the electric field. The inset shows the input waveform
and the recorded output used in the integer charge experiments. Input waveform consists of
the forward injection pulse and a sharp negative pulse for field ionization. During the region
between these two, the IR exposure will create photoexcited electrons, which will be swept
out by the field. The output shows the charge released during the field ionization process at
the correct field strength. A comparison of this released charge with the total charge will give
the photoexcited amount of charge. After Ref. (78).

1. Detection of Fractional Charge Impurities in Si.   The charge storage mode
IR detection method described above has been used as a novel optoelectronic technique to detect unconfined fractional charge ($\pm \frac{1}{3}e$) impurities
in semiconductors (82). In analogy with the well known shallow "donor"
impurities in semiconductors in which an electron is weakly bound to a
positive ion $D = D^+ + e^-$, the existence of shallow fractional charge
donors has been predicted (83):

$$D^{-1/3} = D^{+2/3} + e^-.$$

With reference to the peaks observed for integer charge impurity atoms in
Fig. 25, one would expect the photoionization of the electrons from
fractionally charged impurity atoms to show a similar response in the
corresponding wavelength range (82, 83). This optoelectronic approach
has the advantage of giving an estimate for the concentration of fractional
charges, being repeatable and without mechanical techniques. An upper
limit (95% C.L.) of 2.3×10^{-20} fractional charges per atom is obtained
using a combination of an IR photoionization and a field ionization
technique (82).

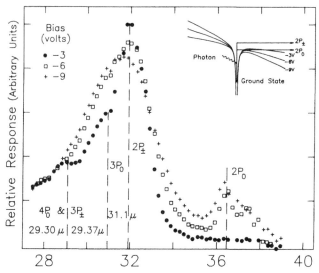

FIG. 25. Response below the zero-field threshold. Peaks correspond to transitions from ground state to excited states. The potential is also shown in the inset for the fields used in this experiment. For -3 V bias, the photoionized electron (from the ground state to $2P_0$) has to tunnel through the barrier. As the bias is changed to -6 V the excited electron at $2P_0$ is free giving rise to a peak in the spectral curve. After Ref. (78).

B. HgCdTe Homojunctions

Although HgCdTe has been studied for many years for short-, medium-, and long-wave infrared ranges (SWIR, MWIR, LWIR), this material has major problems when uniformity is an issue. To make matters worse the material itself is hard to grow. Although ion implantation into p-type material has been the common technique to fabricate n^i–p HgCdTe junctions, liquid phase epitaxy (LPE) and MBE have been used recently. Arias *et al.* have demonstrated LWIR and MWIR p^+–n homojunctions using MBE-grown HgCdTe epilayers (84). Their detector performance was comparable to that of the LPE-grown structures (85). The resistance area product at zero bias, R_0A, is a standard figure of merit for these photovoltaic detectors, which can be used to predict the detectivity limit and the dark current mechanisms by studying the variation with temperature (86). Although theoretical (87) studies have been carried out on p^+–n–n^+ $Hg_{1-y}Cd_yTe$ homojunctions, most of the experiments were performed on $Hg_{1-y}Cd_yTe/Hg_{1-x}Cd_xTe$ heterojunctions (88). HgCdTe p-on-n junctions have higher R_0A products compared to the n-on-p HgCdTe junctions. One more advantage of p-on-n junctions is the controllability of n-type carriers in the range of 10^{14}–10^{15} cm^{-3} using extrinsic doping compared to the p-type.

IV. Spontaneous Spiketrain Generation in Si Homojunctions

Another remarkable feature observed with the Si p^+–n–n^+ structures was their spontaneous firing phenomena at cryogenic temperatures under a constant voltage (7, 89, 90, 91, 92) or current supply (23, 58, 93). Successive pulses are almost identical so that the information transfer can be handled via the modulation of interpulse time intervals (IPTIs). Coon and Perera (68, 94) have reported pulse heights of the order of 40 V and rise times of about 80 ns giving a slew rate of 0.1 V/ns. This offers an advantage in precision timing of IPTIs between pulses. Most of the structures studied consist of Si:P in the lightly doped n-region, where space charge buildup leads to the breakdown, giving rise to the pulse. Although the work described here was carried out at 4.2 K, the structures were able to pulse up to about 10 K. By changing the phosphorus dopant to a deeper (8) level (As, B, Al or Ga), the operating temperature could be increased.

A. DEVICE PHYSICS OF SPIKETRAIN GENERATION

Although the devices will fire either under voltage or current bias, the device physics treatment will be little different in the two situations. We will first discuss the current-driven situation, and the voltage-driven case will be briefly discussed later. The models discussed here have been developed in conjunction with the experimental results, making them more realistic. Hence, the models could be used in device fabrication to optimize the parameters.

1. Spontaneous Pulsing under Constant Current. The p^+–n–n^+ structure is at 4.2 K with a forward current applied to the p side as shown in Fig. 26. The input and load capacitance arc due to the cables. The model assumes that a slow buildup of space charge, characterized by a low electron injection current, is followed by a fast high electron injection current pulse during which some of the space charge is neutralized, resetting the diode for the next pulse. This space charge layer of thickness w and the resulting band diagram are shown in Fig. 26. The voltage across the diode, neglecting the discharge of the load capacitor through the load resistor, is given by

$$V_{diode} = \frac{A}{C} \int_{t_i}^{t_f} (i_T - j) \, dt - \frac{A}{C_L} \int_{t_i}^{t_f} j \, dt, \tag{35}$$

with i_T the input current density (A/m^2), j the injection current density, C_L and C the load and input capacitances, and A the area of the diode. This is valid as $R_L C_L$ ($\sim 50~\mu s$) $\ll$ IPTIs (1–100 ms). Another expression for the voltage across the diode is $V_{diode} - V_0 = ED + V_{bi}$ where E is the field across the n-region, $V_{bi} = new_i^2/(2\epsilon)$ is the built-in potential (40) due to the n-region space charge, $V_0 = 1.1$ V in the flatband voltage for Si which is due to the Fermi level difference between the p- and n-type material, and D is the n-region thickness.

The space charge buildup is given by

$$\frac{dw}{dt} = \frac{D - w}{e} \sigma j, \tag{36}$$

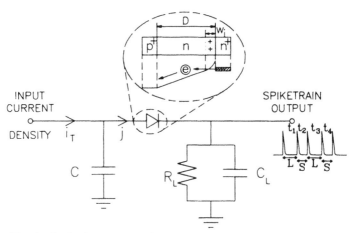

FIG. 26.   The simple circuit used to obtain the spontaneous pulsing. The diode is driven by a dc current at 4.2 K. $C_L \simeq 100$ pF and $C \simeq 100$ pF are due to the cables and $R_L \simeq 300$ kΩ. Interpulse time intervals (IPTIs) are denoted by t_1, t_2, etc. The inset shows the space charge distribution and the resulting band diagram for the diode. D is the n-region thickness, w_i is the initial depletion region, i_T and j denote the input current density and the current density through the device. After Ref. (23).

where σ is the impact ionization cross-section (24). After using Eq. (36) to change the integration variable from t to w in Eq. (35) we integrate Eq. (35). Then using $V_{bi} = w^2 en/2\epsilon$ and the expression for the voltage across the diode with the approximation $w - w_i \ll w_i$ we obtain the following equation for i_T and t:

$$\frac{Ai_T t}{C} - \frac{Ae(w - w_i)(C + C_L)}{(D - w_i)\sigma CC_L} = ED + V_0 + \frac{w^2 en}{2\epsilon}. \qquad (37)$$

The injection current density is given by a modified (94) Richardson–Dushman equation

$$j = A^* T^2 \exp\left(\frac{\Delta - edF}{kT}\right),$$

where Δ is the interfacial workfunction at the n^+-n interface (94) and F is the field at the n^+-n interface. Assuming uniform distribution of impurities ($\rho = ne$), the field at the n^+-n interface was calculated. Combining these results and linearizing the equations with the approximation $\Delta w = |w - w_i| \ll w_i$, the temporal dependence of w could be obtained as

$$\exp[(\beta - \gamma)w] = \theta + R\frac{(\beta - \gamma)}{\alpha}\exp(\alpha t)$$

in terms of the constants

$$\alpha = Aedi_T/CDkT,$$

$$\beta = Ae^2 d/CDkT(D - w_i)\sigma,$$

$$\gamma = ne^2 d(D - w_i)/D\epsilon kT,$$

$$\theta = \exp[(\beta - \gamma)w_i] - R(\beta - \gamma)/\alpha,$$

$$R = \exp(\beta w_i)\exp\left[\frac{ne^2 dw_i^2}{2D\epsilon kT}\right]\eta,$$

where

$$\eta = [(D - w_i)e]\sigma A^* T^2 \exp\left[-\frac{\Delta + edV_0/D}{kT}\right].$$

Depletion will continue efficiently as long as the electric field sweeps carriers out of the n-region, i.e., until $E = 0$. Substituting the expression for temporal dependence on w, in Eq. (37) and setting $E = 0$, we obtain an expression which describes the relationship between the firing time interval and input current density with the device parameters given by

$$\frac{Ai_T t}{C} - (G + M)\ln\left[\theta + \frac{R(\beta - \gamma)}{\alpha}\exp(\alpha t)\right] + G_1 = 0 \qquad (38)$$

where

$$G = Ae(C + C_L)/CC_L(D - w_i)\sigma(\beta - \gamma),$$

$$M = enw_i/2\epsilon(\beta - \gamma),$$

and

$$G_1 = G(\beta - \gamma)w_i - V_0 + enw_i^2/2\epsilon.$$

This was solved numerically for the IPTIs, and the results were compared with the experimental results from a sample consisting of a $D = 150$-μm-thick, 10^{14}-cm^{-3} doped Si:P n-region, sandwiched between 1-μm-thick Si:P and Si:B regions, with an $A = 1$ mm^2 cross-sectional area. The known circuit parameters used were C and $C_L = 150$ pF, $R_L = 300$ kΩ. The data consisted of sets of 2100 IPTIs taken for drive currents between 33 and 72 nA at 1- or 2-nA intervals. A forbidden region in the IPTI distribution allows us to define the IPTIs as short (S) or long (L) and whose mean values are shown in Fig. 27.

a. Effect of Device Parameters on Pulsing This model was also used to study the effect of various device parameters on spontaneous pulsing. All the parameters in the diode and the circuit except for R_L, which does not enter into the final equation, were varied. The results of these variations for currents between 20 and 100 nA are shown in Figs. 28 and 29. All the curves shown had a cutoff value above which pulsing would not occur. In addition, D and n also had a minimum value below which pulsing was not obtainable, probably caused by the inability of the n-region to contain sufficient space charge. The upper limit cutoffs are probably related to the existence of a steady state with high injection current. The limits on pulsing varied with drive current. These cutoffs have been seen in the output of some diodes. The IPTIs varied most for variations of w_i, d, and n, and least for variations of C_L and A. Of significance for fabrication is d, which depends on the fabrication method. Although quite costly compared to diffusion or even implant techniques, epitaxial techniques could control d down to a few monolayers. Fig. 30 shows the variation of the IPTIs with D and d. The sharp drop in this plot shows the cutoff in the pulsing. Both the variations in the cutoff and the changes in the variation of the IPTIs with current are visible.

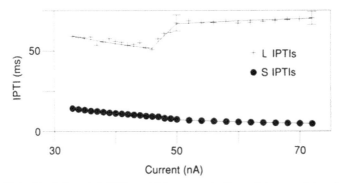

FIG. 27. Plot of both L and S IPTI vs. driving current. The points were determined from averaging the IPTI data. The theoretical fit was obtained by a least squares fit in the low- and high-current regions. The dashed lines indicate the regions used to obtain the fits. The center region values were determined by connecting the endpoints of the d and w_i values determined for the other two regions. The error bars for the L IPTIs vary from 8 to 4 ms. They were not drawn for S, since the values vary only from 0.8 to 0.4 ms. After Ref. (*23*).

2. Spontaneous Pulsing under Constant Voltage. In this situation the device is connected to a constant voltage source, with the current varying with time due to the space charge buildup, which is given by

$$d\rho/dt = n\sigma j,$$

where σ is the impact ionization cross-section and n is the concentration of the n-region. The field at the n^+-n interface can be obtained by Gauss's law and integration by parts as

$$F = \frac{V - V_0}{D} + \frac{1}{D\epsilon} \int_{D-w}^{D} x\rho(x)\, dx,$$

where the same notation as in Sec. IV.A.1 is used. Injection of electrons into the n-region will be inhibited by the workfunction (the gap from the impurity band to the top of the conduction band) at the n^+-n interface. The sharpness of the interfaces, which is measured by a thickness parameter d, can lower the effective inhibition by lowering the interfacial work function Δ by a term edF where e is the electron charge, and F is the field at the n^+-n interface (*20*). Considering this lowering, the injection-current density across the interface is given by the modified (*20*) Richardson–Dushmann thermionic formula. In this simplified approach, the recapture, the charge density of transiting electrons, and the effect of the load

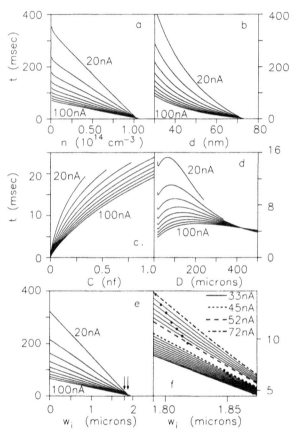

FIG. 28.   The variation of the IPTIs with the various diode parameters and circuit elements for currents between 20 and 100 nA. The parameters varied were (a) n, (b) d, (c) C, (d) D, and (e) w_i. (f) Expansion of the region marked by arrows in (e) with the experimental IPTIs marked as points along 2-nA current interval model outputs. The values for the constant parameters are $D = 150$ μm, $d = 70 - 10i$ nm, $w_i = 1.83$ μm, $n = 1 \times 10^{14}$ cm^{-3}, $A = 1$ mm^2, $C = 150$ pf, and $C_L = 150$ pf. The cutoffs at high parameter values and at low parameter values for D and n represent regions which have no pulsing solutions. After Ref. (23).

impedance have been ignored. Also the impact-ionization cross-section σ is taken as a constant. Combining the above equations gives an expression for the variation of current density with time as

$$j(t) = j(0)/(1 - t/T_0), \tag{39}$$

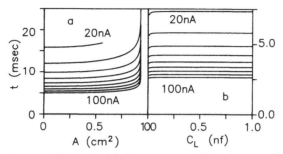

FIG. 29. The variation of IPTI with (a) diode area and (b) load capacitance for the same parameters as in Fig. 28. The large rise in the IPTIs for (a) is caused by the discharge of the input capacitor at low currents. Decreasing the input capacitor decreases the area at which this rise occurs. After Ref. (*93*).

where

$$j(0) = A^* T^2 \exp\left(\frac{-\Delta + (ed/D)(V - V_0)}{kT} \right),$$

and

$$T_0 = 2\epsilon kT / edDn\sigma j(0).$$

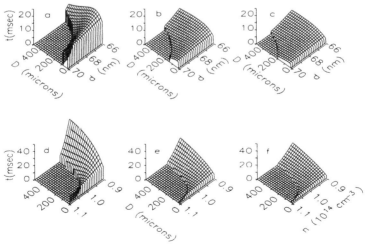

FIG. 30. Plot of IPTIs vs. *d* and *D* for currents of (a) 20 nA, (b) 60 nA, and (c) 100 pA. Plot of IPTIs vs. *n* and *D* for currents of (d) 20 nA, (e) 60 nA, and (f) 100 nA. The rapid drops indicate the limits for pulsing.

A virtue of this simplified model is that one can see how the system evolves slowly for a long time as a result of small $j(0)$, after which the injected current can increase rapidly. A more detailed treatment of this can be found in an article published in *Neural Networks* by Coon and Perera (*95*). Although this simple model predicts the timing of the pulse accurately, the model output of the rapid rise of current lags the experimental current. This may be due to the neglect of tunneling phenomena in the model, which can be significant with the high space charge field.

B. INFRARED DETECTION WITHOUT AMPLIFIERS

The time intervals between these pulses were observed to become shorter as the structure was exposed to IR radiation (*68*). This is due to the photoionization (*96*) of the n-region impurities, which will enhance the space charge buildup, leading to faster breakdown. The simple model described in section IV.A.2 can be expanded to include this additional space charge factor (*89*). This can be done by adding a term $nIe\sigma_p/h\nu$ to the space charge buildup formula given before to obtain

$$d\rho/dt = n\sigma j + nIe\sigma_p/h\nu.$$

Here I is the incident IR flux, $h\nu$ is the energy of the IR photon, and σ_p is the photoionization cross-section. Following the same procedure as in section IV.A.2, an expression for the current density with time can be obtained (*89*) as

$$j(t) = \frac{bc}{n\sigma_i[(c + be^{-a\rho_0})e^{-abt} - c]},$$

where $c = n\sigma_i A^* T^2 \exp[-\Delta + ed(V - V_0)/D]/kT$, $b = \sigma_p nIe/h\nu$ and $a = edD/2\epsilon kT$. From this, an expression for firing was obtained:

$$f = f_{dark} \frac{I/I_0}{\ln(1 + I/I_0)}, \tag{40}$$

where

$$f_{dark} = \frac{edDc}{2\epsilon kT} e^{a\rho_0},$$

and

$$I_0 = \frac{f_{dark}\,dDh\nu}{2\epsilon kTn\sigma_p e^{\Delta/kT}}.$$

A comparison with experimental data found this simple model to be in good agreement (*89*). Current responsivities of the order of 10 A/W (or 10^{10} Hz/W) and specific detectivities of the order of 3×10^{12} cm $Hz^{1/2}$ W^{-1} were obtained for unoptimized commercial samples operated in this pulsing infrared detection mode.

1. Spectral Response Analysis and Interference Patterns. The spectral response measurements on these pulse mode infrared detectors have clearly shown the multiple internal reflection effects inside the detector material leading to Fabry–Perot intensity patterns (*97*). An extremely low-background liquid-helium-cooled custom monochromator was used to measure the spectral response by counting the pulse rates at different wavelengths. The difference of IR-on and IR-off pulse rates (detector response) plotted against the wavelength is shown in Fig. 31, which is qualitatively similar to the response of other extrinsic silicon detectors (*78*) except for the inter-

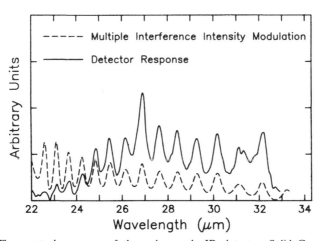

FIG. 31. The spectral response of the pulse mode IR detector. Solid Curve: detector response using an 800-K black-body source with a correction for the transmission of the order sorting filter. Dashed curve: Calculated multiple internal intensity modulation for a detector thickness of 148 μm and empirically determined transmission coefficients in the range 0.1–0.4. The extension of the long wave cutoff due to the high fields (0.3 V/μm in this case) beyond the low field cutoff of 27.3 μm is shown clearly. After Ref. (*68*).

ference pattern. A multiple reflection analysis (taking the absorption of all three regions of the detector and the total reflection at the backside of the structure into account) showed that the detector thickness is very sensitive to the peak positions of the intensity peaks.

This effect might be useful in optical filter development. For example, utilizing high-purity wafers with partially reflective metal layers on both sides, one would expect Fabry–Perot-like interference patterns with sharp maxima in the transmitted radiation. By selecting pairs (or more) of wafers with the correct thickness one could make IR filters. Thus, similar materials and similar technology could be used to construct several optical components operating in the same wavelength range.

C. INTERPULSE TIME SERIES ANALYSIS (NONLINEAR DYNAMICS)

Since the origin of the pulse is related to the space-charge buildup as discussed before, the reset should be connected to the neutralization of the space charge, making the IPTIs dependent on the space-charge neutralization. This leads to different IPTI patterns for different device and circuit parameters and with incident radiation. Nonlinear dynamics have captured the attention of researchers in vastly different areas from anthropology (or arts) to zoology. Some specific examples are the population variation of predator-prey systems, epidemic cycles, weather patterns, stock market performance, landscape design, etc. Here we analyze the IPTIs using nonlinear dynamic tools and briefly describe an application of nonlinear dynamical tools in an infrared detector application using the homojunction structure described before.

1. Nonlinear Dynamics under Constant Bias Voltages. Analyzing up to 6144 IPTIs for a constant voltage-driven (*98*) p^+-n-n^+ structure with nearly periodic patterns, and with period varying from 2 to 9, were found as shown in Fig. 32 for 12.35 V, 12.00 V, 11.91 V and 11.55 V. For 12.35 V bias, the IPTI modulation (frequency modulation) shows a strong period-2 pattern (see Fig. 27) which parallels the large-small (LSLSLS...) amplitude modulation patterns observed in the Belousov–Zhabotinski (BZ) reaction (*99*) with the variation of a dc parameter, the flow rate in the BZ work being parallel to the variation in dc bias in the p^+-n-n^+ structure.

Power spectra obtained from t_r as a function of r are shown in Fig. 32, where the horizontal scale corresponds to frequency f as a fraction of f_{max}. The power spectra show peaks at $f = p/q$ (in units of f_{max} where

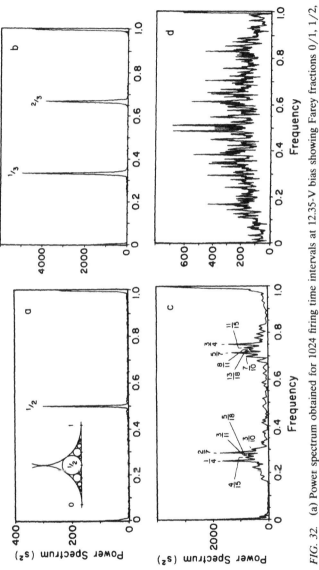

FIG. 32. (a) Power spectrum obtained for 1024 firing time intervals at 12.35-V bias showing Farey fractions 0/1, 1/2, 1/1. Unity on the horizontal frequency scale corresponds to $f_{max} = 1/2$ or period 2. The inset shows the construction of Ford circles. (b) Power spectrum obtained at 12.00-V bias showing fractions at 0/1, 1/3, 2/3, 1/1. The 1/3 and 2/3 components are indicative of period 6. (c) Spectrum at 11.91 V showing peaks at 0/1, 1/4, 4/15, 3/11, 5/18, 2/7, 3/10, 7/10, 5/7, 13/18, 8/11, 11/15, 3/4, 1/1. (d) Power spectrum obtained from 6144 firing time intervals at a bias voltage of 11.55 V. The inset shows the Ford circles showing a 2D connectedness. Identification of 91 peaks at Farey fraction frequencies with order 20 is given in Fig. 33. After Ref. (94).

p/q represents a fraction with p and q being positive, coprime integers). The set F_N of all such fractions with $0 \le p \le q \le N$ is called a Farey sequence of order N (100, 101). When the sequence F_N is ordered by magnitude, consecutive fractions p/q and k/l satisfy the unimodularity condition: $pl - qk = 1$. Here unimodularity is associated with adjacency in a one-dimensional array (102). Each set F_N is symmetric under $f \to 1 - f$. Hence this symmetry is apparent in the power spectra, while it is not apparent in other experimental work (103, 104).

The mediant $(p + k)/(q + l)$ of consecutive Farey fractions is a Farey fraction belonging to a higher order Farey sequence. The structure of a Farey sequence of order 20 is shown in Fig. 33. The lines between fractions indicate unimodular relationships, and pairs of lines descending onto a fraction indicate the parentage of that fraction via the mediant operation.

Another way to illustrate Farey fractions is with Ford circles (100). Each Farey fraction p/q corresponds to a Ford circle of radius $1/2q^2$ centered at $x = p/q$, $y = 1/2q^2$. Ford circles associated with p_1/q_1 and p_2/q_2 are tangent (see the inset in Fig. 32a) if and only if $p_1q_2 - p_2q_1 = 1$. Lines between the centers of adjacent Ford circles are in one-to-one correspondence with lines in the graph in Fig. 33. Unimodularity is the algebraic equivalent of Ford-circle tangents. Thus, the Ford circles construct Farey fractions with a plane geometry nearest-neighbor (adjacency) concept and

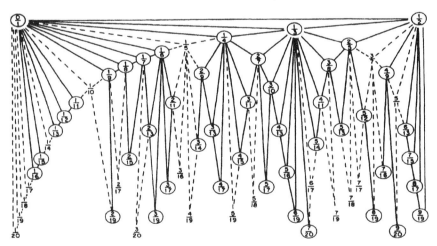

FIG. 33. Farey fractions of order 20 with $f \le 1/2$. Fractions corresponding to peaks in Fig. 32 are circled. Because of the $f \to 1 - f$ symmetry, only fractions on the intervals $(0, 1/2)$ are displayed. After Ref. (94).

a connectedness like the rule for connectedness of sets of stones on a Go board.

a. Nonlinear Dynamics and IR Detection   The detector output, which consists of pulses as described before, can be characterized by a time series t_r with $r = 1, 2, 3, \ldots$, where t_r represents the time intervals between successive pulses. The response to incident IR flux consists of changes in the rate and temporal pattern of pulsing. By adjusting circuit parameters, the pulsing patterns could be locked in to certain periodic patterns under IR illumination, which is termed "mode locking" (*90*). Within the mode locked regime, the pulse rate remains sensitive to changes in IR intensity, and fluctuations in the pulse rate are small. Since the electrical background noise in these pulsing structures is negligibly small compared to the size of the pulses, the relevant noise is contained within the fluctuations in the IPTIs. However, not all of the IPTI fluctuations reflect noise. Recent work on nonlinear dynamics indicates that fluctuations in nonlinear systems can have a deterministic relationship to the parameters of the nonlinear system (*105*). This suggests that for the devices under consideration, a sizable component of the fluctuations in the temporal pattern could be deterministically related to parameters like IR intensity and bias voltage, and unrelated to true performance-limiting noise.

The distinction between deterministic fluctuations and true noise, together with recently achieved understanding of nonlinear systems, could lead to significant advances in IR detector performance. Mathematical models of nonlinear systems provide some guidance and understanding in this area. Many of the models are formulated in terms of iterative maps. A spiketrain can be regarded as a physical representation of a time series $t_1, t_2, t_3, \ldots$. An illustrative example for this pulsing structure is

$$t_{r+1} = f(t_r, \mu), \tag{41}$$

where f is a nonlinear function which relates successive terms in a time series t_r with $r = 1, 2, 3, \ldots$, and μ indicates circuit and device parameters. The explicit form of f is associated with the dynamical system (real or artificial neuron, etc.). These parameters can dramatically affect the temporal pattern of the time series as seen in Fig. 32.

For the system described here, two important parameters are IR intensity and bias voltage. Because the output of these injection-model (*91*) IR detectors is in the form of a time series, it is natural to apply the insight gained from iterative maps, which display a variety of behavior depending on the parameters. In particular, they can generate time series which range from periodic to chaotic even though the maps are deterministic. They thus provide a mathematical example of how fluctuations in time series might not represent noise. If time-series information including deterministic fluctuations were fully utilized, and one had sufficient knowledge about f, one could pin down the parameters of f quite precisely. One can think of this as decoding the time series to obtain information about the parameters of the iterative mapping. The decoding probem is in some sense the inverse of the mapping problem. Since IR intensity is a sought-after parameter, and bias voltage is a known adjustable parameter, one can regard IR detection as a special type of decoding, or inverse mapping problem. In a region of parameter space where mapping generates a chaotic time series the decoding problem is difficult, whereas in a periodic regime the decoding problem is simple. If an injection mode detector output is periodic with period P, i.e., $t_r = t_{r+P}$ for a range of IR intensities, then decoding simply involves empirical determination of the relationship between the IR intensity and P successive time intervals.

In between the extremes of periodicity and chaos lie regimes of varying complexity. Small fluctuations are possible, as are larger fluctuations and intermittency (106). It is an especially interesting result of work on iterative maps that even intermittency can be understood as deterministic behavior (106). From a detector point of view, it seems most convenient to work as closely as possible to the simple periodic regime. Fortunately, work on iterative maps leads one to anticipate the existence of nearly periodic or mode-locked regions of parameter space. Mode locking within certain ranges of the system parameters is a strong indication of deterministic behavior in nonlinear systems. The existence of narrow peaks in the injection-mode-detector power spectra (see Figs. 34 and 35) supports the view that injection mode detectors are deterministic systems. The power spectra are associated with Fourier transforms over the time series index r. Period P is associated with peaks at a frequency $1/P$ and harmonics thereof.

In Figs. 34 and 35, the detector is passing through mode-locked regimes as a function of IR source temperature, with a fixed bias voltage. The large, narrow peaks indicate closeness to the ideal situation of perfectly periodic, noise-free output. For a range of source temperatures, the detectors lock onto nearly periodic modes of pulsing. Figure 36 shows return maps, which also indicate the nearly periodic behavior and display the fluctuations. From Fig. 36, it can be seen that the 107-K data is more nearly periodic than the 102-K data. Similar mode-locking regions are observable as a function of bias voltage (see Fig. 32).

Based on the variations observed for a 5-K IR source, and using an eigenvalue analysis on the periodic IPTI data sets, resulting in a weighted

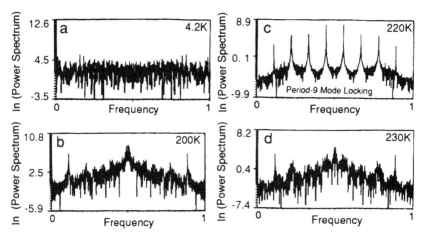

FIG. 34. Power spectra obtained from 1024 measured interpulse time intervals at 9.138-V bias for different black-body source temperatures: (a) 4.2 K, (b) 200 K, (c) 220 K, (d) 230 K. The vertical axis has a logarithmic scale. The frequency variable on the horizontal axis is the inverse of the time series periodicity. After Ref. (*90*).

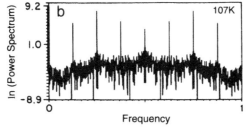

FIG. 35. Power spectra obtained from 1024 interpulse time intervals at 8.886 V for two different black-body temperatures. (a) 102 K, (b) 107 K. After Ref. (*90*).

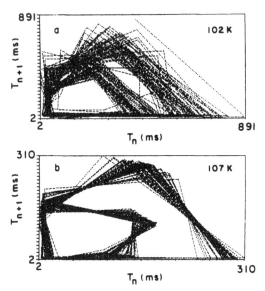

FIG. 36. Return maps for 500 measured interpulse time intervals at (a) 102 K, (b) 107 K for constant bias voltage, for the data shown in Fig. 35. Even though the 2 power spectra are similar the return maps indicate a better resolution for the mode-locking phenomena. After Ref. (*90*).

averaging technique, Coon and Perera calculated (*90*) a minimum resolvable temperature of 0.01 K. The corresponding unweighted averaging resulted in a value of 0.03 K for the same data sets. This technique shows the application of nonlinear dynamic techniques to improve detector performance.

2. Nonlinear Dynamics under Constant Bias Currents. In section IV.A.1, we described a model to explain the spontaneous pulsing in a p^+-n-n^+ homojunction under a constant current source. Here we show some nonlinear behavior observed under current driven situations. Most of the results shown are from a single diode in the circuit shown in Fig. 26 driven under different constant current values.

The data consisted of sets of 2100 IPTIs taken for drive currents between 25 and 72 nA at 1- or 2-nA intervals. The patterns formed by the IPTIs fell into 3 general groups. Examples of these groups with successive IPTIs are shown in Fig. 37. For driving currents less than $\sim$ 33 nA, the IPTIs take on all values between their extremes. In Fig. 37b, 100 IPTIs for

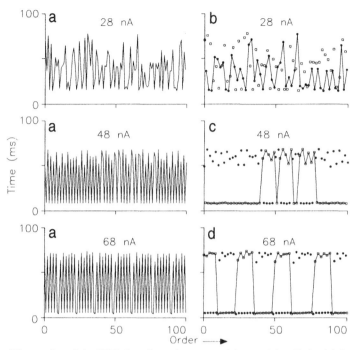

FIG. 37. Time series of the IPTI data for a dc current driven pulsing diode. (a) for 28-nA, 48-nA and 68-nA driving current (b), (c) and (d) up to order 100, showing odd (□) and even (•) points plotted with different symbols and connected with straight and dotted lines respectively. After Ref. (*93*).

a 28-nA driving current are shown with every second point connected. The two groups of points (labeled odd and even) show a similar pattern and give no clear indication of periodicity. Above 33 nA, the presence of a forbidden region in the IPTI distribution allows a definition of the IPTIs as short (S) or long (L) according to whether they are longer or shorter than the forbidden values. The most striking feature of this segment is the strong period 2 component in the series. This is shown in Fig. 37c and d where there is a strong preference for a given IPTI to be of the same type as the IPTI two intervals before it. The period 2 nature is broken intermittently by an extra pulse producing two successive pulses of the same type. Below 50 nA, these are almost exclusively LL groupings that occur at what appear to be random intervals. For 50 nA and above, the extra pulse is always part of an SS pattern. These SS groupings also occur much more periodically than do the LL patterns at lower currents. At

68 nA, the intermittency has developed into a fairly regular pattern of pulses occurring every six or seven LS patterns.

Figure 38 shows the return maps for the first 2048 IPTIs shown in Fig. 37. At a 28-nA driving current, the pattern fills the space almost completely inside a well-defined area, though even here some structure is visible. In particular, the gap (shown with an arrow) is a region where an IPTI can occur only if the previous IPTI is near the minimum; this is an indication of the pulse separation into long and short interval groups, and shows that the pattern is not random. At 48 nA, the spread is reduced and the points are moved toward the boundary giving well defined groups. This can be interpreted as three types of IPTI patterns, one which has a strong period 2 firing pattern, a noisy period 1 pattern caused by an LL pattern, and a single point near the origin, which indicates an SS pattern. As the driving current is increased to 68 nA, the group which gives the noisy period 1 pattern disappears and the spread (and modulation) of the linear period 2 pattern gets reduced. At the same time, the number of SS patterns increases forming a period 1 pattern that is much less noisy. This reduction of the spread gives rise to a finer grouping in the linear segment near the Y axis associated with the fact that the second interval in an SS pattern is shorter than the average for S intervals. The number of LSL (845) and SSL patterns (119) indicates that the extra S IPTI occurs after an average of seven LS patterns or every 14 pulses. This growth in the period 1 pattern at 68 nA indicates that the single SS at 48 nA is not random, but may be due to a higher-order periodic pattern. However, at 48 nA there is no apparent shortening of the L following the SS pattern, so that this pattern may have different causes at the two currents.

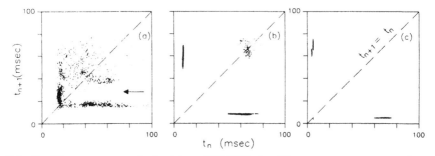

FIG. 38.   The return maps obtained from 2048 IPTIs for different constant current values. (a) Data for 28 nA showing a pattern which fills the space almost completely with a gap. (b) Data for 48 nA shows the LS pattern with some LL and a single SS pattern. (c) Data for 68 nA showing the LS pattern with very sharp SS pattern. Even the SL pattern indicates some structure. After Ref. (93).

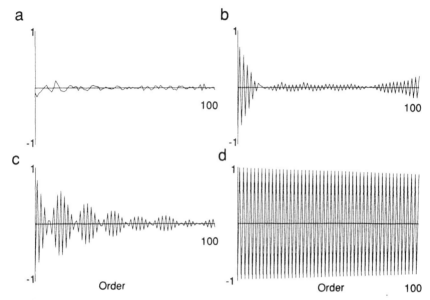

FIG. 39. Autocorrelation function from order 1 to 100 for the time series at (a) 28 nA, (b) 48 nA, (c) 68 nA driving current. The period 14 nature of the envelope in (c) shows the period 14 nature of the extra S IPTI. (d) Autocorrelation function for the time series in (c) modified by dropping the extra short interval at odd/even interchanges to show the basic period 2 nature of the system. After Ref. (*93*).

The autocorrelation (see Fig. 39) also shows the trends toward periodic patterns as the current increases. At 28 nA, no significant correlation is seen. As the current increases, a number of distinctive features appear, the first being an alternating correlation and anticorrelation that indicate the presence of a strong period 2 component in the time series. On top of this period 2 oscillation there is an envelope that modulates the function, which is determined by the pattern of intermittencies (*106*). At 48 nA, this envelope decays rapidly indicating the LL patterns occur at an average interval of 14 pulses but with significant deviations. At 68 nA, the envelope has developed a periodic pattern indicating a periodicity of ~ 14 for the SS patterns.

D. NEURON EMULATION WITH SEMICONDUCTORS

The spontaneous pulsing observed in p^+-n-n^+ structures has led us to compare them to action potentials in biological neurons (*94, 107, 108*). The remarkable similarity has opened new research areas in artificial neural networks where parallel asynchronous multispectral image (signal) processors based on biological vision systems could be achieved (*109*). Since semiconductors, most notiably silicon, form the basis of modern computers, it is hard to avoid the idea of invoking the physical phenomena discussed here as a basis for new types of computers or new strategies for information processing, which would parallel actual intelligent systems at the electrical circuit level. The idea would be to form semiconductor implementations of neural networks in which the primary active circuit components are in one-to-one correspondence with neurons. This is quite distinct from the notion of software/digital computer modeling of neural networks and from the notion of building complex circuits that simulate individual neurons (*110, 111*) or models of neurons (*112*). Neural networks in intelligent systems in nature have more neurons (10^{11} or more) than the number of transistors in a supercomputer ($\sim 10^{10}$), which suggests that fabrication of intelligent system networks would require individual neuron simulators to be simple. That is, for a very large system simple components are needed. Therefore, it would be desirable to simulate neuron electrical processes as closely as possible with analogous microelectronic processes. Thus, the approach would go well beyond the use of hardware instead of software. The approach would rely on adopting new hardware. Ideally, the strategy would be to establish a one-to-one correspondence between neurons and neuron-like semiconductor devices. From a standpoint of microelectronic design, the concepts described here are radical in the sense that the ubiquitous transistor might play no role or possibly only a peripheral role in neuron-like networks based on the aforementioned pulsing phenomena. A transistor simply does not have neuron-like electrical characteristics and, on the other hand, the p^+-n-n^+ diodes, under suitable operating conditions, exhibit the appropriate transient behavior and are also intrinsically fast devices. In their normal use as fiber-optic receivers, response times down to 100 ps are possible. The interest in speed is related to the observation that it might be possible to fabricate neuron-like semiconductor networks that are much faster than neural networks in organisms. Characteristic times for neuron transients' associated ionic processes are of the order of 0.1 ms or more, so that a 100-ns electronic simulation would represent a factor of 1,000 improvement in speed. While silicon technology has reached high device densities, even supercomputers are considerably less complex than neural networks in intelligent organisms. Thus, the most powerful neuron-like semiconductor networks which one could reasonably envision might be orders of magnitude less complex, but also orders of magnitude faster, than neural networks in intelligent organisms. Such networks could form part of

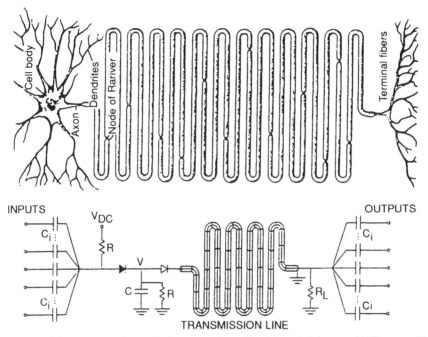

FIG. 40. (a) Features of a typical neuron from Kandel and Schwartz (*113*) page 15 (reproduced by permission of Elsevier, N.Y.), and (b) the artificial neuron, which exhibits the summation over synaptic inputs and fan-out. The input and output capacitive couplings are useful in conjunction with spiketrains. The darkened diode is a p–n junction device used for pulse-height discrimination. The other diode is a p^+–n–n^+ diode used for spiketrain generation. After Ref. (*114*).

large-scale computers or part of specialized systems such as image processing systems which would parallel animal vision systems. The possibility of fast neuron-like networks suggests artificial intelligence applications with the capabilities of a small brain operating at 1000 times normal speed.

1. Similarity with Biological Neurons. The development of this hardware basis for the fabrication of neural networks could be pursued through exploratory work on circuits which exhibit basic features of biological neural networks. Figure 40 shows a typical biological neuron (*113*) and the artificial neuron proposed by Coon and Perera (*114*). Single p^+–n–n^+ diode circuits have been shown to be similar to Hodgkin–Huxley equivalent circuits (*95*). Some of the other features already shown (*107, 108*) with pulsing artificial neurons include action potentials, refractory periods, excitation, inhibition, summation over synaptic inputs, synaptic weights, temporal integration, memory, network connectivity modification based on experience, pacemaker activity, firing thresholds, coupling to sensors with graded signal outputs and the dependence of firing rate on input current. In addition, impedance-matched 50-Ω circuits have been studied (107) to perform logic operations using these p^+-n-n^+ structures. The input signals were added as "AND" gates or "OR" gates and by introducing a dc resting level, the circuit was converted to an "OR" from an "AND."

These homojunction pulsing structures were used in circuits to detect multispectral transient signals (116). The outputs show responses similar to the photo receptors in the lateral eye of a (*Limulus*) horseshoe crab (115) as seen in Fig. 41. It also emulates the behavior of some units in a neural model (117) of the fly visual system, and some components in a motion-tracking model (118) inspired by the fly visual system. A model based on simple circuit theory can explain the outputs, giving important insight into design and optimization. One practical advantage of the spiketrain signals is that the instantaneous power can be very high, while the average power is still kept at a very low level. Whereas the time intervals or the frequency of pulsing represent analog information, the pulse itself can be considered a digital signal since the height remains the same. This mixed character (analog–digital) in biological neural systems was recognized by von Neumann (119). Recently both digital and analog implementation (120) of neural networks have taken place but very few systems with mixed character have been considered. One advantage of the partly digital character of real neural systems is that they can implement pulse-height discrimination in order to get immunity to noise. The pulse-height immunity to noise is evident in biological systems too. The best example is the disease myasenthia gravis (113).

2. Parallel Asynchronous Processing. It has been demonstrated that simple circuits containing only one p^+-n-n^+ diode can code low-level currents from any source into neuron-like spiketrains consisting of fast (< 100-ns risetime) pulses with typical pulse heights (114) (see Fig. 42). Currents (including photocurrents or integrated output from other network elements) down to about 500 fA have been converted into spiketrains with noise (frequency fluctuations), referred to an input, of about 10 fA. These observations were made on millimeter-size devices, and it is expected that the current sensitivity and noise current should scale down with the area and (area)$^{1/2}$, respectively. This scaling would clearly lead to extraordinar-

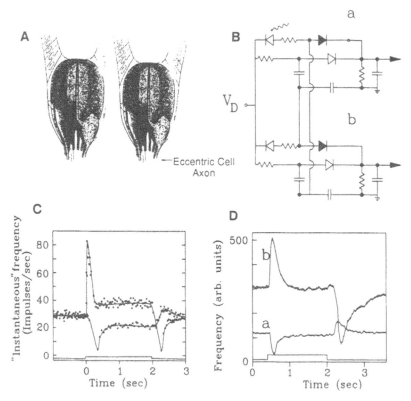

FIG. 41. The schematic of the ommatidium of the *Limulus* (horseshoe crab) (A) showing the eccentric cell axon (*115*) in the ommatidium and its response (C) of adjacent axons, to a step light input to one ommatidium and a common background to the other. The two-channel lateral inhibition transient sensing circuit (B) and its experimental response (D) to inputs identical to (C). (A) and (C) after Ref. (*115*) and (B) and (D) after Ref. (*116*).

ily interesting performance. These spontaneous pulses are of the order of a volt [or higher up to 40 V (*68*)]; there is no need of a preamplifier. The absence of preamplifiers and the electronic simplicity open the way to *much higher parallel processing densities than is possible with other approaches* (*109, 114, 121*). The high dynamic range, optoelectronic capability, and some additional features (like two-dimensional parallel input-output) make this approach especially well suited to sensor–processor applications and hybrid optoelectronic processor applications.

These devices could form the hardware basis for a parallel asynchronous processor in much the same way that transistors form the basis for digital

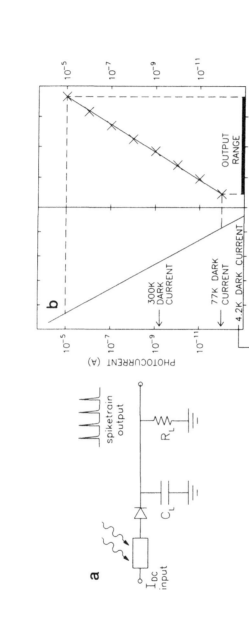

FIG. 42. (a) Interfacing of photodetectors to the input stage of an artificial neural network, which performs spiketrain coding of input current or photocurrent. (b) Experimental output firing rate vs. input current for a particular p^+-n-n^+ diode with a dynamic range of 10^7. The circuit and the pulse-rate curve modified after Ref. (58).

computers. Pulse propagation through silicon chips (parallel fire-through as in Fig. 43) as opposed to the lateral planar propagation in conventional integrated circuits has been proposed (121). This approach could utilize laminar, stacked-wafer architectures (114). Such architectures could eliminate the serial processing limitations of standard processors, which utilize multiplexing and charge transfer.

The power consumption of large-scale analog (112) and digital (122) systems is always a major concern. For example, the power consumption of the CRAY XMP-48 is of the order of 300 kW. For p^+-n-n^+ diodes described here, neuron-like action potential pulses with energy dissipation down to 4 pJ/mm^2/pulse and a quiescent power dissipation of 10 pW/mm^2 are feasible (95). Considering the thermodynamic efficiency of cooling, these numbers correspond to 290 pJ/mm^2/pulse and 710 pW/mm^2 at room temperature. A system with 10^{11} 10×10-μm active elements [comparable to the number of neurons in the brain (113)] all firing with an average pulse rate of 1 kHz [corresponding to a high neuronal firing rate (123)] would have a comparable (114) power dissipation to the human brain (119, 123).

A proposed architecture that has similarities to biological vision systems is depicted in Fig. 44. The primate retina has the same parallel architecture as seen in Fig. 44. An iterative map of the form of Eq. (41) was obtained (126) by an extension of the device model for spontaneous

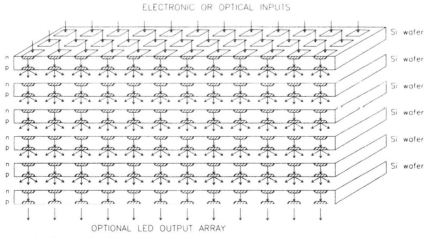

FIG. 43. Schematic of the signal flow pattern through a real time parallel asynchronous processor consisting of a stacked silicon wafers. Modified after Ref. (121).

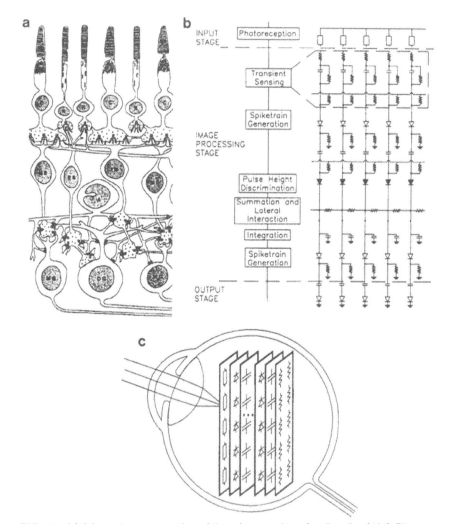

FIG. 44. (a) Schematic representation of the primate retina, after Dowling (*124*). The upper part of the diagram shows photoreceptor rods and cones. Synoptic contacts are made with horizontal cells, along which impulses spread to adjoining parts of the retina, and with centripetal bipolar cells, along which they pass to the inner plexiform layer where synaptic contacts are established with ganglion cells (*125*). (b) The parallel processor circuit, showing different stages and their neuron equivalent functional capabilities. The components inside the dashed area represent the transient sensing circuit components. Most of these functional capabilities have been tested individually, and the full processor system needs to be studied in order to understand the collective behavior. (c) A futuristic processor in an eye illustrating the role it could play in an artificial vision system. (a) after Ref. (*124*) and (b) after Ref. (*126*).

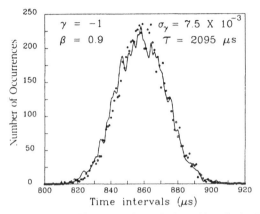

FIG. 45. Interpulse time interval histograms for a single pulsing diode. The circles indicate the experimental points and the line is from the iterative map model. The values for parameters used are also given. The space-charge noise in γ is denoted by σ_γ. After Ref. (*126*).

pulsing as

$$t_{r+1} = f(t_r) = \left(1 - [1 - t_r]^{-\beta} e^\gamma\right)$$

where β is the feedback gain and γ is the dynamical space charge variation noise factor. The time interval from the model is a dimensionless ratio of the real time interval between spikes and a factor τ. The parameters depend on the inputs and device parameters. The results obtained using this model for a single pulsing diode are shown in Fig. 45 as a fit to the experimental data obtained. Allowing inputs as part of the control parameters the iterative map given in Eq. (41) could be extended to give

$$t_{r+1} = f(t_r, \{\text{device parameters}, \langle \text{inputs} \rangle\}). \tag{42}$$

This system, which includes summed and temporally integrated inputs ($\langle \text{inputs} \rangle$), could convert spiketrain inputs (as well as graded signal inputs) from other neurons into an analog signal that can exercise parametric control over the generation of a new time series in a network situation. Using this iterative map, and considering a single channel of the proposed parallel processor as two homojunction pulsing units connected with a filter circuit as in Fig. 46, the input/output of the channel was monitored. This allows one to see some nonlinear behavior for some device and control parameters.

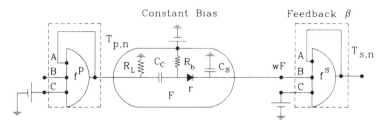

FIG. 46. The circuit diagram used in both the experiment and the model, showing the coupling of two pulsing diodes. The dashed areas represent the pulsing diodes in the experimental circuit. B and C terminals refer to the transient and constant inputs to the circuit. The transfer functions of the secondary and primary pulsing diodes are denoted by f^s and f^p. The transfer function of the filter circuit used in the coupling is F and wF is weighted output of the filter. The filter values are $R_L = 310$ KΩ, $C_c = 0.01$ μF, $R_b = 150$ KΩ, and $C_S = 0.07$ μF. After Ref. (126).

The network output is extremely sensitive to changes in the γ/λ ratio, which is equivalent to the diode bias. This is illustrated in Fig. 47, where a small change in the ratio (in absence of noise) results in the stable fixed point being converted to a limit cycle (Hopf bifurcation). As the control parameter is decreased, the limit cycle grows until only the unstable fixed point at its center is visible in Fig. 47(d). Dynamic noise in a diode can be a stabilizing factor and may produce a limit cycle. The two-diode experimental circuit gives a rich spectrum of nonlinear behavior depending on the driving parameters. For example under a constant subthreshold bias (to C terminal in Fig. 46) and a transient input B [the pulse output t_p through a filter, equivalent to a weighted (w) transfer function F in the model, the 2-D return map shows a stable fixed point as indicated in Fig. 48b. A stable fixed point is seen for the experimental data at a constant bias of 4.7 V in Fig. 48b. As the constant bias is increased to 4.8 V, a Hopf bifurcation (127) occurs producing the limit cycle in Fig. 48a. A similar effect is seen for the model results in Fig. 48c and d.

V. Summary and Future

The evolution and successful integration of any device into practical applications require several prerequisites. In addition to the technical details, the cost and the ease of use with the existing technologies and integrating several components into an integrated circuit (IC) will play a major part. The interest in infrared detector techniques other than HgCdTe

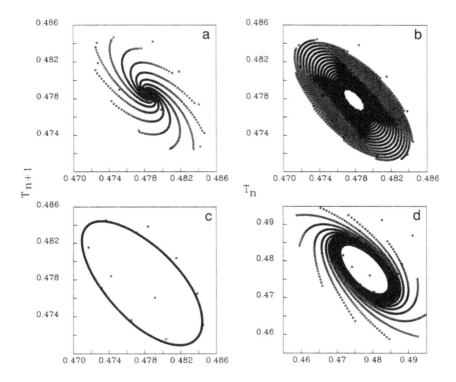

FIG. 47. Model output for γ/λ = (a) 8.18, (b) 8.159, (c) 8.1567, and (d) 8.14. We can observe a Hopf bifurcation from (a) to (c). At (d) the limit cycle has grown so that only the unstable fixed point is visible. The successive points do not move along the spirals seen but instead jump between the branches (skipping three branches) around the pattern moving steadily inward or outward. The other parameters are $\gamma_S = -1$, $\gamma_p = -1$, $\beta_p = 0.125$, $t_{n,1} = 0.5$. After Ref. *(126)*.

is due mainly to these other factors. The photoemissive homojunction IR detectors discussed here will be promising candidates for focal plane array applications which require the high degree of pixel-to-pixel uniformity obtainable in Si. They also offer wavelength tunability in MWIR, LWIR, VLWIR and even longer FIR regions for astronomical applications and the optoelectronic integration of the components such as shift registers. Although the present versions of these detectors have low quantum efficiencies, which are shared by other (high-uniformity) IR detectors such as Schottky barrier detectors, development of more sensitive multilayer

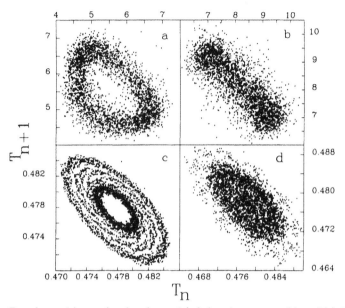

FIG. 48. Experimental interpulse time interval (t_n) data for constant bias of (a) 4.8 V and (b) 4.7 V. The model output from the circuit in Fig. 46, with (c) $\gamma_S/\lambda = 8.1567$ and (d) $\gamma_S/\lambda = 8.17$. We observe a Hopf bifurcation from a stable fixed point (b, d) to a limit cycle (a, c). The common parameters are $\beta_p = 0.125$, $\gamma_S = \gamma_p = -1$, the noise $\sigma_S^\gamma = \sigma_p^\gamma = 10^{-3}$ and the coupling $w = 10^{-4}$. After Ref. (126).

detectors could remedy this situation. The advances in epitaxial fabrication techniques such as MBE and MOCVD could easily achieve the multilayer structures. Using more layers to increase the performance has been successfully implemented in heterostructure multiquantum well detectors. Low-cost, monolithic silicon focal planes with broad-band IR sensitivity could have an enormous impact on imaging applications.

The research on pulsing-junction structures would lead to new hardware components for electronic and hybrid optoelectronic neural networks. These networks would have lower power dissipation, greater electronic simplicity, and greater similarity with biological information coding than any existing adaptive network technology. The approach would provide an ideal basis for silicon device implementation of neural-network and cognitive-science models. The hardware would represent a uniquely advantageous alternative or supplement to existing device technology. One of the major limitations of this p^+-n-n^+ pulsing approach is the low operating temperatures. This is a serious limitation for many applications but probably not for (1) IR image processing, where processing at or near a cooled focal plane detector array would require cooled operation, and (2) large-scale information processing, where work on fundamental limits has already led to the consideration of low-temperature operation because of advantages related to speed, power dissipation and reliability (*122*). In any case, it is possible that higher-temperature operation could be achieved by employing deeper impurity levels or by exploiting the flexibility associated with quantum-well barriers and modulated doping in semiconductor heterostructures. Deeper impurity levels possess smaller impact-ionization cross-sections, but they might also permit higher densities of localized states and smaller devices, as well as higher-temperature operation. In addition to the technological advantages of pulsing structures, they also provide a deeper insight into the physics of these types of devices. Previous work (*93*) has already investigated some of the connections between the underlying physics of the devices and nonlinear dynamics.

Acknowledgments

Some of the work described here was supported in part by the U.S. NSF under Contract ECS-9412248. The author acknowledges the contributions made by students and colleagues while the author was in Pittsburgh and in Georgia, especially S. R. Betarbet, J.-W. Choe, M. H. Francombe, Y.-J. Ko, S. G. Matsik, and H. X. Yuan. In addition, the author thanks J. H. Hadley, Jr., for his careful review of the manuscript and helpful suggestions.

References

1. E. T. Young, M. Scutero, G. Riecke, T. Milner, F. J. Low, P. Hubbard and J. Davis, *Proc. SPIE* **1684**, 63 (1992).

2. A. G. U. Perera, R. E. Sherriff, M. H. Francombe and R. P. Devaty, *Appl. Phys. Lett.* **60**, 3168 (1992).

3. I. C. Wu and E. E. Haller, *in* "Photo-Induced Space Charge Effects in Semiconductors: Electro-Optics, Photoconductivity and the Photorefractive Effect," Materials Research Society, 1992, p. 175.

4. S. B. Stetson, D. B. Reynolds, M. G. Stapelbroek and R. L. Stermer, "Infrared Detectors, Sensors and Focal Plane Arrays," *Proc. SPIE* **686**, 48 (1986).

5. F. J. Low, *Proc. SPIE* **1684**, 168 (1992).

6. J. Huffman, private communication.

7. H. C. Liu, J.-P. Noel, Lujian Li, M. Buchanan and J. G. Simmons, *Appl. Phys. Lett.* **60**, 3298 (1992).

8. P. R. Bratt, *in* "Semiconductors and Semimetals, Volume 12, Infrared Detectors II" (R. K. Willardson and A. C. Beer, eds.), Academic Press, London, 1977, p. 39.

9. N. Sclar, *in* Progress in Quantum Electronics 9, Pergamon, Elmsford, NY, 1984, p. 149.

10. D. D. Coon, *J. Vac. Sci. Technol. Rev. A* **8**, 2950 (1990).
11. B. F. Levine, *J. Appl. Phys.* **74**, R1 (1993).
12. K. M. S. V. Bandara, J.-W. Choe, M. H. Francombe, A. G. U. Perera and Y. F. Lin, *Appl. Phys. Lett.* **60**, 3022 (1992).
13. M. D. Petroff and M. G. Stapelbroek, U. S. Patent No. 4,568,960 (1986).
14. F. D. Shepherd, *Proc. SPIE* **1735**, 250 (1992).
15. W. F. Kosonocky, *Proc. SPIE* **1685**, 2 (1992).
16. T. L. Lin and J. Maserjian, *Appl. Phys. Lett.* **57**, 1422 (1990).
17. R. P. G. Karunasiri, J. S. Park and K. L. Wang, *Appl. Phys. Lett.* **59**, 2588 (1991).
18. D. D. Coon, R. P. Devaty, A. G. U. Perera and R. E. Sherriff, *Appl. Phys. Lett.* **55**, 1738 (1989).
19. N. M. Haegel, *J. Appl. Phys.* **64**, 2153 (1988).
20. Y. N. Yang, D. D. Coon and P. F. Shepard, *Appl. Phys. Lett.* **45**, 752 (1984).
21. P. Lenard, *Ann. Physik* **2**, 359 (1900).
22. A. Einstein, *Ann. Physik* **17**, 132 (1905).
23. A. G. U. Perera and S. Matsik, *Appl. Phys. Lett.* **64**, 878 (1994).
24. N. F. Mott, "Metal–Insulator Transitions," Barnes and Noble, New York, 1974.
25. A. G. U. Perera, H. X. Yuan, M. H. Francombe and J.-W. Choe, *in* "Proc. of the Electrochem. Soc., Sym. on LWIR Detectors and Arrays, New Orleans, LA, Oct. 11–15, 1993," Electrochemical Society, Pennington, New Jersey, 1995, 36–45.
26. A. G. U. Perera, J.-W. Choe, M. H. Francombe, R. E. Sherriff and R. P. Devaty, *Superlattices and Microstructures* **14**, 123 (1993).
27. J.-W. Choe, Byungsung O, K. M. S. V. Bandara and D. D. Coon, *Superlattices and Microstructures* **10**, 1 (1991).
28. T. F. Lee and T. C. McGill, *J. Appl. Phys.* **46**, 373 (1975).
29. G. F. Neumark, *J. Appl. Phys.* **48**, 3618 (1977).
30. S. C. Jain, R. P. Mertens and R. J. Van Overstraeten, *in* "Advances in Electronics and Electron Physics, Vol. 82," (P. W. Hawkes, ed.), Academic Press, Boston, 1991.
31. R. P. Mertens, R. J. van Overstraeten and H. J. de Man, *in* "Advances in Electronics and Electron Physics, Vol. 55" (L. Marton and C. Marton, ed.), Academic Press, Boston, 1981.
32. G. D. Mahan, *J. Appl. Phys.* **51**, 2634 (1980).
33. N. F. Mott, "Metal–Insulator Transition," Barnes and Noble, New York, 1974.
34. J. R. Klauder, *Ann. Phys.* **14**, 43 (1961).
35. J. Serre and A. Ghazali, *Phys. Rev. B* **28**, 4704 (1983).
36. J. R. Lowney, *J. Appl. Phys.* **59**, 2048 (1986).
37. A. G. U. Perera, H. X. Yuan and M. H. Francombe, *J. Appl. Phys.* **77**, 915 (1995).
38. J. S. Escher, *in* "Semiconductors and Semimetals, Vol. 15" (R. K. Willardson and A. C. Beer, eds.), Academic Press, Boston, 1981.
39. W. W. Anderson, *Solid-State Electron.* **18**, 235 (1975).
40. S. M. Sze, "Physics of Semiconductor Devices," 2nd edition, Wiley, New York, 1981.
41. F. Szmulowicz and F. L. Madarsz, *J. Appl. Phys.* **62**, 2533 (1987).
42. Hisashi Hara and Yoshio Nishi, *J. Phys. Soc. Japan* **21**, 1222 (1966).
43. M. Balkanski, A. Aziza and E. Amzallag, *Phys. Status Solidi* **31**, 323 (1969).
44. D. K. Schroder, R. N. Thomas and J. C. Swartz, *IEEE Trans. Electron Devices* **ED-25**, 254 (1978).
45. W. Spitzer and H. Y. Fan, *Physical Review* **108**, 268 (1957).
46. R. A. Smith, "Semiconductors," Cambridge, London, 1978.
47. J. M. Mooney, *J. Appl. Phys.* **65**, 2869 (1989).
48. C. N. Berglund and R. J. Powell, *J. Appl. Phys.* **42**, 573 (1971).

49. V. E. Vickers, *Appl. Opt.* **10**, 2190 (1971).

50. J. M. Mooney and J. Silverman, *IEEE Trans. Electron Devices* **ED-32**, 33 (1985).

51. R. Williams, *in* "Semiconductors and Semimetals, Vol. 6" (R. K. Willardson and A. C. Beer, eds.), Academic Press, 1970.

52. S. M. Sze, "Physics of Semiconductor Devices," Wiley Eastern, New York, 1981.

53. R. H. Fowler and L. W. Nordheim, *Roy. Soc. Proc. A* **119**, 173 (1928).

54. C. Y. Chang and S. M. Sze, *Solid-State Electron.* **13**, 727 (1970).

55. R. J. Keyes, Optical and infrared detectors, Topics in Applied Physics, Vol. 19, Springer-Verlag, Berlin, New York, 1980, p. 41.

56. J. M. Mooney and E. L. Dereniak, *Optical Engineering* **26**, 223 (1987).

57. A. G. U. Perera, R. E. Sherriff, M. H. Francombe and R. P. Devaty, *in* "Photo-Induced Space Charge Effects in Semiconductors," Materials Research Society, 1992, p. 119.

58. D. D. Coon and A. G. U. Perera, *Appl. Phys. Lett.* **55**, 478 (1989).

59. F. A. Riddoch and B. K. Ridley, *J. Phys. C: Solid State Phys.* **16**, 6971 (1983).

60. M. C. Tatham, J. F. Ryan and C. T. Foxon, *Phys. Rev. Lett.* **63**, 1637 (1989).

61. J.-W. Choe, A. G. U. Perera, M. H. Francombe and D. D. Coon, *Appl. Phys. Lett.* **59**, 54 (1991).

62. S. Tohyama, N. Teranishi, K. Konuma, M. Nishimura, K. Asai and E. Oda, *IEEE Trans. Electron Devices* **ED-38**, 1136 (1991).

63. Shigeru Tohyama *et al.*, *in* "IEDM Technical Digest," IEEE, 1988, p. 82.

64. D. D. Coon, R. P. Devaty, A. G. U. Perera and R. E. Sherriff, U.S. Patent No. 5,030,831 (1991).

65. R. J. van Overstraeten and R. P. Mertens, *Solid State Electronics* **30**, 1077 (1987).

66. J. Wagner and J. A. del Alamo, *J. Appl. Phys.* **63**, 425 (1988).

67. Kazuo Kikuchi, *Infrared Phys. Technol.* **35**, 33 (1994).

68. D. D. Coon and A. G. U. Perera, *Solid-State Electron.* **29**, 929 (1986).

69. E. E. Haller, *Infrared Phys.* **35**, 127 (1994).

70. D. M. Watson, M. T. Guptill, J. E. Huffman, T. N. Krabach, S. N. Raines and S. Satyapal, *J. Appl. Phys.* **74**, 4199 (1993).

71. D. M. Watson and J. E. Huffman, *Appl. Phys. Lett.* **52**, 1602 (1988).

72. I. C. Wu and E. E. Haller, *in* "Photo-Induced Space Charge Effects in Semiconductors," Materials Research Society, 1992, p. 175.

73. G. E. Stillman, C. M. Wolfe and J. O. Dimmock, *in* "Semiconductors and Semimetals, Volume 12, Infrared Detectors II," (R. K. Willardson and A. C. Beer, eds.), Academic Press, London, 1977, p. 169.

74. J. R. Banavar, D. D. Coon, and G. E. Derkits, *Phys. Rev. Lett.* **41**, 576 (1978).

75. J. R. Banavar, D. D. Coon, and G. E. Derkits, *Appl. Phys. Lett.* **34**, 94 (1979).

76. M. Lax, *Phys. Rev.* **119**, 1502 (1960).

77. S. Chaudhuri, D. D. Coon and G. E. Derkits, *Appl. Phys. Lett.* **37**, 111 (1980).

78. D. D. Coon, S. D. Gunapala, R. P. G. Karunasiri and H.-M. Muehlhoff, *Proc. SPIE* **430**, 144 (1983).

79. A. K. Ramdas and S. Rodriguez, *Rep. Prog. Phys.* **44**, 1297 (1981).

80. S. D. Gunapala, R. P. G. Karunasiri and D. D. Coon, *Solid State Commun.* **63**, 1165 (1987).

81. D. D. Coon, S. D. Gunapala, R. P. G. Karunasiri and H.-M. Muehlhoff, *Solid State Commun.* **53**, 1144 (1985).

82. A. G. U. Perera, S. R. Betarbet, Byungsung O, and D. D. Coon, *Phys. Rev. Lett.* **70**, 1052 (1993).

83. S. Chaudhuri, D. D. Coon, and G. E. Derkits, *Phys. Rev. Lett.* **45**, (1980).

84. J. M. Arias, S. H. Shin, J. G. Pasko, R. E. DeWames and E. R. Gertner, *J. Appl. Phys.* **65**, 1747 (1989).

85. A. Rogalski, *Infrared Phys.* **28**, 139 (1988).

86. Y. Nemirovsky, D. Rosenfeld, R. Adar and A. Kornfeld, *J. Vac. Sci. Technol. A* **7**, 528 (1990).

87. A. Rogalski, A. Jozwikowska, K. Jozwikowski and J. Rutkowski, *Infrared Phys.* **33**, 463 (1992).

88. J. M. Arias, J. G. Pasko, M. Zandin, S. H. Shin, G. M. Williams, L. O. Bubulac, R. E. DeWames and W. E. Tennant, *Appl. Phys. Lett.* **62**, 976 (1993).

89. D. D. Coon and A. G. U. Perera, *Appl. Phys. Lett.* **51**, 1711 (1987).

90. D. D. Coon and A. G. U. Perera, *Appl. Phys. Lett.* **51**, 1086 (1987).

91. A. G. U. Perera, Ph.D. thesis, "Injection Mode Devices (IMD) for Infrared Detection and Neural Network Emulation," University of Pittsburgh, 1987.

92. L. H. De Vaux, U.S. Patent No. 3,466,448 (1969).

93. A. G. U. Perera and S. Matsik, *Physica D*, **84**, 615 (1995).

94. D. D. Coon, S. N. Ma and A. G. U. Perera, *Phys. Rev. Lett.* **58**, 1139 (1987).

95. D. D. Coon and A. G. U. Perera, *Neural Networks* **2**, 143 (1989).

96. D. D. Coon and R. P. G. Karunasiri, *Solid-State Electron.* **26**, 1151 (1983).

97. D. D. Coon and H. C. Liu, *J. Appl. Phys.* **60**, 445 (1986).

98. D. D. Coon and H. C. Liu, *Superlattices and Microstructures* **3**, 95 (1987).

99. J. Maselko and H. L. Swinney, *Phys. Scr.* **T9**, 35 (1985).

100. H. Rademacher, *Lectures on Elementary Number Theory*, Krieger, Huntington, NY, 1977.

101. P. Bak, T. Bohr, and M. H. Jensen, *Physica Scripta* **T9**, 50 (1985).

102. D. L. Gonzalez and O. Piro, *Phys. Rev. Lett.* **50**, 870 (1983).

103. J. Maselko and H. L. Swinney, *Physica Scripta* **T9**, 35 (1985).

104. L. D. Harmon, *Kybernetik* **1**, 89 (1961).

105. R. M. May, *Nature* **216**, 459 (1976).

106. J. Hirsch, B. Huberman, and D. Scalapino, *Physical Review A* **25**, 519 (1982).

107. D. D. Coon and A. G. U. Perera, *Solid-State Electronics* **31**, 851 (1988).

108. D. D. Coon and A. G. U. Perera, *Int. J. Electronics* **63**, 61 (1987).

109. D. D. Coon and A. G. U. Perera, *Proc. SPIE* **1360**, 1620 (1990).

110. L. D. Harmon, *Science* **129**, 962 (1959).

111. L. D. Harmon, *Kybernetik* **1**, 89 (1961).

112. M. A. Sivilotti, M. R. Emerling and C. A. Mead, *in* "Neural Networks for Computing," Am. Inst. of Phys., New York, 1986, p. 408.

113. E. R. Kandel and J. H. Schwartz, "Principles of Neural Science," Elsevier, New York, 1985.

114. D. D. Coon and A. G. U. Perera, *in* "Neural Information Processing Systems," Am. Inst. of Phys. New York, 1988, p. 201.

115. F. Ratliff, H. K. Hartline and W. H. Miller, *J. Opt. Soc. of America* **53**, 110 (1963).

116. A. G. U. Perera, S. Betarbet and M. H. Francombe, *in* "World Congress on Neural Networks—Portland," Erlbaum, Hillsdale, NJ, 1993, p. IV 835.

117. H. Ogmen and S. Gagne, *Neural Networks* **3**, 487 (1990).

118. James M. Missler and Farhad A. Kamangar, *in* "World Congress on Neural Networks—Portland, Erlbaum," Hillsdale, NJ, 1993, p. I 62.

119. J. von Neumann, *in* "Collected Works Vol. 5" (A. H. Taub, ed.), Pergamon Press, New York, 1961.

120. "DARPA Neural Network Study," (Study Director: Bernard Widrow) AFCEA International Press, Fairfax, Virginia, 1988.

121. D. D. Coon and A. G. U. Perera, *Int. J. Infrared and Millimeter Waves* **7**, 1571 (1986).
122. R. W. Keyes, *Proc. IEEE* **63**, 740 (1975).
123. J. von Neumann, "The Computer and the Brain," Yale University Press, New Haven and London, 1958.
124. J. E. Dowling and B. B. Boycott, *Proc. Royal Soc. B* **166**, 80 (1967).
125. S. Polyak, The Vertebrate Visual System, Univ. of Chicago Press, Chicago, 1957.
126. A. G. U. Perera, S. R. Betarbet and S. G. Matsik, Bifurcations and Chaos in Pulsing Si Neurons, *in World Congress on Neural Networks-San Diego*, Erlbaum, Hillsdale, NJ, 1994, p. IV 704.
127. J. Hale and H. Koçak, "Dynamics and Bifurcations," Springer-Verlag, New York, 1991.

THIN FILMS, VOLUME 21

Progress of SiGe / Si Quantum Wells for Infrared Detection

R. P. Gamani Karunasiri

Department of Electrical Engineering, National University of Singapore, Singapore 0511

J. S. Park

Jet Propulsion Laboratory, Pasadena, California

AND

K. L. Wang

*Department of Electrical Engineering, University of California, Los Angeles,
Los Angeles, California*

I. Introduction

Intersubband transitions in quantum wells and superlattices have received considerable attention because of their potential use in infrared detection and imaging. The interesting features of these transitions, those that distinguish them from interband ones, include unique polarization selection rules and strong oscillator strength. Until recently, most of the research on this subject was carried out using III–V-based material systems. However, SiGe/Si systems, in addition to having the advantage of potential versatility of Si based technology, also exhibit intersubband transitions at normal incidence. The infrared absorption process in quantum wells is based on populating the lowest bound state in the well by either doping the well region, or doping the barrier region near the well. The photoexcitation occurs to a higher subband or to the continuum states above the barrier. Intersubband transition in the SiGe/Si system was first observed between heavy hole subbands using p-type multiple quantum wells (1). Recently, in addition to intersubband transition, transitions between different hole bands (for example, between the heavy and split-off hole subbands) were also observed (2). The origin of the intervalence band transition can be traced to the coupling of the conduction band with the valence band (2). These infrared transitions in the valence band of a SiGe/Si quantum well are illustrated schematically in the Fig. 1. One of the key advantages of the intervalence band transition is the possibility of normal incidence detection, which is forbidden in the case of the intersubband transition at the Γ-point (3). More recently, intersubband transitions in the conduction band have also been observed (4) including normal incidence absorption due to the off-diagonal effective masses for structures grown on substrates of specific orientations (5). In this chapter, experimental studies of infrared transitions in SiGe/Si quantum well structures will be reviewed. In addition to intersubband transitions, two

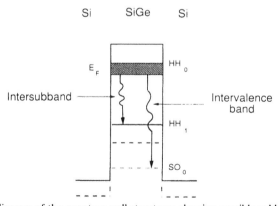

FIG. 1.   Band diagram of the quantum well structures showing possible subband transitions.

normal incident absorption processes, intervalence band transition and internal photoemission from two-dimensional hole gas in the quantum well, will also be discussed. Furthermore, we will illustrate the importance of many-body effects on the optical transitions due to heavy doping in the quantum well. Finally, the progress of application of these transitions for the fabrication of infrared detectors will be discussed. We will begin with a brief discussion on material issues that are important in the design of SiGe/Si quantum well structures, in particular, for infrared detector applications. The emphasis will be given to the valence band due to the large band offset compared to the conduction band when the structures are grown pseudomorphically on Si substrate.

II. Material Considerations

The important parameters in the design of p-type SiGe/Si multiple quantum well structures are the critical thickness, the band offset, the relative positions of the three hole bands (i.e., light, heavy, and split-off), and the respective effective masses. The critical thickness of a strained SiGe layer on Si is strongly dependent on the growth parameters, especially the substrate temperature. For a typical growth temperature ($\sim$ 500–600°C) used in the heteroepitaxy of SiGe on Si, the values of the critical thickness as a function of the Ge composition has been a subject of a great discussion and the model proposed by People and Bean seems to agree well with experimental data (6). For example, a SiGe layer with a 50% Ge composition has a critical thickness of about 100 Å. In the case of multiple-layer growth, the critical thickness is obtained using the average Ge composition of a single period of the structure (i.e., the average Ge content of a period, $x_{Ge} = (x_1 d_1 + x_2 d_2)/(d_1 + d_2)$, where the x's and d's are the Ge content and thickness of each constituent layer, respectively) (7). In order to avoid the generation of misfit dislocations as a result of strain relaxation, the individual layer thicknesses as well as the multiple-layer thickness must be kept below their respective critical thicknesses. One of the other important parameters in determining the energy of quantized states (or detector response) is the band offset. This is obtained by interpolation using the band offsets of Si and Ge, which were originally obtained by Van de Walle and Martin (8). Figure 2 shows the valence band offset for the three hole bands of a SiGe layer grown on a Si substrate. For practical purposes, the conduction band offset can be neglected if the structure is grown psedomorphically on an Si substrate. For the detectors based on internal photoemission (2D free-carrier absorp-

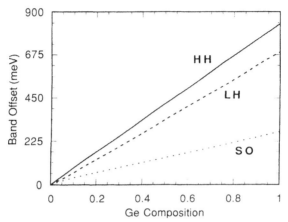

FIG. 2.   Band offsets of the three hole band as a function of Ge composition for a strained SiGe layer grown on Si.

tion) from quantum wells, the heavy hole band offset mainly determines the cutoff wavelength of the photoresponse. On the other hand, for the detectors based on intervalence band absorption, the detector response is strongly dependent on the strain-induced splitting of the three valence bands. Such splittings can be calculated as a function of Ge composition using deformation potentials (9). Figure 3 shows the positions of the three valence bands as a function of the Ge composition of a SiGe layer grown pseudomorphically on a Si substrate. It is also important to know the effective masses of the three valence bands for determining the energies of quantized states. The values of the effective masses, including strain effects, have been calculated using the $\mathbf{k} \cdot \mathbf{p}$ approximation derived by Chun and Wang (10). Figures 4a and 4b show the in-plane and growth direction effective masses, respectively, for the three hole bands of a SiGe layer pseudomorphically grown on a (100)-oriented Si substrate. Orientation dependence of the effective mass is due to the deformation of the energy bands as a result of the strain in the SiGe layer. These parameters can be readily used for the design of structures for the study of quantum size effects in SiGe/Si heterostructures. Before we move on to the discussion of infrared transitions, we will first give the evolution of quantum size effects in SiGe/Si heterostructures, which led to the observation of intersubband transition.

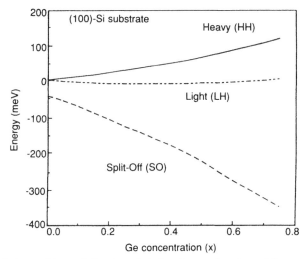

FIG. 3. Relative positions of the three hole bands as a function of Ge composition for a strained SiGe layer grown on Si.

III. Quantum Size Effects

A. RESONANT TUNNELING

The successful demonstration of tunneling is the first step toward the exploitation of the quantum effects for device applications. Such observations in the III–V-based quantum wells (*11*) led to the observation of many novel effects. In SiGe/Si heterostructures, the large band offset in the valence band and the small light-hole effective mass compared to that of the electron effective mass in the growth direction favor hole tunneling devices (*12–14*). The schematic band diagram of a typical structure used in the study of resonant tunneling of holes is shown in Fig. 5. A detailed description of the structure and the growth can be found in Ref. (*13*). The current–voltage (*I–V*) of the tunneling diode at 4.2, 77, and 300 K are shown in Fig. 6. The negative differential resistance is clearly seen at 4.2 and 77 K; however, at room temperature NDR is not visible. Inset of the Fig. 5 shows the *I–V* and conductance voltage (*G–V*) for an additional peak in the high-voltage (> 500 mV) regime. The experimental peak voltage positions are in good agreement with the estimated values taking into account the bound-state energy and voltage drops across the spacer

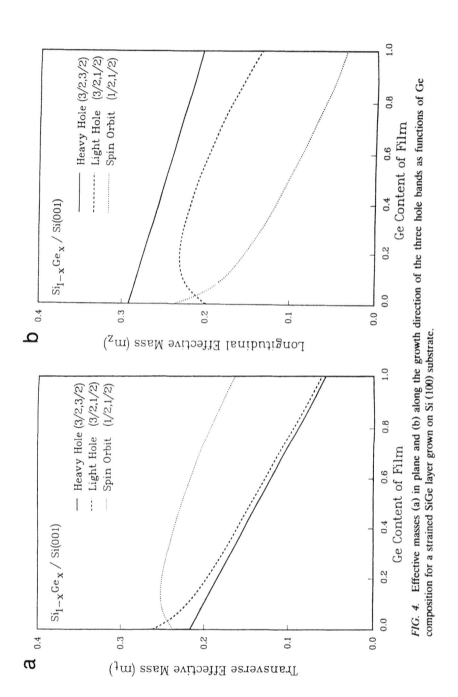

FIG. 4. Effective masses (a) in plane and (b) along the growth direction of the three hole bands as functions of Ge composition for a strained SiGe layer grown on Si (100) substrate.

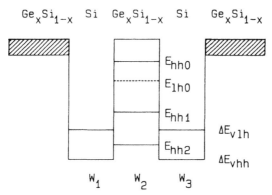

FIG. 5.   Schematic diagram of the double-barrier diode. The device parameters are $W_{1,3}$ = 50 Å, W_2 = 40 Å, and x = 0.4.

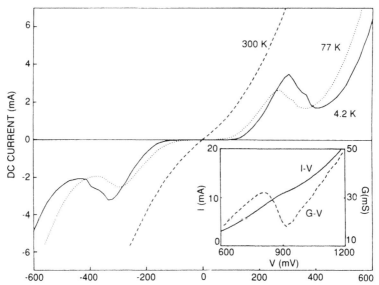

FIG. 6.   Observed I–V characteristics for the structure shown in Fig. 5 at 300 K, 77 K, and 4.2 K. Inset shows the I–V and conductance G–V measurement of an additional peak for high bias at 77 K.

layers (15). The peak at 300 mV has a peak-to-valley current ratio of 1.6 at 77 K. The predominant tunneling mechanism is confirmed to be due to the light hole. The latter is verified by investigating the tunneling under a strong magnetic field parallel to the interface. As the magnetic field is applied parallel to the interface, the motion of holes passing through the double barrier structure is detoured. This can be considered a decrease of the kinetic energy in the tunneling direction or, in other words, an effective increase of the tunneling barrier. In this case, the additional voltage needed for resonant tunneling through can be estimated (16) using WKB approximation to be about $eB^2(2s^2 + 2b^2 + w^2)/3m^*$, where s, b, and w are the thickness of the spacer, barrier and well, respectively. Figure 7 shows the magnetic field dependence of the voltage shift for the light and heavy hole peaks (solid lines) and from these dependencies, the light and heavy hole peaks can be clearly discriminated. The dashed lines represent the results of Gennser et al. (17) for a similar resonant tunneling diode with a slightly different Ge composition for comparison. The data of Ref. (17) have been scaled using the geometrical factor $(2s^2 + 2b^2 + w^2)$, such that the slope of lines are inversely proportional to the respective effective masses. Next, we will discuss the transport properties involving SiGe/Si superlattices.

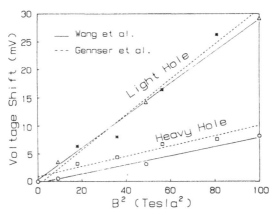

FIG. 7.   Measured peak voltage shift vs. B^2 for the light and heavy ground states (solid lines), when the magnetic field is parallel to the interfaces. The dashed lines show the data of Gennser et al. (17).

B. SUPERLATTICE MINIBANDS

1. Transport Properties. The successful demonstrations of resonant tunneling in the SiGe/Si heterostructures reveals the coherent nature of the carrier transport through the double barrier structure. This indicates that the study can be extended to structures involving multiple quantum wells and superlattices. Here, the results of the transport measurement of a symmetrically strained, $Si_{1-x}Ge_x/Si$ superlattice grown on a $Si_{1-x/2}Ge_{x/2}/Si$ buffer layer are presented. The superlattice structure is grown on a 2-μm-thick unstrained p^+ $Si_{0.8}Ge_{0.2}$ buffer layer doped to 5×10^{18} cm^{-3}. The active layers of the structure consist of a 15-period $Si_{0.6}Ge_{0.4}/Si$ (each layer 50 Å thick) superlattice of p-doped (1×10^{17} cm^{-3}) and 150 Å undoped contact layers on each side of the superlattice. The schematic band diagram of the superlattice and the conductance process are shown in the Figs. 8a and b, respectively. Figure 9 shows the measured I–V and dI/dV at 4.2, 35, and 77 K. Two peaks at 1.1 V and 2.5 V (77 K) are present in the I–V and dI/dV. The peak at 2.5 V shows a clear negative differential resistance. As the temperature of the sample decreases, the peak-to-valley ratio of the peak at 2.5 V increases, and the peak position shifts toward a higher voltage while the peak at 1.1 V

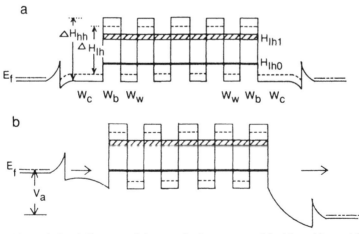

FIG. 8. Schematic band diagrams of the superlattice structure (a) without bias and (b) with miniband conduction under bias. The device parameters are $n = 15$, $x = 0.4$, $W_c = 150$ Å, $W_b = 50$ Å, $W_w = 50$ Å, ΔH_{hh} (heavy-hole barrier height) = 302 meV, ΔH_{lh} (light-hole barrier height) = 255 meV, and heavy-hole and light-hole band splitting = 39 meV.

becomes more apparent from the dI/dV curve. Above 100 K, the NDR is no longer clearly observed. The observed process is a result of hole transport through the minibands. When the Fermi level in the emitter region is aligned with one of the minibands as shown in Fig. 8b the carriers flow through the miniband. The NDR shows up when the Fermi level is moved away from the miniband. Although the NDR due to conduction through the second miniband is clearly observed as shown in Fig. 9, no NDR is seen for the transport through the first miniband, probably due to its narrow bandwidth. The energies of the minibands are also investigated experimentally using the I-V-T measurement. Thermionic emission, which is sensitive to the miniband energies of the superlattice, is the predominant component of the current when the bias across the structure is below the value required for direct injection into the minibands. The thermionic emission current through a miniband of bandwidth ΔH in a superlattice is described by the equation (14)

$$ J = \eta(E)A^*T^2\left[\exp\left(-\frac{\phi(V)}{kT}\right) - \exp\left(-\frac{\phi(V) + \Delta H}{kT}\right)\right], \qquad (1) $$

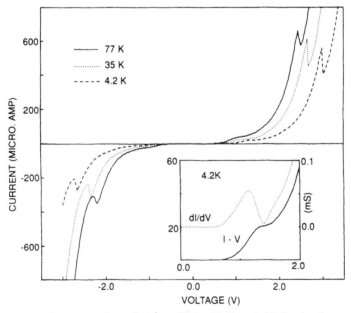

FIG. 9. Observed current-voltage (I–V) at 77 K, 35 K and 4.2 K, showing conduction through minibands in the superlattice. Inset of the figure shows the magnified I–V, and dI(V)/dV for the first peak.

where A^* is the effective Richardson constant, T is the temperature, k is the Boltzmann constant, $\phi(V)$ is the potential difference between the Fermi level at the emitter and the miniband level in the superlattice, V is the bias voltage, and $\eta(E)$ is the tunneling transmission coefficient. The second exponential term in Eq. (1) accounts for the current component due to the carriers with energies above the miniband, which are blocked by the barrier. The total current through the structure is the sum of currents through each miniband. If the ΔH is smaller than kT, the above equation can be simplified to

$$J = \frac{\eta(E)A^*T\Delta H}{k}\exp\left(-\frac{\phi(V)}{kT}\right),\qquad(2)$$

and $\phi(0)$ can be estimated from the experimental I–V–T data by taking the slope of $\log(I/T)$ vs. $1/T$ at biases greater than several kT, and then extrapolating to zero bias (14). The plot of $\log(I/T)$ vs. $1/T$ at different biases for a typical sample is shown in Fig. 10. For a given bias, two regions with distinct slopes are observed. One appears in the relatively high-temperature region (200–300 K), whereas the other appears in the lower-tem-

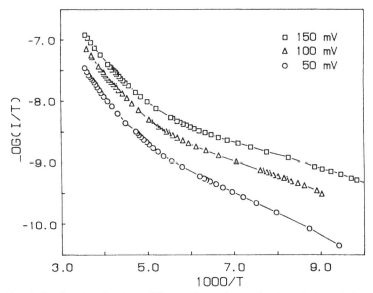

FIG. 10. Arrhenius plots for three different biases due to the thermionic emission current through minibands.

perature region (100–170 K). This observation indicates that there are two different thermionic current components (due to the first and the second minibands) flowing through the superlattice, as expected. In the low-temperature region, most of the thermionic current is from the first miniband (ground state). The barrier energy of the first miniband can be estimated from the low-temperature thermionic data using Eq. (2). On the other hand, at higher temperatures, the increased thermal energy allows the carriers to flow through the second miniband as well. In this case, the current due to the second miniband can be estimated by subtracting the current through the first miniband extrapolated to the high-temperature region. Thus, the barrier energy for the second miniband can be obtained using Eq. (2), similar to the case of first miniband. The experimental values for the first and second miniband positions are 95 meV and 250 meV, respectively. These values are in good agreement with the calculated values of 91 meV and 220 meV from the emitter band edge including the heavy and light-hole band splitting. We have also probed the optical transitions between minibands by measuring the $I–V$ characteristics under optical excitation as described in the following section.

2. *Photoassisted Miniband Transport.* The effect of hole transport through minibands under an optical excitation is investigated using a tunable CO_2 laser as excitation source. The structure of this sample is the same as the previous sample, except that 60-Å wells and 40-Å barriers are used to obtain light-hole miniband separation near 10 μm to match the laser wavelength. The samples are grown on p^- substrate (100 $\Omega \cdot$ cm) to reduce the free-carrier absorption. The ohmic contact is made to the $Si_{0.8}Ge_{0.2}$ buffer layer instead of the backside of the wafer.

Figure 11 shows the measured photocurrent using a 1-kΩ load resistor as a function of applied bias across the device for the polarization of the infrared beam parallel and perpendicular to the interfaces of the superlattice. The details of the measurement can be found in Ref. (*18*). Two clear peaks at 1.5 and 5 V are observed when the polarization of the incident beam has a component perpendicular to the interfaces (*z*-polarization). In the case of polarization parallel to the interfaces (*xy*-polarization), the peak at 5 V completely disappears while the one at 1.5 V becomes smaller. The polarization dependence is a well-known characteristic of the intersubband transition. It is interesting to note that the peaks in the photoresponse appear very close to the high conductivity regions in the conductance–voltage ($dI/dV–V$) characteristics as shown by dashed line in

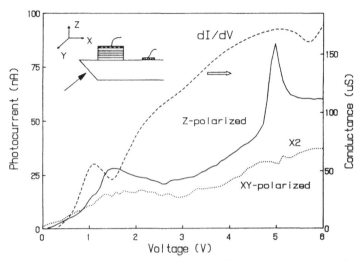

FIG. 11.   Measured photocurrent as a function of applied bias for xy- and z-polarizations. The peaks at 1.5 V and 5 V in the z-polarization data are due to intraband absorption between $LH_0 \rightarrow LH_1$ and $LH_1 \rightarrow LH_2$ minibands. The small photoresponse in the xy-polarization data at 1.5 V may be due to the interband absorption between $LH_0 \rightarrow SO_0$ minibands. The conductance-voltage characteristic (dashed curve) shows the enhancement of conductivity when the minibands are aligned with the emitter. The positions of the conductance peaks appear close to those of the photocurrent peaks.

Fig. 11. This confirms that, when the photoresponse peaks, the Fermi level of the emitter aligns with one of the minibands.

The photoabsorption process can be understood by considering the minibands in the superlattice, which are separated by $10.6\text{-}\mu$m photon energy, as well as the alignment of minibands with the emitter. Figure 12 shows the detailed energy level diagram of the superlattice including the minibands of the light hole (LH) and the split-off hole (SO) states. For the device parameters used, only localized bound states are possible for the heavy holes due to the large barrier height as well as the large effective mass. The peak in the photoresponse at 1.5 V appears when the Fermi level of the emitter is aligned with the LH_0 miniband and the infrared absorption occurs for the transitions from the LH_0 to LH_1 (intraband) and SO_0 (interband) minibands. The peak at 5 V occurs when the Fermi level of the emitter is aligned with the LH_1 miniband and the only allowed transition is to the LH_2 miniband. Such transitions have a strong polarization dependence since only the z-component of polarization contributes to the matrix element. The details of the polarization dependence and

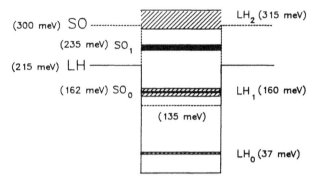

FIG. 12.   Band structure of the superlattice showing the minibands due to both light and split-off bands. The light hole and split-off band edges are shown in solid and broken lines, respectively. Hole energy is assumed to be positive.

transitions between different hole bands will be discussed in the following section. This initial observation led to the probe of intersubband transitions in SiGe/Si quantum wells and fabrication of infrared-detector operation in the mid to long wavelengths.

IV. Intersubband Transition in p-Type SiGe / Si Quantum Wells

We mainly focus on subband transitions in the valence band due the large band offset and the possibility of many transitions as a result of multiple band structure. The structures used in the experiment were grown by MBE on high-resistivity (100 $\Omega \cdot$ cm) (100)-Si wafers to ensure low free-carrier absorption in the substrate. The growth temperature is kept as high as possible (in the 500–600°C range, depending on the Ge composition) in order to improve the quality of the Si barrier, where most of the photoexcited carrier transport occurs. Several $Si_{1-x}Ge_x$/Si multiple quantum well structures with different Ge compositions, $x = 0.15, 0.3, 0.4,$ and 0.6 in the quantum well, were employed to study the strain dependence of the subband transitions. One period of the SiGe/Si multiple quantum well structure consists of a 40-Å $Si_{1-x}Ge_x$ well and a 300-Å Si barrier. The middle 30 Å of the $Si_{1-x}Ge_x$ wells is boron doped to about 5×10^{19} cm^{-3} while the Si barriers are undoped. For the $Si_{0.85}Ge_{0.15}$/Si, $Si_{0.7}Ge_{0.3}$/Si, and $Si_{0.6}Ge_{0.4}$/Si samples, 10 periods of multiple quantum wells are grown, whereas for the $Si_{0.4}Ge_{0.6}$/Si sample, only five periods are grown because of the critical thickness limitation of the SiGe strained layers. The absorption spectra of the samples were taken at room temperature using a Fourier transform infrared (FTIR) spectrometer. The different transitions were identified using the characteristic polarization dependence of the absorption (2). Figure 13 shows the measured absorption spectra of the $Si_{0.4}Ge_{0.6}$/Si quantum well sample at two different polarization angles using a waveguide structure as shown in the inset. At the 0° polarization angle (electric field having a component in the growth direction as depicted in the inset), an absorption peak occurs at 5.3 μm. At the 90° angle (beam polarized parallel to the plane), this peak vanishes, but another peak appears at a shorter wavelength, 2.3 μm. The peak at the 0° polarization is due to the intersubband transition between two heavy-hole states (19). This is confirmed by the polarization dependence of the peak shown in Fig. 14a. The peak at 2.3 μm shows a polarization dependence opposite that of the intersubband transition as illustrated in Fig. 14b. At normal incidence (90° polarization), the absorption reaches the maximum and it drops as the angle decreases. Two other samples, $Si_{0.7}Ge_{0.3}$/Si and $Si_{0.6}Ge_{0.4}$/Si, show absorption spectra similar to that of the $Si_{0.4}Ge_{0.6}$/Si sample previously discussed (20). For the sample with 15% Ge concentra-

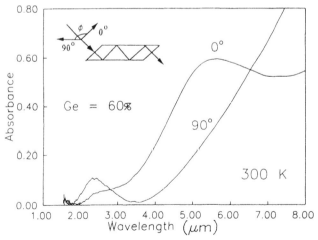

FIG. 13. Measured room-temperature absorption spectra of a $Si_{0.6}Ge_{0.4}$/Si multiple quantum well sample at 0 and 90° polarization angles. At 0°, a clear peak due to intersubband transition is shown at near 5.6 μm. However, at 90° this peak vanishes, and another peak appears at 2.3 μm, which is due to intervalence band transition. The inset shows the waveguide structure used for the measurement.

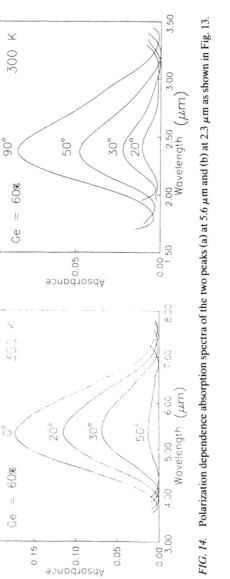

FIG. 14. Polarization dependence absorption spectra of the two peaks (a) at 5.6 μm and (b) at 2.3 μm as shown in Fig. 13.

tion, the absorption spectra reveal a clear peak at 0°, due to the heavy-hole intersubband transition; however, no obvious peaks observed at 90°. The origin of the different absorption processes can be understood qualitatively by considering the matrix element of optical transition including band mixing. The initial Ψ_i and final Ψ_f state wavefunctions can be expressed using a complete set of subband wave functions of all the bands (21).

$$|\Psi_i\rangle = \sum_{n\nu} |\phi_{n,\nu} U_\nu\rangle \qquad (3)$$

and

$$|\Psi_f\rangle = \sum_{n'\nu'} |\phi_{n'\nu'} U_{\nu'}\rangle \qquad (4)$$

where ϕ's are the envelope functions of subbands calculated using effective mass approximation and U's are Bloch functions. The n's and ν's are the quantum number of the subband and band index of the corresponding subband, respectively. The wavefunction in the plane of the quantum well ensures the conservation of momentum in the optical transition. The optical matrix element (M) in the dipole approximation can be written as

$$M = \langle \Psi_f | \hat{\epsilon} \cdot \vec{p} | \Psi_i \rangle \qquad (5)$$

where $\hat{\epsilon}$ is the photon polarization vector; $\vec{p}$ is the momentum of the hole. Using the above wavefunctions and assuming the envelope functions are slowly varying compared to Bloch functions, the matrix element can be expanded into

$$M = \sum_{nn'\nu} \langle U_\nu | U_\nu \rangle \langle \phi_{n'\nu} | \epsilon_z \cdot p_z | \phi_{n\nu} \rangle$$
$$+ \sum_{n\nu\nu'} \langle U_{\nu'} | \hat{\epsilon} \cdot \vec{p} | U_\nu \rangle \langle \phi_n | \phi_n \rangle \qquad (6)$$

where z indicates the growth direction. The first term of the Eq. (6) is responsible for transitions within a single band. Such transitions have a strong polarization dependence because the matrix element occurs between envelope functions, which depend only on z. This behavior is clearly evident in the experimental data shown in Fig. 14a. This absorption disappears for normal incident light because there is no polarization component of light along the z direction. On the other hand, the second term in the Eq. (6) is responsible for transitions between different bands where the matrix element occurs between the Bloch states. For the transitions involving two different valence bands, the matrix element vanishes to the first order since the Bloch functions are p-like at the

Γ-point assuming no band mixing (22). If, however, the light and spin-orbit split-off Bloch functions have significant band mixing from the s-like conduction band, a substantially large matrix element can result. The band mixing can be significant if the (Γ) bandgap is small enough and if the light hole and spin-orbit split-off bands involved in the transition that have a large crystal momentum in the plane ($k_{\parallel}$) (21, 23).

For Si, the coupling between the conduction and valence bands is weak due to the large direct bandgap at the Γ point (4.2 eV). On the other hand, for Ge, significantly strong coupling is expected since the bandgap at the Γ point is small (0.89 eV). Due to this coupling, the valence band wavefunctions can have a component of the s-like conduction band state, particularly for large $k_{\parallel}$. For $Si_{1-x}Ge_x$, the bandgap at the Γ-point decreases as the Ge content increases, and, therefore, with a sufficient Ge content, the intervalence band transition will become stronger. One earlier report on intervalence band transition in p-type bulk Ge supports this assessment (24). It is also possible that a similar mechanism is responsible for the observed normal-incident photocurrent spectrum in a p-type AlGaAs/GaAs multiple quantum well structure (25). In this case, the direct bandgap is 1.42 eV and is much smaller than that of Si (4.2 eV).

To estimate the absorption strength qualitatively, the coupling of the valence band Bloch states with the conduction band can be expressed by using the first order perturbation theory (22),

$$U_i = U_i^0 + \frac{\hbar}{m_0} \frac{\langle U_c^0 | \mathbf{k} \cdot \mathbf{p} | U_i^0 \rangle}{(E_g + |E_i|)} U_c^0, \tag{7}$$

where the index i ($= h, l, s$) refers to the heavy, light and spin-orbit split-off holes, respectively, and U_c is the Bloch state of the conduction band. E_g is the Γ energy gap and E_i is the energy of the light or split-off hole band edge measured from the heavy hole band edge. It can be seen that as the energy gap decreases, coupling with the conduction band increases. This explains the fact that intervalence band absorption was observed only for high Ge concentrations. From Eq. (7), it is expected that the absorption strength will be proportional to $(E_g + |E_i|)^{-2}$. Indeed such a dependence is observed experimentally as illustrated in Fig. 15.

Another interesting observation from Fig. 14b is that the strength of the intervalence band absorption peaks when the polarization of the light is in the plane of the quantum well. Such a behavior can be understood by considering the optical matrix element between Bloch functions. The matrix element involved in this case is proportional to $\langle U_c^0 | \hat{\epsilon} \cdot \mathbf{p} | U_{HH} \rangle$ where the Bloch function U_{HH} is only a function of the (x, y) plane (22).

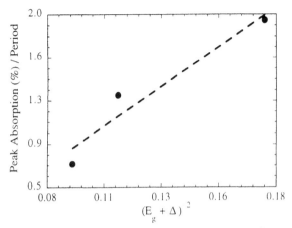

FIG. 15. Measured absorption strength as a function of $(E_g + |E_i|)^2$, where E_g is the Γ energy gap and E_i is the band-edge energy of the light or spin-orbit split-off hole bands measured from the heavy hole band edge.

Thus, the largest absorption is expected when the field is polarized along the (x, y) plane. Next, we will discuss formation δ-doped quantum wells in Si and optical transitions between subbands of such a quantum well. We will also illustrate the importance of many-body effects in determining the transition energy as the doping concentration in the quantum well is large.

V. Intersubband Transition in *p*-Type δ-Doped Si Quantum Wells

In this section, we will illustrate intersubband transition in quantum wells formed by selectively narrow doping of Si layers. In this case, the transition energy is strongly dependent on the amount of doping used in the layer compared to the Ge composition for SiGe/Si quantum well structures. The sample structure used in this study consists of 10 periods of 35-Å-thick heavily boron-doped (or δ-doped) Si layers separated by 300-Å-thick undoped Si barriers. Four samples with doping concentrations of about 0.7, 2.7, 4.0, and 5.7×10^{20} cm^{-3} were prepared in order to study the effect of doping on optical transition (*20, 26*). The absorption spectra of the samples were taken at room temperature using a waveguide structure similar to that of SiGe/Si quantum wells described above. The measured absorption spectra of the four samples as a function of photon

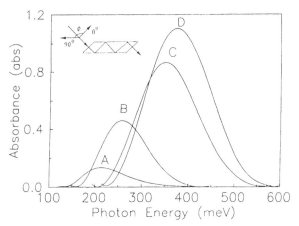

FIG. 16. Absorption spectra for the four δ-doped samples in Si as a function of photon energy at 300 K. The set of curves represent different doping concentrations. (A) 0.7×10^{20}, (B) 2.7×10^{20}, (C) 4.0×10^{20}, and (D) 5.7×10^{20}.

energy are shown in Fig. 16. It can be clearly seen that the absorption spectra shift toward the high-energy regime with increasing absorption as the doping density is increased. The shift of the absorption peak is due mainly to the increase of the potential well depth as well as plasma shifts at high doping densities. The widths of the absorption peaks are more than one order of magnitude larger than those observed in GaAs/AlGaAs quantum well structures, which are typically about 10 meV. The non-parabolicity of the hole bands plays an important role in determining the peak width; in particular, at high doping densities, the hole bands can be filled up to a few hundreds of meV (27, 28). Other possible mechanisms include ionized impurity scattering, which was recently demonstrated by Dupont et al. (29) for intersubband transition in GaAs/AlGaAs quantum wells. We have examined the polarization dependent absorption as shown in Fig. 17, in order to understand the nature of the transition. The lack of the absorption at 90° polarization is due to the disappearance of the photon electric field along the quantum well similar to that of SiGe/Si quantum wells described in the previous section. The intervalence band transition is not seen in this case due to the large direct bandgap of Si as previously discussed. Next, we will discuss the estimation of the transition energy positions using a three-band self-consistent calculation including many-body effects.

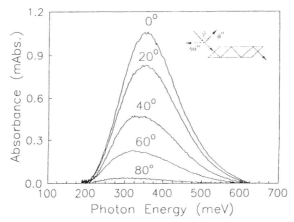

FIG. 17. Polarization-dependent absorption spectra of sample with 4×10^{20} cm^{-3} doping density at 300 K. The decrease of the absorbance with increasing polarization angle is due to the reduction of the component of photon electric field along the growth direction.

In order to understand the experimental data, in particular the variation of peak positions as a function of the 2-D hole density in the wells, the subband energies are estimated using a multiband self-consistent calculation (26). Many-body effects are incorporated using an exchange–correlation potential derived from the 3-D local-density-functional approximation (30). The energy position is also corrected for depolarization and exciton-like shifts (31). In this calculation, we assume that the doping is uniformly distributed over the doped layer. In the case of Si, the light- and heavy-hole bands are assumed to be degenerate at the Γ-point and the split-off band is 44 meV above the heavy-hole band. The effective masses along the growth direction are used for the subband energy calculation and the average in-plane masses are used for the estimation of the populations of subbands. The effective masses of the three hole bands as a function of Ge composition (10) are shown in Figs. 4a and 4b. In the following, we will briefly discuss the steps that are used in the self-consistent calculation. The Schrodinger equation including the many-body effects can be written as

$$\left[-\frac{\hbar^2}{2m^*} \nabla^2 + V_H(z) + V_{xc}(z) \right] \psi_n = E_n \psi_n, \tag{8}$$

where V_H and V_{xc} are the Hartree and exchange–correlation potentials, respectively. The Hartree potential is given by

$$V_H(z) = \frac{e}{\epsilon\epsilon_0} \int (z - z')\rho(z')\,dz',$$ (9)

where $\rho(z) = e[N_D - n(z)]$, ϵ is the dielectric constant, N_D is the dopant density in the doped region and assumed to be zero elsewhere, $n(z) = n^{2D}\sum_n|\psi_n|^2$ is the carrier concentration at z and n^{2D} is the 2-D density of carriers. The exchange–correlation potential in the local-density-functional approximation using the Hedin and Lundqvist parameterization (32) can be expressed as

$$V_{xc}(z) = -\left[1 + \frac{0.7734r_s}{21}\ln\left(1 + \frac{21}{r_s}\right)\right]\left(\frac{2}{\pi\alpha r_s}\right)Ry^*$$ (10)

where $r_s = [4\pi a^{*3}n(z)/3]^{-1/3}$, the effective Bohr radius $a^* = 4\pi\epsilon\hbar^2/m^*e^2$, $\alpha = [4/(9\pi)]^{1/3}$ and the effective Rydberg constant $Ry^* = e^2/(8\pi\epsilon a^*)$. The validity of this approximation for a narrow sheet of charge has been discussed by Hautman and Sander (33). The condition for the 3-D approximation to be valid is that the interparticle distance (r_s) should be smaller than the width of the charged sheet. For the doping densities used in this study, the largest value of r_s is about 15 Å and the width of the charged sheet is more than 30 Å. This justifies the use of the 3D approximation in this analysis. It is also known that the photon induced many-body contributions such as depolarization (34) and exciton-like interaction (31) between the ground and excited states shift the calculated energy position by a substantial amount. In the case of multiple quantum wells, the coupling of wells can also be shown to introduce an additional shift of the transition energy (35). However, in our case the large barrier thickness keeps the coupling of wavefunctions between adjacent wells that participate in the transition small enough so that we may neglect this contribution. The peak position including the shift due to depolarization and exciton-like effects can be expressed as

$$\tilde{E}_{10}^2 = E_{10}^2(1 + \alpha - \beta),$$ (11)

where $\tilde{E}_{10}$ is the shifted energy and E_{10} is the difference ($E_1 - E_0$) of the subband energy levels calculated by self-consistently using Eq. (8). The quantities α and β for the transition between the ground and excited states are given by (*31*)

$$\alpha = \frac{2e^2 n_0^{2D}}{\epsilon\epsilon_0}\left(\frac{\hbar^2}{2m^*E_{10}}\right)^2$$

$$\times \int_{-\infty}^{+\infty} dz\left[\psi_1(z)\frac{d\psi_0(z)}{dz} - \psi_0(z)\frac{d\psi_1(z)}{dz}\right]^2 \quad (12)$$

and

$$\beta = \frac{2n_0^{2D}}{E_{10}}\int_{-\infty}^{+\infty} dz\,\psi_1(z)^2\psi_0(z)^2\frac{\partial V_{xc}[n(z)]}{\partial n(z)}. \quad (13)$$

We will first illustrate the effect of the exchange–correlation potential on the potential well. Figure 18 shows the calculated potential well and the wavefunctions of the bound states using the above formalism for a Si δ-doped layer. We have also included the potential profile shown by the dashed line without many-body effects for comparison. It can be seen that the many-body effects increase the depth of the potential well. This is a direct consequence of Pauli's exclusion principle causing the holes with the same spin to repel each other, which effectively increases the net positive charge at the center of the doped region. In order to compare experimental peak positions, we have first calculated the transition energy without

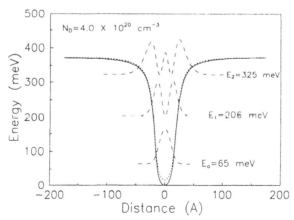

FIG. 18. Calculated potential well and wavefunctions for the sample with a doping density of 4×10^{20} cm^{-3}. The dotted curve shows the potential profile without exchange–correlation effects. The calculation was carried out using the 3-band self-consistent approach discussed in the text.

the many-body effects (Hartree approximation). The calculated peak positions (dashed curve) along with the experimental data (solid circles) for the δ-doped structures in Si are shown in Fig. 19. It can be seen that the calculated values are considerably smaller than the measured peak positions. This discrepancy is due mainly to the omission of the many-body effects, depolarization (34), and exciton-like (31) interactions in the calculation. This is particularly important in the case of boron (B) doping; due to its high solid solubility in Si, a considerably large doping density can be achieved (36). The many-body effects are estimated using the wavefunctions and bound-state energies calculated self-consistently using the Hartree and exchange–correlation potentials. The dotted curve in Fig. 19 shows the calculated peak positions including exchange–correlation potentials. These values in fact are a few meV above those obtained using the Hartree approximation even though the potential well depth increases due to the exchange–correlation, which is in the range of 20–40 meV for the doping densities used. This is due mainly to the fact that the energy-level separation is relatively insensitive to the change of the potential well depth. The solid curve in Fig. 19 shows the results including depolarization and exciton-like shifts. This brought the calculated and experimental peak positions into reasonably close agreement. The slightly lower values obtained in the calculation may be due to the omission of nonparabolic effects of the hole band structure in the analysis. Other possible causes

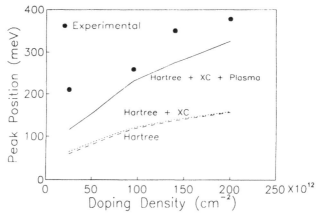

FIG. 19. Subband separation as a function of doping density for δ-doped layers in Si: experimental data (solid circles), using Hartree approximation (dashed curve), Hartree with the exchange–correlation potential (dotted curve), and including depolarization and exciton-like effects (solid curve).

include the contribution of multiple bands in depolarization and exciton-like effects and uncertainty in determining the exact peak position, when transitions involving other bands occur. Next, we will discuss the application of intersubband transitions in SiGe/Si quantum wells for infrared detection.

VI. Detector Design

For attaining optimum sensitivity, a quantum-well structure with a single bound state and an excited extended state close to the barrier is desirable (3, 37). The operating wavelength of the detector is usually obtained by calculating the energy levels of the quantum well using the effective mass approximation. The energy of the extended state is taken to be the maximum of the transmission coefficient. The photon energy required for the transition is usually higher than the energy difference between the bound and extended states due to plasma shifts as discussed above. The doping is typically done in the quantum well in order to minimize the scattering in the barriers, where the photoexcited carriers spend most of their transit time before reaching the contacts. However, doping in the well can reduce the lifetime of the excited carriers, which can be captured back into the well without contributing to the photocurrent. This can have a strong effect on the response of the detector. Therefore, when considering the amount of dopants in the well, a compromise must be made between quantum efficiency and the leakage current (mainly due to thermionic emission) that can be tolerated for a given application. In the case of p-type SiGe/Si detectors, one order-of-magnitude larger doping density is required to obtain quantum efficiencies comparable to that of n-type GaAs/AlGaAs detectors. This is because the hole effective mass is larger than the electron effective mass in GaAs. The absorption strength (I_A) for a given doping density can be estimated using the matrix element given in Eq. (6) as

$$I_A = \rho_s N_T \frac{e^2 h}{4 c \varepsilon_0 n_r m^*} f \frac{\cos^2\theta}{\sin\theta} \cos^2\phi, \tag{14}$$

where N_T is the total number of quantum wells that the infrared beam passes through due to multiple reflections, ρ_s is the 2-D density of carriers in the well, I_A is the integrated absorption strength, f is the oscillator strength of transition, θ is the angle of incidence extending from the normal to the plane, ϕ is the angle between the photon electric vector and the plane of incidence, and m^* is the effective mass taken along the (100)

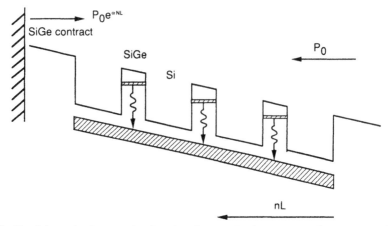

FIG. 20. Schematic diagram showing the photoconduction process of a *p*-type multiple quantum well detector.

direction. Here, we ignore the nonparabolic effects on the intersubband transition due to the quantization and strain present in the quantum well.

Two other important quantities which describe the detector performance are the responsivity and detectivity. Figure 20 schematically shows the photoconduction process of a *p*-type multiple quantum well detector. The dependence of the responsivity of the detector on transport parameters can be estimated by calculating the number of photogenerated carriers at a given quantum well and multiplying it by the probability of their reaching the contact

$$I_p(F) = \sum_{n=1}^{N} \frac{eP_0}{h\nu} (e^{-\alpha nl} + e^{-\alpha(2N-n)l}) \alpha l\, e^{-nL/v(F)\tau}, \qquad (15)$$

where P_0 is the incident infrared power, $h\nu$ is the photon energy, α is the absorption coefficient assuming the light absorption occurs only in the quantum well, l is the well width, L is the period of the multiple quantum wells, N is the number of quantum wells, $v(F)$ is the drift velocity under an applied electric field F, and τ is the excited-carrier lifetime. The second term in the parenthesis is due to the absorption by the reflected light from the metal contact as illustrated in Fig. 20. Since, typically, for quantum well detectors $\alpha l \ll 1$, the above equation can be simplified to

$$I_p(F) \approx \frac{2eP_0}{h\nu} \alpha l \sum_{n=1}^{N} e^{-nL/v(F)\tau}. \qquad (16)$$

Since, for moderate electric fields, the mean free path is considerably larger than the period of the multiple quantum wells (i.e., $v(F)\tau > L$), the responsivity, $R(F) = I_p(F)/P_0$, after we evaluate the summation, can be approximated as

$$R(F) \approx \frac{2e}{h\nu} \alpha l \frac{v(F)\tau}{L}.$$ (17)

It is interesting to note that the responsivity is directly proportional to the drift velocity. This indicates that the bias dependence of the photoresponse is strongly dependent on the bias dependence of the drift velocity. In the case of holes, the drift velocity is given by (38)

$$v(F) = \frac{\mu F}{1 + \mu F/v_s},$$ (18)

where μ is the mobility and v_s is the saturated drift velocity. Then the field dependence of the responsivity of a p-type detector can be written as

$$R(F) \approx \frac{2e\alpha\tau}{h\nu} \frac{l}{L} \frac{\mu F}{1 + \mu F/v_s}.$$ (19)

This expression can be used to determine the mobility and lifetime of the photoexcited carriers from the measured bias-dependent photoresponse, which will be discussed later. This formalism can be readily applied to n-type devices using the proper form of the field dependence of the velocity. For example, detectors based on AlGaAs/GaAs quantum wells, the photoresponse as a function of field indicate velocity overshooting behavior (39). The detectivity of a quantum well detector operating in the photoconductive mode can be estimated by using the relationship (40)

$$D^* = \eta(e \cos \theta/I_d)^{1/2}/2h\nu,$$ (20)

where η ($= 1 - e^{2\alpha l}$) is the quantum efficiency, θ is the incident angle, $h\nu$ is the photon energy and I_d is the dark current mainly due to thermionic emission from the quantum well. Since the quantum well detector is a majority carrier device, the main contribution to the leakage current comes from the thermionic emission of trapped carriers in the quantum well. As an attempt to reduce the leakage current, the use of bound-to-miniband transitions was proposed (41) and experimented (42). Such a configuration can improve the blocking of the thermionic current while the photoexcited carriers can be extracted through the miniband. In this case, the spectral extent of the photoresponse depends strongly on the width of the miniband. In the following sections, we will discuss the experimental results of infrared detection using p-type SiGe/Si quantum wells.

VII. Detector Photoresponse

For the photoresponse measurement, mesa diodes of 200 μm in diameter were fabricated using the above structures. The spectral dependence of the photocurrent was measured using a glowbar source and a grating monochromator with lock-in detection. Infrared light was illuminated on the mesa at either an angle or from the backside of the wafer depending on the transition process to be examined. The measured leakage currents for the detectors with 15 and 60% Ge composition at 77 K as a function of bias across the device are shown in Fig. 21. It can be seen that there is a drastic reduction of the leakage current as the Ge composition is increased. This is due mainly to the increased of the barrier height at higher Ge compositions, which suppresses the leakage current arising from the thermionic emission. Figure 22 shows the photoresponse of the detector with 15% Ge composition at 0 and 90° polarizations (43). The data were taken at 77 K using a 7-kΩ load resistor with a 2-V bias across the detector. In the 0° polarization case, a peak is found near 8.6 μm, which is in agreement with the FTIR absorption spectra (43). The full width at half-maximum (FWHM) is about 80 meV whereas, for the 90° polarization,

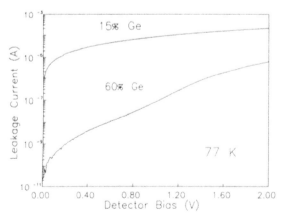

FIG. 21.   Current-voltage and differential resistance measured at 77 K for the 15% and 60% Ge composition device. Positive bias indicates a positive top contact.

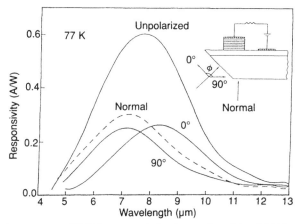

FIG. 22. Photoresponse at 77 K for two polarization angles with a 2-V bias applied across the detector. Infrared light is illuminated on the facet at the normal such that the incident angle on the multiple quantum well structure is 45° as shown in the inset. The dashed curve shows the normal incidence photoresponse, when the light is illuminated normal from the backside of the detector.

the peak near 7.2 μm with a similar FWHM is observed. For the latter, the shift of the peak to a shorter wavelength may be due to the sharing of phonon energy with momentum conservation processes such as impurity, alloy, and phonon scatterings. The photoresponsivity for both cases is about the same, 0.3 A/W. In the case of unpolarized light incident on the structure, a peak is found near 7.5 μm and the photoresponsivity is about 0.6 A/W, which is approximately the sum of the two polarization cases. The photoresponse is also measured by illuminating the light normal from the backside of the device. The photoresponse with a similar peak responsivity and width to the 90° polarization case is observed, as shown by the dashed curve in Fig. 22.

It is clear that for the 0° polarization case, the photoresponse is due to intersubband transitions between two heavy-hole subbands (*1*) with some contributions from the internal photoemission of holes, which are excited via free-carrier absorption. For the 90° polarization case, the intersubband transition is forbidden but the free-carrier absorption is stronger than that of the 0° polarization. This is because the entire photon electric field lies in the *xy*-plane. The photoresponse in this case is due to the internal photoemission of free carriers because the mixing of the hole bands due to the *s*-like conduction band and/or nonparabolicity is too small to cause significant absorption. The following explains the normal incident photoconduction process. If the energy of the incoming photon is large enough to create holes having an energy larger than the barrier height, the photoemitted holes can travel above the barriers and be collected under an applied electric field (internal photoemission). This gives rise to the photocurrent. As the photon energy decreases (wavelength increases), the photocurrent increases due to the large free-carrier absorption at longer wavelengths as shown in Fig. 13. When the photon energy is further decreased such that the energy for generating holes is lower than the barrier height, the photocurrent vanishes since the flow of holes is blocked by the potential barriers. This results in the cutoff of the photocurrent. This type of photoemission is similar to that observed by Lin and Maserjian for a SiGe/Si heterojunction (44). The difference is that for their case the free-carrier absorption is three dimensional. In our case, all multiple layers contribute to the photoemission and thus a larger quantum efficiency is expected. The internal quantum efficiency (η) of the detector for either polarization is given by $\eta = (1 - e^{-2\alpha l})$ (45). For the present detector, a quantum efficiency of $\eta \sim 14\%$ is obtained.

In order to estimate the mobility and the excited carrier lifetime, the bias dependence of responsivity (R) of the detector is measured using the glowbar–monochromator setup at 9.5 μm wavelength. The measured responsivity as a function of bias across the device is shown in Fig. 23. The responsivity rapidly increases as the bias is increased until about 2.5 V, where it saturates. The maximum value of R for the present device is

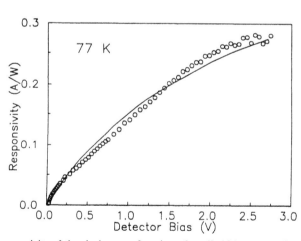

FIG. 23. Responsivity of the device as a function of applied bias across the device at 77 K. The solid curve is calculated values using Eq. (19).

about 0.3 A/W, and is comparable with that typically observed in III–V-based quantum well detectors. The experimental data is fitted to Eq. (19) using a saturated drift velocity (38) of about 1×10^7 cm s^{-1} and the measured absorption coefficient at 9.5 μm. The mobility of holes at 1000 cm^2/Vs and the excited carrier lifetime of 15 ps seems to fit the experimental data well as shown in the solid line of Fig. 23. The values of mobility and lifetime are comparable to that obtained for a typical GaAs/AlGaAs detector (45, 39). The nonpolar nature of Si/Ge should result in a longer intersubband lifetime compared with polar GaAs; however, the alloy and impurity scatterings can have a strong influence on the carrier lifetime. Since the responsivity is inversely proportional to the barrier width, it may be possible to further increase the responsivity by reducing the barrier width without degrading the detectivity because the large heavy-hole effective mass keeps the tunneling leakage current low. Next, we will discuss the photoresponse of the detectors based on the intervalence band transition.

The photoresponse of the structures, which showed intervalence band transition, has also been measured using mesa diodes similar to those already described except that the infrared light is illuminated from the backside of the substrate as schematically shown in the inset of Fig. 24.

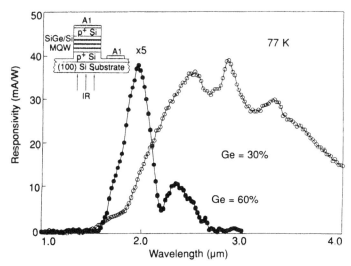

FIG. 24. Measured responsivity at 77 K as a function of wavelength for 30% and 60% samples. Infrared light is incident from the backside of the wafer (normal to the substrate plane) as shown in the inset.

Responsivity spectra as a function of wavelength at 77 K for the samples with 30 and 60% Ge compositions are shown in Fig. 24. It can be clearly seen from the measured result that, as the Ge composition is increased, the peak photoresponse moves toward a shorter wavelength due to the large splitting of the heavy-hole and split-off hole bands. In comparison with the absorption data at room temperature, the photoresponse shows several peaks, whereas the absorption spectrum has only one broad peak. The existence of several peaks is due to the transitions to both bound and continuum states, which can be discriminated in the photocurrent measurement (46). In order to better understand the detailed structure of the photoresponse, we have studied the bias dependence of the photocurrent at 77 K. For this measurement, the structure with 60% Ge was used because the larger barrier height keeps the leakage current low ($\sim$ 1 nA) even at a moderate bias ($\sim$ 0.5 V) across the device. The responsivity shown in Fig. 25 was obtained by measuring the voltage across a 4-MΩ load resistor used in the detector biasing circuit. At low bias, only the peak at 2.0 μm is clearly visible, indicating the excitation of holes into the continuum where the holes can easily reach the contact without having to tunnel through a barrier (45). However, as the bias across the device is increased, the peak at 2.4 μm grows rapidly as compared with the peak at 2.0 μm, and at about 0.5 V, both peaks become almost equal as can be

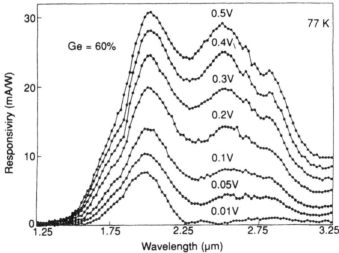

FIG. 25.   Bias dependence of responsivity spectra for the sample with 60% Ge in the well at 77 K.

seen from Fig. 25. This indicates that the 2.4-μm transition is to a bound state and, as the bias is increased, the photoexcited holes tunnel out easily, giving rise to a larger photocurrent. At about 0.5 V across the device, the two peaks nearly merge and look similar to that observed in the absorption measurement. For identification of the origin of these transitions, we have estimated the bound state energies for the three hole bands using the multiband self-consistent scheme described above including the potential due to heterostructure and the splitting of the hole bands due to strain. The potential well, calculated ground state energies, and wavefunctions of the three valence bands for the 60% Ge sample are shown in Fig. 26. For the doping concentration used, only the ground states of the heavy and light holes are populated as indicated by the Fermi level of Fig. 26. This indicates that for normal incidence, light transitions can occur between the heavy-hole ground state and the ground states of the light and split-off bands and the continuum states (2). The excited bound states mainly participate in the intersubband transition, which is forbidden for normally incident light (2). The observed peak at 2.0 μm is most likely due to the transition from the heavy-hole ground states to the continuum states (i.e., a mixture of the light, heavy, and split-off subbands) while the peak at 2.4 μm may be attributed to the transition from the heavy-hole ground state to the ground state of the split-off band. In the latter case, photoexcited carriers must tunnel through the barrier to reach the contact and should

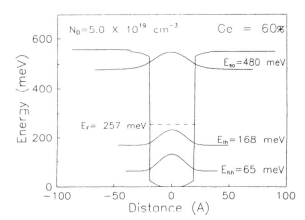

FIG. 26. Calculated wavefunctions and energies for the three hole bands for the sample with 60% Ge. In this plot, the hole energy is taken to be positive. For the doping used in this sample, only the ground states of the light and heavy holes are occupied as indicated by the Fermi energy. Many-body effects are also included in the calculation.

give a strong dependence on the external bias, which is in agreement with the measured bias-dependent data. The calculated positions for these transitions are 438 and 523 meV, respectively. These energies have been corrected for plasma shifts associated with subband transition as described earlier. From our data, the experimentally observed energies for the same transitions are 516 and 620 meV, respectively. The discrepancy between the experimental and the calculated peak positions arises mainly from the uncertainty in the determination of the exact doping concentration and to some extent the exclusion of nonparabolic effects in the analysis.

The detectivity of the detectors with different Ge compositions was estimated using Eq. (20) with measured dark current and the quantum efficiency. For the detector with 15% Ge, the estimated detectivity at 77 K is about $D^*(9 \ \mu m) = 1 \times 10^9 \ cm \ \sqrt{Hz}/W$ while for the one with 60% Ge is about $D^*(3 \ \mu m) = 4 \times 10^{10} \ cm \ \sqrt{Hz}/W$. These values are one order of magnitude lower than that of comparable GaAs/AlGaAs detectors. This is due mainly to the larger doping density required for attaining appreciable quantum efficiency.

VIII. Summary

In summary, quantum size effects in SiGe/Si heterostructures are discussed. A detailed discussion on infrared absorption between different subbands of $Si_{1-x}Ge_x$/Si multiple quantum wells and δ-doped structures is presented. The origins of different transitions were identified using the polarization-dependence measurements. The importance of many-body effects in determining transition energy is also demonstrated. Infrared detectors based on intersubband and intervalence band transitions have been demonstrated. The photoresponse was shown to cover both the 3–5- and 8–14-μm atmospheric windows. A possible avenue for improvement of the responsivity as well as the detectivity of the device was outlined. The continued progress in the study of SiGe intersubband transitions indicates that monolithic integration of SiGe/Si multiple quantum well infrared detectors with Si signal processing circuits should be realizable for focal plane applications.

Acknowledgments

The author would like to thank S. K. Chun, S. S. Rhee, Y. J. Mii, and C. H. Chern for collaboration in this work. This project was supported in part by the Army Research Office (Dr. John Zavada) and the Air Force Office of Scientific Research (Dr. Gerald Witt). One of us (R.P.G.K.) would like to thank Electrical Engineering Department of the National University of Singapore for the support during the preparation of the manuscript.

References

1. R. P. G. Karunasiri, J. S. Park and K. L. Wang, *Appl. Phys. Lett.* **59**, 2588 (1991).
2. J. S. Park, R. P. G. Karunasiri and K. L. Wang, *Appl. Phys. Lett.* **61**, 681 (1992).
3. D. D. Coon and R. P. G. Karunasiri, *Appl. Phys. Lett.* **45**, 649 (1984).
4. H. Hertle, G. Schuberth, E. Gornik, G. Abstreiter and F. Schaffler, *Appl. Phys. Lett.* **59**, 2977 (1991).
5. Chanho Lee and K. L. Wang, *Appl. Phys. Lett.* **60**, 2264 (1992).
6. R. People and J. C. Bean, *Appl. Phys. Lett.* **39**, 538 (1986).
7. R. Hull, J. C. Bean, F. Cerdeira, A. T. Fiory and J. M. Gibson, *Appl. Phys. Lett.* **48**, 56 (1986).
8. C. G. Van de Walle and R. Martin, *J. Vac Sci. Technol.* B **3**, 1256 (1985).
9. R. People, *Phys. Rev.* B **32**(2), 1405 (1985).
10. S. K. Chun and K. L. Wang, *IEEE Trans. Electron Devices* **39**, 2153 (1992).
11. L. L. Chang, L. Esaki and R. Tsu, *Appl. Phys. Lett.* **24**, 593 (1974).
12. H. C. Liu, D. Landheer, M. Buchanan and D. C. Houghton, *Appl. Phys. Lett.* **52**, 1809 (1988).
13. S. S. Rhee, J. S. Park, R. P. G. Karunasiri, Q. Ye and K. L. Wang, *Appl. Phys. Lett.* **53**, 204 (1988).
14. J. S. Park, R. P. G. Karunasiri, K. L. Wang, C. H. Chern and S. S. Rhee, *Appl. Phys. Lett.* **54**, 1564 (1989).
15. K. L. Wang, R. P. Karunasiri, J. Park, S. S. Rhee and C. H. Chern, *Superlattices and Microstructures* **5**(2), 201 (1989).
16. K. L. Wang and C. H. Chern, *Resonant tunneling of holes in strain layer structures*, "Proc. Workshop on Resonant Tunneling Physics and Applications, El Escorial, Spain, May 14–18, 1990."
17. Ulf Gennser, V. P. Kesan, S. S. Iyer, T. J. Bucelot and E. S. Yang, *J. Vac. Sci. Technol.* B **8**, 210 (1990).
18. R. P. G. Karunasiri, J. S. Park, K. L. Wang and Li-Jen Cheng, *Appl. Phys. Lett.* **56**(14), 1342 (1990).
19. R. P. G. Karunasiri, J. S. Park, Y. J. Mii and K. L. Wang, *Appl. Phys. Lett.* **57**, 2585 (1990).
20. J. S. Park, R. P. G. Karunasiri and K. L. Wang, *Appl. Phys. Lett.* **58**, 1083 (1991).
21. G. Bastard and J. A. Brum, *IEEE J. Quantum Electronics*, **22**, 1625 (1986).
22. E. O. Kane, *J. Phys. Chem. Solids* **1**, 82 (1956).
23. P. Man and D. S. Pan, *Appl. Phys. Lett.* **61**, 2799 (1992).
24. W. Kaiser, R. J. Collins and H. Y. Fan, *Phys. Rev.* **91**, 1380 (1953).
25. B. F. Levine, S. D. Gunapala, J. M. Kuo, S. S. Pei and S. Hui, *Appl. Phys. Lett.* **59**(15), 1864 (1991).
26. R. P. G. Karunasiri, K. L. Wang and J. S. Park, *in* "Semiconductor Interfaces and Microstructures (Z. C. Feng, ed.), World Scientific, New Jersey, 1992, p. 252.
27. A. D. Wieck, E. Batke, D. Heitmann and J. P. Kotthaus, *Phys. Rev.* B **30**, 4653 (1984).
28. Y. C. Chang and R. B. James, *Phys. Rev.* B **39**(17), 12672 (1989).
29. E. B. Doupont, D. Delacourt, D. Papillon, J. P. Schnell and M. Papuchon, *Appl. Phys. Lett.* **60**, 2121 (1992).
30. W. Kohn and L. J. Sham, *Phys. Rev.* **140**, 1133 (1965).
31. T. Ando, *Z. Phys.* B **26**, 263 (1977).
32. L. Hedin and B. I. Lundqvist, *J. Phys.* C **4**, 2064 (1971).
33. J. Hautman and L. M. Sander, *Phys. Rev.* B **30**(8), 7000 (1984).

34. S. J. Allen, D. C. Tsui and B. Vionter, *Solid State Commun.* **20**, 425 (1976).

35. W. L. Bloss, *Solid State Commun.* **46**, 143 (1983).

36. S. S. Rhee, R. P. G. Karunasiri, C. H. Chern, J. S. Park and K. L. Wang, *J. Vac. Sci. Technol. B*, **7**, 327 (1989).

37. A. G. Steele, H. C. Liu, M. Buchanan and Z. R. Wasilewski, *Appl. Phys. Lett.* **59**, 3625 (1991).

38. D. M. Caugley and R. E. Thomas, *Proc. IEEE* **55**, 2192 (1967).

39. C. G. Bethea, B. F. Levine, V. O. Shen, R. R. Abbott and S. J. Hseih, *IEEE Trans. Electron Devices* **38**, 1118 (1991).

40. M. A. Kinch and A. Yariv, *Appl. Phys. Lett.* **55**(20), 2093 (1989).

41. D. D. Coon, R. P. G. Karunasiri and L. Z. Liu, *Appl. Phys. Lett.* **47**, 289 (1985).

42. L. S. Yu and S. S. Li, *Appl. Phys. Lett.* **59**, 1332 (1991).

43. J. S. Park, R. P. G. Karunasiri and K. L. Wang, *Appl. Phys. Lett.* **60**(1), 103 (1992).

44. T. L. Lin and J. Maserjian, *Appl. Phys. Lett.* **57**, 1142 (1990).

45. B. F. Levine, C. G. Bethea, G. Hasnain, V. O. Shen, E. Pelve and P. R. Abbott, *Appl. Phys. Lett.* **56**(9), 851 (1990).

46. R. P. G. Karunasiri, J. S. Park and K. L. Wang, *Appl. Phys. Lett.* **61**, 2434 (1992).

Recent Developments in Quantum-Well Infrared Photodetectors

S. D. GUNAPALA AND K. M. S. V. BANDARA*

Center for Space Microelectronics Technology, Jet Propulsion Laboratory, California Institute of Technology, Pasadena, California

Permanent address: Department of Physics, University of Peradeniya, Peradeniya, Sri Lanka.

I. Introduction

Intrinsic infrared (IR) detectors in the long wavelength range (8–20 μm) are based on an optically excited interband transition, which promotes an electron across the band gap (E_g) from the valence band to the conduction band as shown in Fig. 1. These photoelectrons can be collected efficiently, thereby producing a photocurrent in the external circuit. Since the incoming photon has to promote an electron from the valence band to the conduction band, the energy of the photon ($h\nu$) must be higher than the E_g of the photosensitive material. Therefore, the spectral response of the detectors can be controlled by controlling the E_g of the photosensitive material. Examples for such materials are $Hg_{1-x}Cd_xTe$, and $Pb_{1-x}Sn_xTe$, in which the energy gap can be controlled by varying x. This means detection of very-long-wavelength IR radiation up to 20 μm requires small band gaps down to 62 meV. It is well known that these low band gap materials, characterized by weak bonding and low melting points, are more difficult to grow and process than large-band gap semiconductors such as GaAs. These difficulties motivate the exploration of utilizing the intersubband transitions in multiquantum well (MQW) structures made of more refractory large-band gap semiconductors (Fig. 2).

The idea of using MQW structures to detect IR radiation can be explained by using the basic principles of quantum mechanics. The quantum well is equivalent to the well-known particle in a box problem in quantum mechanics, which can be solved by the time independent Schrö-

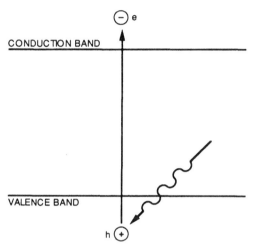

FIG. 1. Band diagram of conventional intrinsic infrared photodetector.

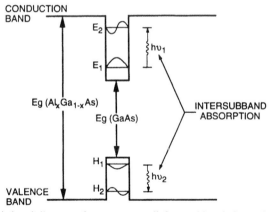

FIG. 2. Schematic band diagram of a quantum well. Intersubband absorption can take place between the energy levels of a quantum well associated with the conduction band (*n*-doped) or the valence band (*p*-doped).

dinger equation. The solutions to this problem are the eigenvalues that describe the energy levels inside the well in which the particle is allowed to exist. The position of the energy levels is primarily determined by the well dimensions (height and width). For infinitely high barriers and parabolic bands, the energy levels in the well are given by

$$E_j = \left(\frac{\hbar^2 \pi^2}{2m^* L_w^2} \right) j^2, \tag{1}$$

where L_w is the width of the quantum well, m^* is the effective mass of the carrier in the well, and j is an integer. Thus, the intersubband energy between the ground and the first excited state is

$$(E_2 - E_1) = (3\hbar^2 \pi^2 / 2m^* L_w^2). \tag{2}$$

The quantum well infrared photodetectors (QWIPs) discussed in this article utilize the photoexcitation of an electron (hole) between ground and first excited state in the conduction (valance) band quantum well (see Fig. 2). The quantum-well structure is designed so that these photoexcited carriers can escape from the well and be collected as photocurrent. In addition to larger intersubband oscillator strength, these detectors afford greater flexibility than conventional extrinsically doped semiconductor IR detectors because the wavelength of the peak response and cutoff can be continuously tailored by varying layer thickness (well width) and barrier composition (barrier height).

The lattice-matched $GaAs/Al_xGa_{1-x}As$ materials system is a very good candidate for creating such a potential well, because the band gap of $Al_xGa_{1-x}As$ can be changed continuously by varying x (and hence the height of the well). Thus, by changing the quantum well width L_w and the barrier height (Al molar ratio of $Al_xGa_{1-x}As$ alloy), this intersubband transition energy can be varied over a wide range from short wavelength IR (SWIR; 1–3 μm), through the midinfrared (MWIR; 3–5 μm), to long-wavelength infrared (LWIR; 8–12 μm), and into the very-long-wavelength (VLWIR; > 12 μm). It is important to note that unlike intrinsic detectors, which utilize interband transition, quantum wells of these detectors must be doped because the photon energy is not sufficient to create photocarriers ($h\nu < E_g$). Carriers can be introduced into the well by doping the GaAs well.

The possibility of using $GaAs/Al_xGa_{1-x}As$ MQW structures to detect IR radiation was first suggested by Esaki et al. (1) experimentally investigated by Smith et al. (2) and theoretically analyzed by Coon et al. (3). The first experimental observation of a strong intersubband absorption was performed by West et al. (4) and the first QWIP was demonstrated by

Levine *et al.* (*5*) (discussed in Section III) at Bell Laboratories. Levine *et al.* also introduced QWIP structures involving bound-to-continuum inter-subband transitions with wider $Al_xGa_{1-x}As$ barriers (*6*) and demonstrated dramatically improved detectivity. Recent developments in these detectors have already led to the demonstration of large 128×128 high-sensitivity starting arrays by several groups (7–9).

II. Intersubband Absorption

Unlike the interband transition, in intersubband transitions the optical matrix element is associated with the envelope functions only (*3*) i.e., between confined electron states or hole states of the quantum well. Thus, the optical dipole moment M can be expressed as

$$M \sim \int \phi_F(z)\vec{\varepsilon} \cdot \vec{r}\phi_I(z)dr \tag{3}$$

where ϕ_I and ϕ_F are initial and final envelope wavefunctions, $\vec{\varepsilon}$ is the polarization vector of the incident photons, and z is the growth direction of the quantum well. This yields a dipole matrix element in the order of the size of the quantum well for intersubband transitions compared to the order of the atomic size for interband transitions. For an infinite well, the value of the dipole matrix element $\langle z \rangle$ between the ground and first excited state is $16L/9\pi^2$ ($\sim 0.18\ L_w$). Since the envelope wavefunctions in Eq. (3) are orthogonal, M is nonzero due to the component of $\vec{\varepsilon} \cdot \vec{r}$ perpendicular to the quantum well (along growth direction). Therefore, the optical electric field must also have a component along this direction in order to induce an intersubband transition; thus normal incidence radiation will not be absorbed.

Because of this, West *et al.* (*4*) have used a Brewster's angle ($\theta_b = 73°$) geometry in order to give a large electric field component of incident light along the growth direction. Using this technique, a 5% absorption was measured at wavelength $\lambda = 8\ \mu m$, corresponding to a near unity oscillator strength. Levine *et al.* (*10*) have measured the IR intersubband absorption at 8.2 μm in a doped $GaAs/Al_xGa_{1-x}As$ quantum-well super-lattice using a multipass waveguide geometry as shown in Fig. 3. This multipass waveguide geometry increased the net intersubband absorption by approximately two orders of magnitude, hence allowing accurate mea-surements of the oscillator strength, the polarization selection rule, and the line shape even for single quantum well (*11–13*). Levine *et al.* (*10*)

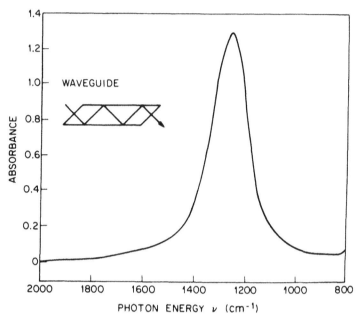

FIG. 3.   Measured QWIP absorbance vs. photon energy for the multipass waveguide geometry shown in the insert (*129*).

have also shown that the polarization selection rule $\alpha \propto \cos^2\theta$ (where θ is the angle between electric field and the z direction) approximately holds as shown in Fig. 4. Also it clearly shows that the absorption vanishes when the light polarization is in the plane of quantum wells (i.e., $\theta = 90°$).

Although these quantum wells have large optical dipole moments, this by itself is not useful for detection unless photoexcited carriers readily escape from the excited state. For bound-to-bound transitions, where both ground and excited states are classically bound within the well, escaping of the photoexcited electrons will be partly suppressed due to the barrier height seen by the electrons in the excited state. By decreasing the width or barrier height of a quantum well that contains two bound states, the excited bound state can be pushed into the continuum, resulting in a larger escape probability of photoexcited electrons. Using this bound-to-continuum design and by increasing the barrier thickness to reduce the tunneling dark current, Levine *et al.* (*14*) dramatically improve the QWIP performances.

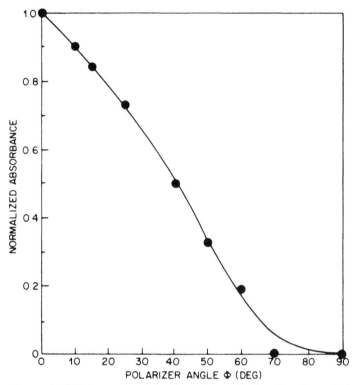

FIG. 4. Measured QWIP intersubband absorbance (normalized to $\phi = 0$) vs. polarizer angle at Brewster's angle $\theta_B = 73°$. The solid line is drawn through the points as a guide to the eye (129).

Absorption peak wavelength (λ_p) of bound-to-bound intersubband transitions is simply determined by the ground (E_0) and excited (E_1) state energy difference.

$$\lambda_p = \frac{2\pi\hbar c}{E_1 - E_0} \tag{4}$$

Typical calculations of E_0 and E_1 require only basic quantum mechanics and boundary conditions linking envelope wave functions of the well and barrier sides of the heterointerface (15). The absorption coefficient $\alpha(\hbar\omega)$ associated with an optical transition of an electron promoted from the ground state E_0 to excited state E_1 by absorbing a photon $\hbar\omega$ can be expressed as

$$\alpha(\hbar\omega) = \frac{\rho_S}{L}\frac{\pi e^2 \hbar}{2n_w \varepsilon_0 m^* c}\frac{\sin^2\theta}{\cos\theta}f(E_1)g(E_1),\tag{5}$$

where ρ_S is the two dimensional carrier density in the quantum well, L is the period length of MQW structure, m^* is the effective mass of the electrons in the well, n_w is the index of refraction, θ is the angle between the direction of the optical beam and the normal to the detector surface, $f(E_1)$ is the oscillator strength, and $g(E_1)$ is the one-dimensional density of the final state. When the scattering effects are neglected, the density of final state $g(E_1)$ in the continuum is given simply by

$$g(E_1) = \frac{L}{2\pi\hbar}\left(\frac{m_b^*}{2}\right)^{1/2}\frac{1}{\sqrt{E_1 - H}},\tag{6}$$

where m_b^* is the effective mass of the electrons in the barrier and H is the barrier height. For dipole transition between two energy states E_0 and E_1 is given by (16)

$$f(E_1) = \frac{2\hbar^2}{m^*(E_1 - E_0)}\left|\left\langle 1\left|\frac{\partial}{\partial x}\right|0\right\rangle_w + \frac{m^* n_w}{m_b^* n_b}\left\langle 1\left|\frac{\partial}{\partial x}\right|0\right\rangle_b\right|^2\tag{7}$$

where n_b is the refractive index of the barrier.

For a quantum well with finite barriers, one can write exact wavefunctions for ground and excited states in both well and barrier regions and evaluate the oscillator strength. [See Ref. (16) for a detailed calculation.] According to Eq. (5), the photon energy at the absorption peak is determined by the product of $f(E_1)$ and $g(E_1)$; however, due to the singularity of $g(E_1)$ at $E_1 = H$ [see Eq. (6)], the absorption strongly peaks close to the barrier height. In reality, the density of states in the continuum is strongly modified by the presence of the well and is broadened by impurity scattering and both will tend to smooth out the singularity. After considering the effect of the well on the density of states, it is safe to assume (16) that local density of states in the continuum varies relatively slowly compared to $f(E_1)$ when E_1 is close to H. Under this assumption, the absorption peak is approximately determined by the energy dependence of the oscillator strength f. Hence, the peak absorption wavelength λ_p can be obtained by computing the value of $E_1 = E_m$ where f is maximized:

$$\lambda_p = \frac{2\pi\hbar c}{E_m - E_1}. \tag{8}$$

Numerical calculations for λ_p based on the above equations were carried out by Choi; Fig. 5 shows the energy of the final state E_m at which f is maximum as a function of well width for different values of the Al molar ratio x of the barrier. For a given value of x, there is a maximum value of E_m, which is associated with a resonant state above the barrier where the wave amplitude in the well region is maximized. Figure 6 shows the calculated oscillator strength f at $E_1 = E_m$ as a function of well width. Figure 7 shows the absorption peak wavelength λ_p as a function of well width. Within the detector parameters shown, λ_p can be varied from 5 to over 25 μm. Although there is a lower limit for λ_p at a given x, there is·no apparent upper limit as long as the quantum well width is technologically predictable.

Detailed theoretical and experimental analyses of intersubband absorption peak wavelength λ_p and spectral band width have been done as a function of quantum-well width, barrier height, temperature, and doping density N_D in the well (17–25). Bandara et al. (26) have shown that, for high doping densities ($N_D > 10^{18}$ cm^{-3}), exchange interaction can significantly lower the ground-state subband energy and that the direct Coulomb shift can increase the excited-state energy; consequently the peak absorp-

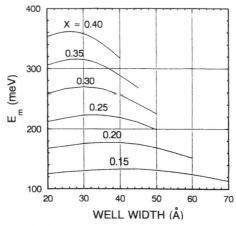

FIG. 5.   The energy of the final state E_m at which f is a maximum for a given Al molar ratio x as a function of well width, (16).

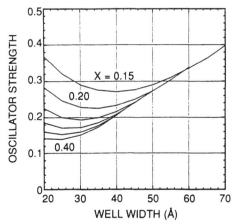

FIG. 6. The maximum oscillator strength f for a given Al molar ratio x as a function of well width for a barrier thickness $L_B = 300$ Å; x changes 0.05 in steps between each curve (16).

tion wavelength shifts to higher energy. In addition to absorption peak shift at high doping densities, the absorption line width broadens and the oscillator strength increases linearly with the doping density. Ando et al. (27) have theoretically analyzed this and shown that most of the shift in intersubband energy is due to the collective plasma oscillation and a

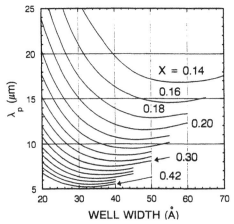

FIG. 7. The position of the absorption peak λ_p for a given Al molar ratio x as a function of well width; x changes 0.02 in teps between each curve (16).

smaller contribution from the exciton-like shift. Ramsteiner *et al.* (*17*) have experimentally confirmed this by measuring the Raman scattering from both the collective plasma mode and the single particle excitation.

Furthermore, there are shifts in the peak absorption wavelength and the absorption linewidth with the temperature (*18–24*). When the temperature is lowered from room temperature to cryogenic temperatures, there is a small decrease in the position of the peak absorption wavelength λ_p and the absorption linewidth Δv. This shift with temperature was experimentally analyzed by Hasnain *et al.* (*21*), who has observed an increase in the peak absorption by a factor of 1.3. This increase in peak absorption is due to the decrease in linewidth and the unchanging oscillator strength. Manasreh *et al.* (*23*) have explained this temperature shift by including the collective plasma, exciton-like, Coulomb and exchange interactions, nonparabolicity, and the temperature dependence of the band gaps and effective mass.

III. Bound-to-Bound State QWIPs

As mentioned previously, the first *bound-to-bound* state QWIP was demonstrated by Levine *et al.* (*5*), which consisted of 50 periods of $L_w = 65$-Å GaAs and $L_b = 95$-Å $Al_{0.25}Ga_{0.75}As$ barriers sandwiched between top (0.5 μm thick) and bottom (1 μm thick) GaAs contact layers. The center 50 Å of the GaAs wells were doped to $N_D = 1.4 \times 10^{18}$ cm^{-3} and the contact layers were doped to $N_D = 4 \times 10^{18}$ cm^{-3}. This structure was grown by molecular-beam epitaxy (MBE). These thicknesses and compositions were chosen to produce only two states in the well with energy spacing close to 10 μm. The measured (*5*) absorption spectrum is shown in Fig. 8, which peaked at 920 cm^{-1} ($\lambda_p = 10.9$ μm) with a full width at half maximum of $\Delta v = 97$ cm^{-1} corresponding to an excited state lifetime of $\tau_2 = (\pi\Delta v)^{-1} = 1.1 \times 10^{-13}$ s. The peak absorbance $A = -\log(\text{transmission}) = 2.2 \times 10^{-2}$ corresponds to net absorption of 5% (i.e., $\alpha = 600$ cm^{-1}). The oscillator strength f was calculated using Eq. (3), which yields $f = 0.6$ is in good agreement with the theoretical value $f = 0.8$.

After the absorption of IR photons, the photoexcited carriers can be transported either along the plane of the quantum wells (with an electric field along the epitaxial layers) or perpendicular to the wells (with an electric field perpendicular to the epitaxial layers). As far as the IR detection is concerned, perpendicular transport is superior to parallel

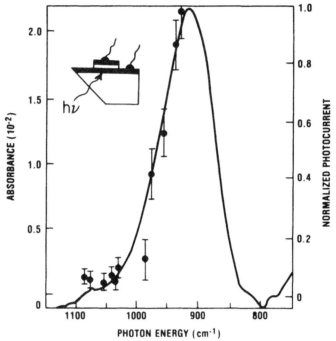

FIG. 8. Absorption spectrum of a bound-to-bound QWIP. The solid points are photocurrent vs. photon energy (normalized to the peak absorbance). The inset shows the device geometry (*129*).

transport (*28*) since the difference between the excited state and ground state mobilities is much larger in the latter case; consequently, transport perpendicular to the wells (i.e., growth direction) gives a substantially higher photocurrent. In addition, the heterobarriers block the transport of ground-state carriers in the wells, and thus lower dark current. For these reasons, QWIPs are based on escape and perpendicular transport of photoexcited carriers as shown in Fig. 9.

In order to measure (*5*) the optical responsivity of the detector, 50-μm-diameter mesas were chemically etched through the superlattice. Ge/Au ohmic contacts were made to the top and bottom *n*-type contact layers. In addition, as shown in Fig. 10 a 45° angle was polished on the substrate to allow the IR radiation to back illuminate the detector at 45° angle of incidence. This allows for a large optical electric field along the growth direction. Levine *et al.* (*5*) have obtained the 10-μm IR radiation

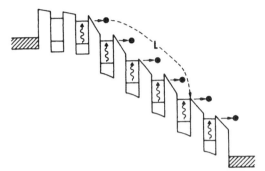

FIG. 9. Photoconductivity produced by absorption of infrared radiation (intersubband transition) followed by tunneling out of well (*129*).

to test this detector, from a grating CO_2 laser, which was tunable from 9.2 to 10.8 μm. The CO_2 laser was normalized to a reference detector and also highly attenuated so that power levels of < 10 μW were incident on the active device area. The photocurrent signal was found to be independent of temperature from 15 to 80 K. A strongly resonant character of the photocurrent has also been observed; as expected it closely agreed with the measured absorption spectra (Fig. 8). Figure 11 shows the normalized photoresponse versus polarizer angle, which clearly indicates that the optical transition dipole moment is along the growth direction. The peak responsivity and the hot-electron mean free path of this first bound-to-bound QWIP were determined to be (*5*) 0.52 A/W and 2500-Å, respectively. It is worth noting that this responsivity and hot-electron mean free

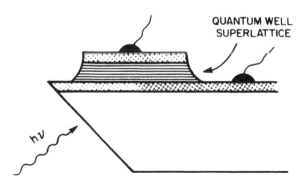

FIG. 10. The infrared radiation incident on a 45° polished substrate face. The cross-hatched layers are the n^+ contact layers with the active multiquantum-well layers between them.

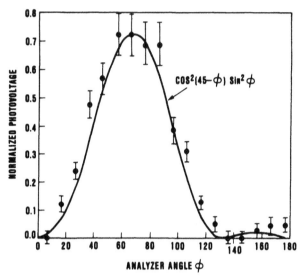

FIG. 11. Polarization dependence of QWIP photo signal. The solid points are experimental data and the curve is obtained by theory (*129*).

path are quite comparable to those of the present state-of-the-art *bound-to-continuum* state QWIPs (see Section IV). However, the dark current (mostly due to tunneling) associated with these thin barrier detectors were much worse than the *bound-to-continuum* QWIPs.

In the latter versions of the bound-to-bound state QWIPs, Choi *et al.* (*29*) have used slightly thicker and higher barriers to reduce this huge tunneling dark current. When they increased the barrier thickness from $L_b = 95$ to 140 Å and $Al_xGa_{1-x}As$ barrier height from $x = 0.25$ to 0.36, the dark current (also the photocurrent) was significantly reduced (Fig. 12). The nonlinear behavior of the responsivity and the dark current vs. bias voltage is due to the complex tunneling process associated with the high-field domain formation (*29*).

IV. Bound-to-Continuum State QWIPs

A. *n*-Doped QWIPs

1. Structure. In the previous section, we mentioned the QWIP containing two bound states. By reducing the quantum-well width, it is possible to

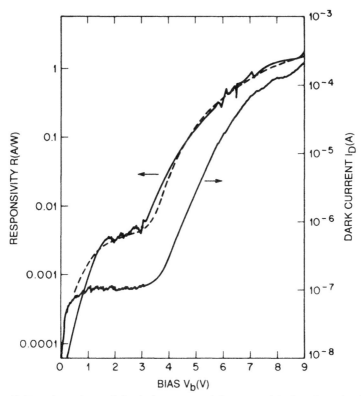

FIG. 12. Voltage dependence of the dark current and the responsivity for a bound-to-bound QWIP. The dashed curve is theory (*129*).

push the bound excited state into the continuum resulting in a strong *bound-to-continuum* intersubband absorption. The major advantage of the bound-to-continuum QWIP is that the photoexcited electron can escape from the quantum-well to the continuum-transport states without tunneling through the barrier, as shown in Fig. 13. As a result, the bias required to efficiently collect the photoelectrons can be reduced dramatically, and hence lower the dark current. Due to the fact that the photoelectrons do not have to tunnel through the barriers, the $Al_xGa_{1-x}As$ barrier thickness of bound-to-continuum QWIP can now be increased without reducing the photoelectron collection efficiency. Increasing the barrier width from a few hundred to 500-Å can reduce the ground-state sequential tunneling by an order of magnitude. By making use of these improvements, Levine *et al.*

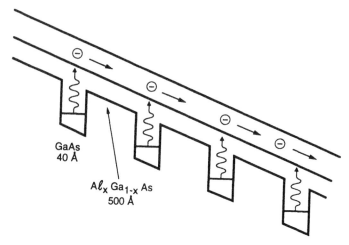

FIG. 13.   Conduction-band diagram for a bound-to-continuum QWIP, showing the photoexcitation and hot-electron transport process.

(14) have successfully demonstrated the first *bound-to-continuum* QWIP with a dramatic improvement in the performance (i.e., detectivity 3×10^{10} cm$\sqrt{\text{Hz}}$/W at 68 K for a QWIP, which had cutoff wavelength at 10 μm). Due to this high performance and the excellent uniformity of GaAs-based QWIPs, several groups have demonstrated large (i.e., 128 × 128) imaging focal plane arrays (7–9).

Now we will discuss and compare the optical and transport properties of bound-to-continuum QWIPs with bound-to-bound QWIPs. The structures of the six samples to be discussed are listed in Table I. These *n*-doped QWIPs were grown using MBE and the wells and contact layers were doped with Si. The quantum-well widths L_w range from 40 to 70 Å, whereas the barrier widths are approximately constant at $L_b = 500$ Å. The Al molar fraction in the Al$_x$Ga$_{1-x}$As barriers varies from $x = 0.10$ to 0.31 (corresponding to cutoff wavelengths of $\lambda_c = 7.9$–19 μm). The photosensitive doped MQW region (doping density N_D containing 25 to 50 periods) is sandwiched between similarly doped top (0.5 μm) and bottom (1 μm) ohmic contact layers. These structural parameters have been chosen to give a very wide variation in the QWIP absorption and transport properties (30). In particular, samples A through D are *n*-doped, with intersubband IR transition occurring between a single localized *bound* state in the well and a delocalized state in the *continuum* (10, 14, 31–38) (denoted B–C in Table I). Sample E has a high Al concentration $x = 0.26$ coupled with a

TABLE I
STRUCTURE PARAMETERS FOR SAMPLES A–F

Sample	L_w (Å)	L_b (Å)	x	N_D $(10^{18}\ cm^{-3})$	Doping type	Periods	Intersubband Transition
A	40	500	0.26	1	n	50	B–C
B	40	500	0.25	1.6	n	50	B–C
C	60	500	0.15	0.5	n	50	B–C
D	70	500	0.10	0.3	n	50	B–C
E	50	500	0.26	0.42	n	25	B–B
F	50	50	0.30	0.42	n	25	B–QC
		500	0.26				

Note. These parameters include quantum-well width L_w; barrier width L_b, $Al_xGa_{1-x}As$ composition x; doping density N_D; doping type; number of multiquantum-well periods, and type of intersubband transition, *bound-to-continuum* (B–C); *bound-to-bound* (B–B), and *bound-to-quasicontinuum* (B–QC) (*30*).

wide well $L_w = 50$ Å, yielding two bound states in the well. Thus, the intersubband transition from the *bound* ground state to the *bound* first excited state (denoted B–B in Table I) and therefore requires electric-field-assisted tunneling for the photoexcited carrier to escape into the continuum as discussed in the previous section (*5, 39, 40*). Sample F was designed to have a *quasicontinuum* excited state (*41*) (designated B–QC in Table I), which is intermediate between a strongly bound excited state and a weakly bound continuum state. It consists of a $L_W = 50$-Å doped quantum well surrounded by 50 Å of a high barrier $x = 0.30$ and 500 Å of a low barrier $x = 0.26$. These quantum-well parameters result in an excited state bound by the high barriers but in the continuum of the low barriers, and are thus expected to have an intermediate behavior. (See the inserts in Fig. 15 for a schematic conduction band diagram of all three types of QWIPs.)

2. Absorption Spectra. The infrared absorption spectra for samples A–F were measured at room temperature, using a 45° multipass waveguide geometry as described in the previous section (except for sample D, which was designed for such a long wavelength that the substrate multiphonon absorption obscured the intersubband transition). As can be readily seen in Fig. 14, the spectra of the *bound-to-continuum* QWIPs (samples A, B, and C) are much broader than the *bound-to-bound* or *bound-to-quasicontinuum* QWIPs (samples E and F or the QWIPs discussed in the previous section). Correspondingly, the magnitude of the absorption coefficient α

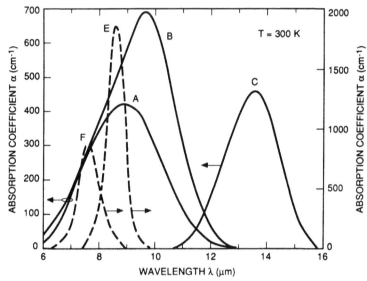

FIG. 14. Absorption coefficient spectra vs. wavelength measured at $T = 300$ K for samples
A, B, C, E, and F (30).

for the continuum QWIPs (left-hand scale), is significantly lower than that
for the bound or quasicontinuum QWIPs (right-hand scale), due to the
conservation of oscillator strength. That is, $\alpha_p(\Delta\lambda/\lambda)N_D$ is highly constant, as was found previously (42). The values of the peak room-tempera-
ture absorption coefficient α_p, peak wavelength λ_p, cutoff wavelength λ_c
(long wavelength λ for which α drops to half α_p) and spectral width $\Delta\lambda$
(full width at half α_p) are given in Table II. The room-temperature
absorption quantum efficiency η_a (300 K) evaluated from α_p (300 K) using

$$\eta_a = \frac{1}{2}(1 - e^{-2\alpha_p l}), \qquad (9)$$

where η_a is the unpolarized double-pass absorption quantum efficiency; l
is the length of the photosensitive region and the factor of 2 in the
denominator is a result of the quantum–mechanical selection rules, which
only allow the absorption of radiation polarized in the growth direction.
The low-temperature quantum efficiency η_α (77 K) was obtained by using
α_p (77 K) $\approx 1.3\ \alpha_p$ (300 K) as discussed previously. The last column
containing η_{max} will be discussed later.

   In order to more clearly compare the line shapes of the *bound, quasicontinuum*, and *continuum* QWIPs, the absorption coefficients for samples

TABLE II

OPTICAL ABSORPTION PARAMETERS FOR SAMPLES A–C, E, AND F

Sample	λ_p (μm)	λ_c (μm)	$\Delta\lambda$ (μm)	$\Delta\lambda/\lambda$ (%)	α_p (300 K) (cm^{-1})	η_a (300 K) (%)	η_a (77 K) (%)	η_{max} (%)
A	9.0	10.3	3.0	33	410	10	13	16
B	9.7	10.9	2.9	30	670	15	19	25
C	13.5	14.5	2.1	16	450	11	14	18
E	8.6	9.0	0.75	9	1820	17	20	20.5
F	7.75	8.15	0.85	11	875	11	14	20.5

Note. These parameters include peak absorption wavelength λ_p, long-wavelength cutoff λ_c; spectral width $\Delta\lambda$; fractional spectral width $\Delta\lambda/\lambda$; peak room-temperature absorption coefficient α_p (300 K); peak room-temperature absorption quantum efficiency η_a (300 K); $T = 77$ K absorption quantum efficiency η_a (77 K), and maximum high bias net quantum efficiency η_{max} (*30*).

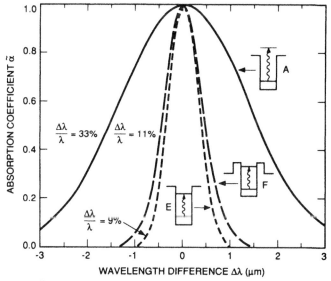

FIG. 15. Normalized absorption spectra vs. wavelength difference $\Delta\lambda = (\lambda - \lambda_p)$. The spectral width $\Delta\lambda/\lambda$ are also given. The insert show the schematic conduction band diagram for sample A (*bound-to-continuum*), sample E (*bound-to-bound*), and sample F (*bound-to-quasicontinuum*) (*30*).

A, E, and F have been normalized to unity and plotted as $\bar{\alpha}$ in Fig. 15 and the wavelength scale has been normalized by plotting the spectra against $\Delta\lambda \equiv (\lambda - \lambda_p)$, where λ_p is the wavelength at the absorption peak. The very large difference in spectral width is apparent with the bound excited-state transitions ($\Delta\lambda/\lambda = 9\%-11\%$) being three to four times narrower than for the continuum excited-state QWIPs ($\Delta\lambda/\lambda = 33\%$).

3. Dark Current. In order to measure the dark current–voltage curves, 200-μm-diameter mesas were fabricated as described in the previous section and the results are shown in Fig. 16 for $T = 77$ K. Note that the asymmetry in the dark current (*43*) with I_d being larger for positive bias (i.e., mesa top positive) than for negative bias. This can be attributed to the dopant migration in the growth direction, which lowers the barrier height of the quantum wells in the growth direction compared to the quantum-well barriers in the other direction (which are unaffected). Note that as expected, the dark current I_d increases as the cutoff wavelength λ_c increases. At very low bias, however, the curves for samples E and F cross. This is due to the fact that even though sample E has a longer-wavelength cutoff than sample F, it has a *bound* excited state so that it is difficult for the excited electrons to tunnel out at low bias. In contrast, sample F has a

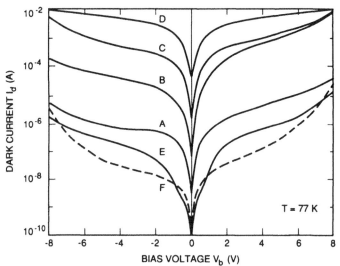

FIG. 16. Dark current I_d as a function of bias voltage V_b at $T = 77$ K for samples A–F (*30*).

quasicontinuum excited state, which is above the $L_b = 500$-Å-thick barrier and can thus tunnel out and escape at zero bias.

Levine et al. (14, 42-45) have analyzed the origin of the dark current in detail and shown that thermionic-assisted tunneling is a major source of dark current. In that analysis, they determined the effective number of electrons $n(V)$, which are thermally excited out of the well into the continuum transport states, as a function of bias voltage V,

$$n(V) = \left(\frac{m^*}{\pi \hbar^2 L_p} \right) \int_{E_0}^{\infty} f(E) T(E, V) \, dE, \qquad (10)$$

where the first factor containing the effective mass m^* is obtained by dividing the two-dimensional density of states by the superlattice period L_p (to convert it into an average three-dimensional density), and where $f(E)$ is the Fermi factor $f(E) = [1 + \exp(E - E_0 - E_F)/kT]^{-1}$, E_0 is the ground state energy, E_F is the two-dimensional Fermi level, and $T(E, V)$ is the bias-dependent tunneling current transmission factor for a single barrier, which can be calculated using Wentzel–Kramers–Brillouin (WKB) approximation to a biased quantum well. Equation (10) accounts for both thermionic emission above the energy barrier E_b (for $E > E_b$) and thermionically assisted tunneling (for $E < E_b$). The bias dependent dark current $I_d(V)$ has been calculated using $I_d(V) = n(V) e v(V) A$, where e is the electronic charge, A is the area of the detector, and v is the average transport velocity given by

$$v(V) = \mu F \left[1 + (\mu F / v_s)^2 \right]^{1/2}, \qquad (11)$$

where μ is the mobility, F is the average electric field, and v_s is the saturated drift velocity. Good agreement is achieved as a function of both bias voltage and temperature over a range of eight orders of magnitude in dark current (14).

4. Responsivity. The responsivity wavelength spectra $R(\lambda)$ were measured (43) on 200-μm-diameter mesa detectors using a polished 45° incident facet on the detector as shown in Fig. 10, together with a globar source and a monochromator. A dual lock-in ratio system with a spectrally flat pyroelectric detector was used to normalize the system spectral response due to wavelength dependence of the blackbody, spectrometer, filters, etc. The absolute magnitude of the responsivity was accurately determined by measuring the photocurrent I_p with a calibrated black-body source. This

photocurrent is given by

$$I_p = \int_{\lambda_1}^{\lambda_2} R(\lambda)P(\lambda)\, d\lambda, \qquad (12)$$

where λ_1 and λ_2 are the integration limits that extend over the responsivity spectrum, and $P(\lambda)$ is the black-body power per unit wavelength incident on the detector, which is given by

$$P(\lambda) = W(\lambda)\sin^2(\theta/2)AF \cos \phi, \qquad (13)$$

where A is the detector area, ϕ is the angle of incidence, θ is the optical field-of-view angle [i.e., $\sin^2(\theta/2) = (4f^2 + 1)^{-1}$, where f is the f-number of the optical system; in this case, θ is defined by the radius ρ of the black-body opening at a distance D from the detector, so that $\tan(\theta/2) = \rho/D$], F represents all coupling factors and $F = T_f(1 - r)C$, where T_f is the transmission of the filters and windows, $r = 28\%$ is the reflectivity of the GaAs detector surface, and C is the optical-beam chopper factor ($C = 0.5$ in an ideal optical-beam chopper), and $W(\lambda)$ is the black-body spectral density given by the following equation (i.e., the power radiated per unit wavelength interval at wavelength λ by a unit area of a blackbody at temperature T_B):

$$W(\lambda) = (2\pi c^2 h/\lambda^5)(e^{hc/\lambda k T_B} - 1)^{-1}. \qquad (14)$$

By combining Eqs. (12) and (13), and using $R(\lambda) = R_p \tilde{R}(\lambda)$, where R_p is the peak responsivity and $\tilde{R}(\lambda)$ is normalized (at peak wavelength λ_p) experimental spectral responsivity, we can rewrite the photocurrent I_p as

$$I_p = R_p G \int_{\lambda_1}^{\lambda_2} \tilde{R}(\lambda)W(\lambda)\, d\lambda, \qquad (15)$$

where G represents all the coupling factors and is given by $G = \sin^2(\theta/2)AF \cos \phi$. Thus, by measuring the $T_B = 1000$-K black-body photocurrent, R_p can be accurately determined.

The normalized responsivity spectra $\tilde{R}(\lambda)$ are given in Fig. 17 for samples A–F, where we again see that the *bound* and *quasicontinuum* excited state QWIPs (samples E and F) are much narrower $\Delta\lambda/\lambda = 10\%$–$11\%$ than the *continuum* QWIPs $\Delta\lambda/\lambda = 19\%$–$28\%$ (samples A–D). Table III gives the peak responsivity wavelengths λ_p and cutoff wavelengths λ_c as well as the responsivity spectral width $\Delta\lambda$. These responsivity spectral parameters given in Table III are similar to the corresponding absorption values listed in Table II.

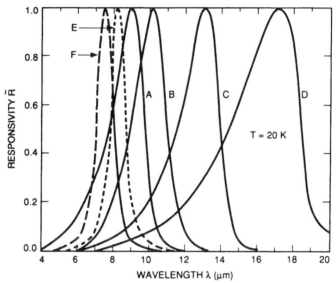

FIG. 17. Normalized responsivity spectra vs. wavelength measured at $T = 20$ K for samples A–F (*30*).

TABLE III
RESPONSIVITY SPECTRAL PARAMETER FOR SAMPLES A–F

Sample	λ_p (μm)	λ_c (μm)	$\Delta\lambda$ (μm)	$\Delta\lambda/\lambda$ (%)
A	8.95	9.8	2.25	25
B	9.8	10.7	2.0	20
C	13.2	14.0	2.5	19
D	16.6	19	4.6	28
E	8.1	8.5	0.8	10
F	7.5	7.9	0.85	11

Note. These parameters include peak responsivity wavelength λ_p; long-wavelength cutoff λ_c; spectral width $\Delta\lambda$, and fractional spectral width $\Delta\lambda/\lambda$ (*30*).

The absolute peak responsivity R_p can be written in terms of quantum efficiency η and photoconductive gain g as

$$R_p = (e/h\nu)\eta g. \tag{16}$$

The bias dependence of R_p is shown in Fig. 18. Note that at low bias the responsivity is nearly linearly dependent on bias and it saturates at high bias. This saturation occurs due to the saturation of drift velocity. For the longest-wavelength sample D, where $\lambda_c = 19$ μm, the dark current becomes too large at high bias to observe the saturation in R_p.

Responsivity versus bias voltage plot for the *bound* and *quasicontinuum* sample is shown in Fig. 19. Whereas the *quasicontinuum* QWIP (sample F) behaves quite similarly to the *continuum* QWIPs of Fig. 18, the fully *bound* sample E has a significantly different shape. The responsivity does not start out linearly with bias but is in fact zero for finite bias. That is, there is a zero bias offset of more than 1 V, due to the necessity of field-assisted tunneling for the photoexcited carriers to escape from the well (*5, 39, 40, 46*).

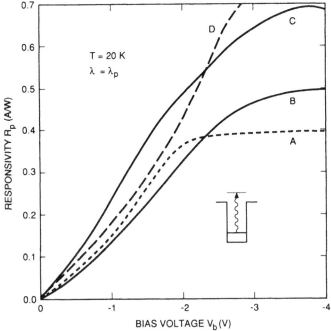

FIG. 18. Bias-dependent peak ($\lambda = \lambda_p$) responsivity R_p measured at $T = 20$ K for samples A–D. The insert shows the conduction band diagram (*30*).

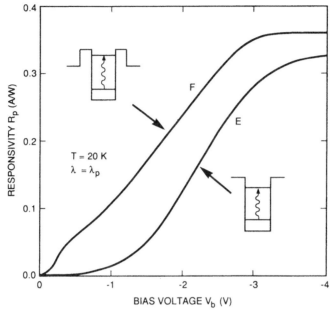

FIG. 19. Bias dependent peak ($\lambda = \lambda_p$) responsivity R_p measured at $T = 20$ K for samples E and F. The inserts show the conduction band diagram (*30*).

5. Dark Current Noise. The dark current noise i_n was measured on a spectrum analyzer for all of the samples at $T = 77$ K as a function of bias voltage (*43*). The result for sample B is shown in Fig. 20. The solid circles were measured for negative bias (mesa top negative), whereas the open circles are for positive bias. The smooth curves are drawn through the experimental data. Note that the current shot noise for positive bias is much larger than that for negative bias (e.g., at $V_b = 3.5$ V it is 4 times larger), and also that near $V_b = 4$ V there is a sudden increase in the noise due to a different mechanism [possibly due to the avalanche gain (*47*) process]. This asymmetry in the dark current noise is due to the previously mentioned asymmetry in I_d. The photoconductive gain g can now be obtained using the current shot noise expression (*14, 43,48*).

$$i_n = \sqrt{4eI_d g\Delta f},\tag{17}$$

where Δf is the band width (taken as $\Delta f = 1$ Hz). This expression is valid for small quantum well capture probabilities (i.e., $p_c \ll 1$). QWIPs satisfy this condition at usual operating bias (i.e., 2–3 V). A more general formula $i_n^2 = 4eI_d g\Delta f(1 - p_c/2)$ was derived by Beck (*49*), which can apply even in

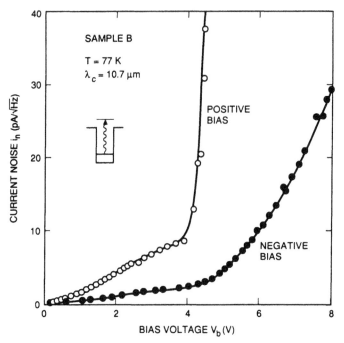

FIG. 20.   Dark current noise i_n (at $T = 77$ K) vs. bias voltage V_b for sample B. Both positive (open circles) and negative (solid circles) bias are shown. The smooth curves are drawn through the measured data. The insert shows the conduction band diagram (*30*).

low-bias conditions where capture probabilities for carriers traversing the wells are high.

6. Photoconductive Gain. Combining i_n from Fig. 20 and I_d from Fig. 16 allows the experimental determination of g as shown in Fig. 21. The solid circles are for negative bias, whereas the open circles are for positive bias, and the smooth curves are drawn through the experimental points. As shown in Fig. 21, the photoconductive gain increases approximately linearly with the bias at low voltage and saturates near $V_b = 2$ V (due to velocity saturation) at $g \sim 0.3$. It is worth noting that in spite of the large difference between the noise current i_n for positive and negative bias (as shown in Fig. 20), the photoconductive gains are quite similar. This demonstrates that the asymmetry in i_n is due quantitatively to the asymmetry in I_d. It further shows that, although the number of carriers which

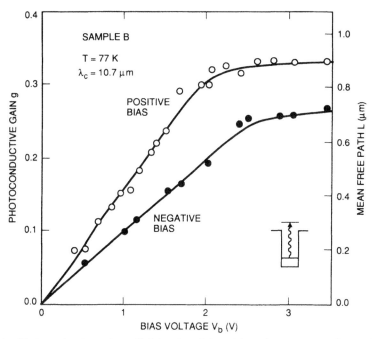

FIG. 21. Photoconductive gain g (left-hand scale) and hot electron mean free path L (right-hand scale) vs. bias voltage V_b for sample B at $T = 77$ K. Both positive (open circles) and negative (solid circles) bias are shown. The smooth curves are drawn through the measured data. The insert shows the conduction band diagram (30).

escape from the well (with probability p_e) and enter the continuum is strongly dependent on bias direction (due to the asymmetrical growth interfaces), the continuum transport, i.e., photoconductive gain, is less sensitive to the direction of carrier motion. The reason for this difference is that the escape probability p_e (and hence I_d), depends exponentially on the bias (30), whereas photoconductive gain is only linearly dependent on bias V_b. The photoconductive gain of QWIPs can be written as (48)

$$g = L/l, \qquad (18)$$

where L is the hot-carrier mean free path and l is the superlattice length ($l = 2.7$ μm for sample B). Therefore, we can evaluate g as shown on the right hand scale of Fig. 21. Thus, for this device L saturates at ~ 1 μm. As mentioned above, the dramatic increase in i_n near $V_b = 4$ V in Fig. 20

is due to additional noise mechanisms and therefore should not be attributed to a striking increase in g.

The photoconductive gains for the other bound-to-continuum n-doped QWIP samples are shown in Figs. 22–24. Note that, unlike sample B, the gain for sample C is just beginning to saturate, whereas sample D shows no sign of saturation. This is similar to the responsivity behavior shown in Fig. 18. From the strong saturation of R_p for sample A (in Fig. 18) we would also expect the photoconductive gain to be completely saturated for $|V_b| > 2$ V, which is how we have drawn the smooth curve in Fig. 24. However, if one obtains g by simply substituting the measured i_n into Eq. (17), the result is the dotted line in Fig. 24. Just as interpreting the large excess noise i_n above 4 V in Fig. 20 as I_d shot noise would lead to an incorrectly large gain, interpreting the low i_n above $V_b = -2$ V in Fig. 24 as due to a low gain would also be incorrect. Levine *et al.* (*30*) have explained this by attributing the excess noise at high bias to the ground-state

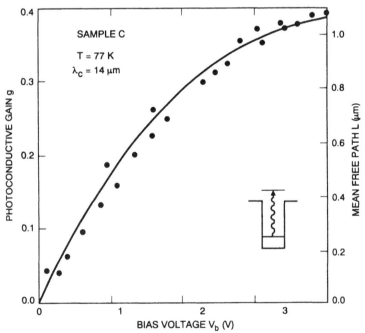

FIG. 22. Photoconductive gain g (left-hand scale) and hot-electron mean free path L (right-hand scale) vs. bias voltage V_b for sample C at $T = 77$ K. The smooth curve is drawn through the measured data. The insert shows the conduction band diagram (*30*).

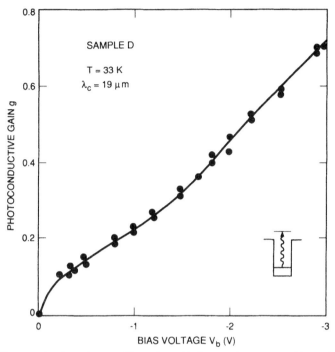

FIG. 23. Photoconductive gain g vs. bias voltage V_b for sample D at $T = 33$ K. The smooth curve is drawn through the measured data. The insert shows the conduction band diagram (*30*).

sequential tunneling, which is increasing the dark current I_d above that due to thermionic emission and thermionically assisted tunneling through the tip of the barriers. That is

$$I_{d,m} = I_{d,th} + I_{d,tu},\tag{19}$$

where $I_{d,m}$ is the total measured current, $I_{d,th}$ is the usual thermionic contribution, and $I_{d,tu}$ is the ground-state tunneling current. It should be noted (*14*) that electrons near the top of the well, which contribute to the dark current $I_{d,th}$, are the same as those which can transport in the continuum and thus contribute to the photocurrent I_p. In contrast, the electrons which contribute to the ground-state tunneling current $I_{d,tu}$ do not enter the continuum but sequentially tunnel from one well to the next (*50*). The gain associated with this process is $g_{tu} = L_p/l$ (where L_p is the superlattice period) and is very small compared with the usual continuum

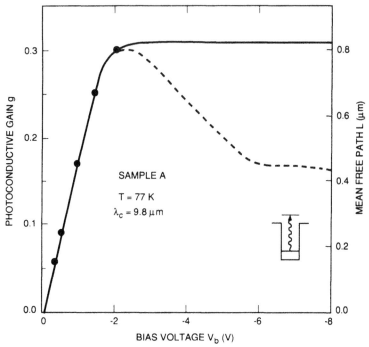

FIG. 24. Photoconductive gain g (left-hand scale) and hot-electron mean free path L (right-hand scale) vs. bias voltage V_b for sample A at $T = 77$ K. The solid curve drawn through the points is the correct interpretation; the dashed line is not. The insert shows the conduction band diagram (30).

transport gain $g = L/l$. These two current processes thus lead to two contributions to the shot noise:

$$i_{d,m}^2 = i_{d,th}^2 + i_{d,tu}^2. \tag{20}$$

By combining Eqs. (17) and (20) and using the fact that $g_{tu} \ll g$ we can write

$$g_m = (1 - f)g, \tag{21}$$

where g_m is the measured gain and $f \equiv I_{d,tu}/I_{d,m}$. Therefore, at high bias when the current contribution from sequential tunneling increases, the measured gain decreases, exactly as found in Fig. 24.

7. *Well Capture Probability.* The photoconductive gain can also be viewed in terms of a well-capture probability p_c (51). A total photocurrent I_p, which is flowing across a quantum well, will have a portion $p_c I_p$ captured by the well and a remaining fraction $(1 - p_c)I_p$ transmitted to the next period, as indicated in Fig. 25. There is also a net photocurrent i_p emitted by the well, which also has a fraction $(1 - p_c)i_p$ transmitted to the next period. By current continuity at steady state,

$$p_c I_p = i_p. \tag{22}$$

The photocurrent emitted by the well is given by

$$i_p = eF\eta_w, \tag{23}$$

where F is the optical flux, and η_w is the net quantum efficiency for a single quantum well. Therefore, now we can write the total net photocurrent as

$$I_p = eF\eta, \tag{24}$$

where η is the total net quantum efficiency of the MQW structure consisting of N wells. By combining Eqs. (22)–(24) and assuming $\eta = N\eta_w$ yields

$$g = 1/Np_c. \tag{25}$$

It is worth noting that at very high well-capture probability (i.e., $p_c = 1$), N photoelectrons are needed to move one electron across the photosensitive MQW region (i.e., $g = 1/N$). Furthermore, a short mean free path (i.e., $L = L_p$) is associated with a high ($p_c = 1$) capture probability, (as will turn out to be the case for *p*-doped QWIPs discussed in Section IV.B). In order to show the dependence of p_c on bias, we have obtained p_c from

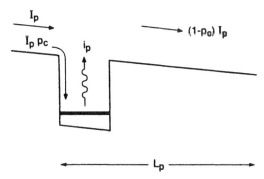

FIG. 25.   Schematic band diagram of one period (L_p) of a QWIP (30).

Eq. (25) and plotted the result for samples B, E, and F in Fig. 26 (samples A, C, and D are similar and not shown to avoid cluttering the figure). Also note from Fig. 26, that $p_c = 1$ at zero bias and then rapidly drops with increasing bias. This is due to the efficient removal of the carriers from the vicinity of the well at high electric fields.

8. Quantum Efficiency. By using Eq. (16), the bias-dependent photoconductive gain (Figs. 21–24) and the responsivity (Figs. 18 and 19), we can now

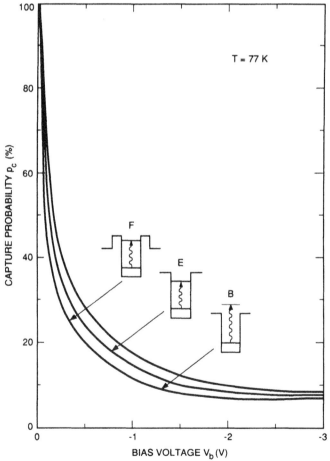

FIG. 26. Plot of the well capture probability p_c at $T = 77$ K vs. the bias voltage V_b for *bound, continuum,* and *quasicontinuum* n-doped QWIPs (*30*).

determine the total measured quantum efficiency (52) η. The results of the continuum samples A–D are shown in Fig. 27. It is important to note that the total quantum efficiency does not vanish at zero bias but has a substantial value ranging from $\eta_0 = 3.2\%–13\%$, corresponding to a finite probability of escaping from the quantum well. As the bias is increased quantum efficiency increases approximately linearly and then saturates at high bias reaching maximum values of $\eta_{max} = 8–25\%$. The saturation values of the total quantum efficiencies are listed in Table II where they can be seen to be in good agreement with the values obtained from the zero-bias absorption measurements η_a (77 K). This expected agreement $\eta_{max} = \eta_a$ (77 K) confirms our analysis.

We now consider the *bound* and *quasicontinuum* QWIPs (samples E and F). The photoconductive gains are plotted in Fig. 28, where they can be seen to be quite similar to the continuum QWIPs shown in Figs. 21–24, and the quantum efficiencies are shown in Fig. 29. For the *quasicontinuum*

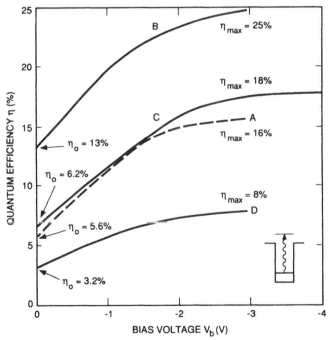

FIG. 27. Quantum efficiency η vs. bias voltage V_b (negative) for samples A–D. The zero-bias quantum efficiencies η_0 and the maximum quantum efficiencies η_{max} are shown. The insert shows the conduction band diagram (30).

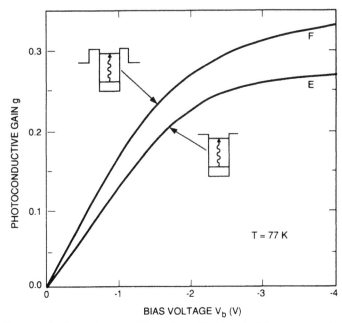

FIG. 28. Photoconductive gain g vs. bias voltage V_b (negative) for samples E and F at $T = 77$ K. The inserts show the conduction band diagram (30).

QWIP (sample F), η is quite similar to that of the continuum QWIPs (Figs. 27 and 30), having a significant value at zero bias ($\eta_0 = 5.8\%$). However, the *bound-to-bound* state QWIP (sample E) is quite different, having an order of magnitude lower value of zero bias quantum efficiency $\eta_0 = 0.6\%$. This is due to the necessity of field-assisted tunneling in order for the bound-state excited photoelectrons to escape from the quantum well.

9. *Escape Probability.* As Levine *et al.* have discussed (30) the behavior of η (shown in Fig. 27) results from the fact that the net quantum efficiency is composed of two components. One is the optical absorption quantum efficiency η_a (i.e., as discussed before, the fraction of the unpolarized incident photons which are absorbed by the intersubband transition). The other factor is the probability p_e that a photoexcited electron will escape from the quantum well and contribute to the photocurrent, rather than being recaptured by the originating well (5, 39, 40). That is, we can

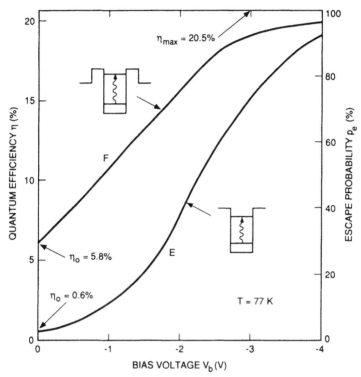

FIG. 29. Quantum efficiency η (left-hand scale) and escape probability p_c (right-hand scale) vs. bias voltage V_b (negative) for samples E and F. The zero-bias quantum efficiencies η_0 and the maximum quantum efficiencies η_{max} are shown. The inserts show the conduction band diagram (*30*).

express the total (net) quantum efficiency η as (*52*)

$$\eta = \eta_a p_c = \eta_{max} p_c \tag{26}$$

Using Eq. (26) together with Fig. 27 allows the determination of p_c as shown in Fig. 30. Note that although the curves for η in Fig. 27 are quite different, the escape probabilities are similar, with p_c varying from 34 to 52% at low bias and increasing toward unity at high bias. This is to be expected because all of these QWIPs are based on *bound-to-continuum* transitions, where the photoexcited carriers are above the top of the barriers and can thus readily escape before being recaptured.

We now evaluate escape probabilities for the *bound-to-bound* and *bound-to-quasicontinuum* QWIPs (samples E and F). For the *quasicontin-*

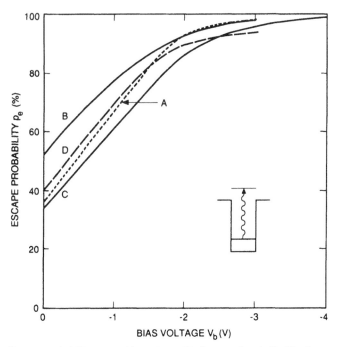

FIG. 30.   Escape probability p_c vs. bias voltage V_b for samples A–D. The insert shows the conduction band diagram (30).

uum QWIP, p_c is quite similar to that of the *continuum* sample (Fig. 30), having significant value at zero bias ($p_c = 28\%$). However, the *bound state* QWIP (sample E) is quite different, having an order of magnitude lower value of escape probability $p_c = 2.9\%$. This is due to the necessity of field-assisted tunneling in order for the bound-state excited photoelectrons to escape from the quantum well. Since the shape of the bias-dependent gain curves for all the QWIPs (sample A–F) are quite similar (i.e., increasing linearly with bias), it means that once the photoexcited electrons escape from the quantum well (whether or not by tunneling) the hot carrier transport processes are nearly identical.

The reason that p_c is so large at zero bias for the *continuum* and *quasicontinuum* samples is that being in the continuum above the barriers the carriers have a short escape time τ_c, relative to the time τ_r required to be recaptured (before contributing to the photocurrent) into the same quantum well from which they originated. The reverse is true for the *bound-to-bound* state QWIP (5, 39, 52, 53). For a more quantitative analysis, see Ref. 30.

10. *Detectivity.* We can now determine the peak detectivity D_λ^* defined as (14, 43)

$$D_\lambda^* = R_p \frac{\sqrt{A\Delta f}}{i_n},\tag{27}$$

where A is the detector area and $\Delta f = 1$ Hz. This is done as a function of bias for a *continuum* (A), a *bound* (E), and a *quasicontinuum* (F) QWIP in Fig. 31. (The dashed lines near the origin are extrapolations.) For all three samples, D^* has a maximum value at a bias between $V_b = -2$ to -3 V. Since these QWIPs all have different cutoff wavelengths, these maximum D^* values cannot be simply compared. In order to facilitate this comparison, we note that the dark current has been demonstrated to

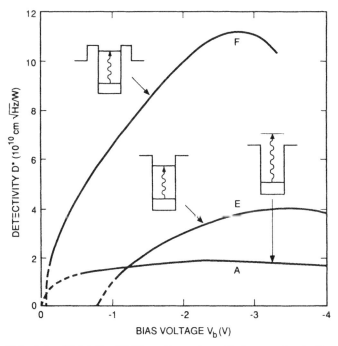

FIG. 31. Detectivity D^* (at $T = 77$ K) vs. bias voltage V_b for samples A, E and F. The inserts show the conduction band diagram (30).

follow an exponential law (*14, 43, 52*) $I_d \propto e^{-(E_c - E_F)/kT}$ (where E_c is the cutoff energy $E_c = hc/\lambda_c$) over a wide range of both temperature and cutoff wavelength. Thus using $D^* \propto (R_p/i_n)$, we have

$$D^* = D_0^\lambda e^{E_c/2kT}. \qquad (28)$$

In order to compare the performance of these different QWIPs, Levine *et al.* have plotted D^* against E_c on a log scale as shown in Fig. 32. [In addition to the data discussed in this article, other measurements they have taken are also included (*14, 43, 54, 55*)]. The straight line fits the data very well, which is satisfying considering the fact that the samples have different doping densities, N_d, different methods of crystal growth, different spectral widths, $\Delta \lambda$, different excited states (*bound, quasicontinuum*, and *continuum*) and even in one case a different materials system (InGaAs) (*54*). The best fit for $T = 77$ K detectivities of *n*-doped QWIPs is

$$D_c^* = 1.1 \times 10^6 e^{E_c/2kT} \text{cm}\sqrt{\text{Hz}}/\text{W}. \qquad (29)$$

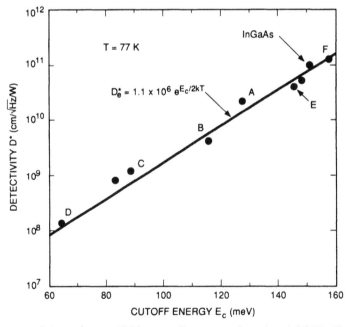

FIG. 32. Detectivity D^* (at $T = 77$ K) vs. cutoff energy E_c for *n*-doped QWIPs. The straight line is the best fit to the measured data (*30*).

Another useful figure of merit is black-body responsivity and detectivity R_B and D_B^*, written as

$$D_B^* = R_B \frac{\sqrt{A\Delta f}}{i_n} \tag{30}$$

with

$$R_B = \frac{\int_{\lambda_1}^{\lambda_2} R(\lambda) W(\lambda) \, d\lambda}{\int_{\lambda_1}^{\lambda_2} W(\lambda) \, d\lambda}. \tag{31}$$

It is worth noting that, for most applications, the black-body responsivity R_B is reduced only a relatively small amount from the peak value R_p. Also note that, since QWIP dark current is mostly due to the thermionic emission and thermionically assisted tunneling, unlike other detectors, QWIP detectivity increases nearly exponentially with the decreasing temperature as shown in Fig. 33.

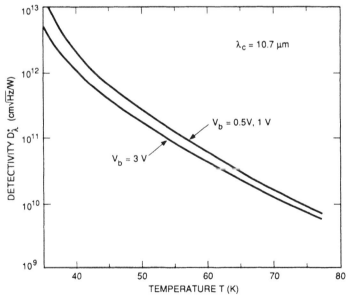

FIG. 33. Peak detectivity D_λ^* for a QWIP having cutoff wavelength of $\lambda_c = 10.7 \ \mu m$ as a function of temperature T for several bias voltages V_b (42).

11. Dependence on Doping, Bias, and Barrier Thickness. In this section we will discuss the dependence of detectivity on doping density of the quantum wells and optimization of doping density in the wells for best performance. The integrated absorption strength (i.e., the area under the absorption spectrum), which is proportional to α_p times the fractional (56) spectral width $\Delta\lambda/\lambda$ is expected to scale with doping density N_D. That is,

$$\alpha_p(\Delta\lambda/\lambda) \propto N_D. \tag{32}$$

Gunapala *et al.* have shown that $\alpha_p(\Delta\lambda/\lambda)/N_D$ is indeed constant to within a factor of two as the doping varied by a factor of 30, in agreement with previous work (13, 17). Thus, by combining Eqs. (9), (16), and (32) and assuming low η, we can show that

$$R_p \propto N_D(\lambda/\Delta\lambda). \tag{33}$$

Also, we have experimentally verified Eq. (33) (42). Since the current noise $i_n \propto \sqrt{n^*}$ and using the expression that black-body detectivity $D_B^* \equiv D_\lambda^*(\Delta\lambda/\lambda)$ we can easily show that [by using Eqs. (27) and (33)]

$$D_B^* \propto N_D/\sqrt{n^*}, \tag{34}$$

where n^* is the number of electrons thermally excited into the continuum at low bias [i.e., $T(E) = 0$ for $E < E_b$ and $T(E) = 1$ for $E > E_b$] and given by

$$n^* = \left(m^*kT/\pi\hbar^2 L_p\right)e^{-(E_c-E_F)/kT}. \tag{35}$$

By combining Eqs. (34) and (35), and using expression $N_D = n_0\ln(1 + e^{E_F/kT})$ where $n_0 \equiv (m^*kT/\pi\hbar^2 L_w)$ we obtain the following direct relation between the detectivity D^* and the doping density N_D

$$D_B^* \propto \rho/\sqrt{e^\rho - 1} \tag{36}$$

where $\rho \equiv N_D/n_0$. This relation shows that the detectivity has a maximum value at $T = 77$ K of $\rho = 1.6$ corresponding to $N_D = 7.4 \times 10^{17}$ cm^{-3} (where we have used $n_0 = 4.6 \times 10^{17}$ cm^{-3} at $T = 77$ K). It is worth noting, however, that this is an extremely *broad* maximum having low- and high-density half-heights of $\rho = 0.18$ (i.e., $N_D = 8.35 \times 10^{16}$ cm^{-3}) and $\rho = 5.1$ (i.e., $N_D = 2.4 \times 10^{18}$ cm^{-3}), respectively. As is shown in Fig. 34, D^* varies by a factor of two for a factor of 30 variation in N_D.

In order to verify this, three standard bound-to-continuum QWIP samples were grown by MBE (several months apart for the high and low doping levels). The device structures of these three samples were identical except for their well doping densities. The well doping densities of these

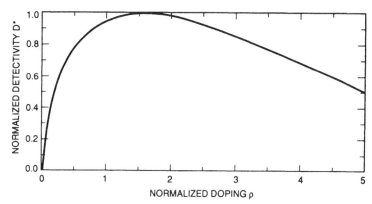

FIG. 34. Plot of normalized QWIP detectivity against normalized doping ρ, showing the very broad maximum (42).

samples were $N_D = 4.7 \times 10^{16}$ cm^{-3}, 1.2×10^{17} cm^{-3} and 1.5×10^{18} cm^{-3}. Lower values of doping were not used because, if N_D is below the metal insulator transition density of 1.4×10^{16} cm^{-3}, the electrons freeze out onto their impurity sites and will therefore not contribute to the intersubband absorption.

In agreement with this theoretical expectation, note that the experimental black-body detectivity D_B^*, shown in Table IV, is essentially independent of doping from $N_D = 4.7 \times 10^{16}$ cm^{-3} to 1.5×10^{18} cm^{-3} (a factor of 30). These results are in agreement with Kane et al. (13). In addition, note that the variations in the absorption and responsivity (approximately an order of magnitude), the variation in dark current (approximately three orders of magnitude), and the spectral line width ($\Delta\lambda/\lambda$) have been correctly normalized out.

This doping independence is advantageous from both a system and fabrication point of view since it means that D^* will remain highly uniform across a large 2-D array even with doping variations across the wafer. Since MBE is *extremely uniform* ($\sim 1\%$) to start with, this shows that doping uniformity should not limit the performance of large QWIP arrays (57). Another conclusion worth mentioning is that in some situations it can be advantageous to use lower doped detectors, which have essentially the *same* detectivity D^* (and hence the same noise equivalent temperature difference NEΔT), but which have substantially *lower* dark current I_d. This can be especially advantageous for large 2-D arrays where filling of the charge storage wells in the multiplexer circuit by the dark current must be avoided.

TABLE IV

COMPARISON OF MEASURED AND CALCULATED PARAMETERS FOR THREE QWIP DETECTORS
HAVING DOPING DENSITIES OF $N_D = 4.7 \times 10^{16}$, 1.2×10^{17}, AND 1.5×10^{18} cm^{-3}

N_D (cm^{-3})	4.7×10^{16}	1.2×10^{17}	1.5×10^{18}
λ_p (μm)	8.0	8.2	8.35
λ_c (μm)	8.6	9.4	9.7
$\Delta\lambda/\lambda$ (%)	19	31	44
α_p (cm^{-1})	110	220	830
η (%)	3	5.6	18
I_D (A)	2.2×10^{-8}	4.5×10^{-7}	1.4×10^{-5}
R_p (mA/W)	74	100	630
g	0.38	0.27	0.52
D_λ^* (cm$\sqrt{\text{Hz}}$ /W)	1.8×10^{10}	6.4×10^{9}	5.2×10^{9}
n^* (cm^{-3})	1.4×10^{6}	2.5×10^{7}	3.0×10^{9}
I_D/n^*	3.4	3.9	$\equiv 1$
α_p $(\Delta\lambda/\lambda)/N_D$	1.8	2.3	$\equiv 1$
R_p $(\Delta\lambda/\lambda)/N_D$	1.6	1.4	$\equiv 1$
$\tilde{D}_\lambda^*$ (exp)	3.5	1.2	$\equiv 1$
$\tilde{D}_\lambda^*$ (cal)	3.4	1.2	$\equiv 1$
$\tilde{D}_B^*$ (exp)	1.5	0.85	$\equiv 1$
$\tilde{D}_B^*$ (cal)	1.5	0.88	$\equiv 1$

Note. The quantities listed include the peak responsivity wavelength λ_p; the long-wavelength cutoff λ_c; the fractional spectral width $\Delta\lambda/\lambda$; the peak absorption coefficient α_p; the peak quantum efficiency η; the dark current I_D (measured at $V_b = -3$ V and $T = 77$ K); the absolute responsivity R_p (measured at $V_b = -3$ V); the photoconductive gain g; the calculated number of thermally excited electrons at $T = 77$ K in the conduction band n^*; the absolute peak detectivity D_λ^* and the normalized peak and blackbody detectivities $\tilde{D}_\lambda^*$ and $\tilde{D}_B^*$ (42).

There is a similarly advantageous insensitivity of D^* to the bias voltage V_b, as can be seen from the following argument (48). First we note that the (transport) velocity v is proportional to V_b (for biases less than few volts), that the responsivity R, photoconductive gain g, and the dark current I_d are all proportional to V_b. Therefore the noise (Eq. 17), which is given by $i_n \propto \sqrt{gI_D} \propto V_b$ is also proportional to the bias. This demonstrates that $D^* \propto R/i_n$ is indeed expected to be independent of V_b, for biases less than a few volts where R and I_d are linear in V_b. Thus, as the bias is reduced, D^* will remain constant until Johnson noise (or system amplifier noise) becomes important. To test this, we have measured the bias and temperature dependence on a $N_D = 1.2 \times 10^{18}$ cm^{-3} detector having a cutoff wavelength $\lambda_c = 10.7$ μm as shown in Fig. 33. Note that not only D_λ^* is *independent* of bias between 0.5 and 1 V, but that in fact the lower bias detectivity is actually *higher* than the D_λ^* at $V_b = 3$ V, because the dark current noise decreases more rapidly than the responsivity. Below $V_b = 0.5$ V, Johnson noise starts to become important. We calculate that at $V_b = 0.2$-V Johnson noise and dark current shot noise become equal, so that operation below $V_b = 0.2$ V is not advantageous.

As was the case when we discussed doping uniformity, the independence of D^* on V_b means that bias nonuniformities will not adversely affect the uniformity of the NEΔT for large 2D arrays. It also indicates that low biases can be advantageously used, which will significantly reduce the dark current and hence reduce multiplexer well filling. It is worth noting from Fig. 33 that, as the temperature is reduced, D^* increases dramatically reaching $D_\lambda^* = 10^{13}$ cm $\sqrt{\text{Hz}}$ /W at $T = 35$ K and even larger values at lower temperatures. This demonstrates the high quality of the $Ga_x Al_{1-x} As$ barriers and the absence of tunneling through defects. In contrast, HgCdTe devices can have significant defect tunneling (58) leading to a saturation in D^* as the temperature is lowered.

As we discussed in Section IV.A.1 Levine et al. have achieved a dramatic improvement in the QWIP performance as the $Ga_x Al_{1-x} As$ barrier thickness increased from 95 to 500 Å. Most of this significant improvement was due to the reduction of tunneling related dark current (sequential and thermally assisted). In order to test whether even thicker barriers can further reduce the dark current, we have grown a 50-period QWIP sample with very wide barriers ($L_b = 1000$-Å). The other device parameters of this QWIP were $L_W = 40$-Å, $N_D = 1 \times 10^{18}$ cm^{-3}, and $x = 0.27$. The responsivity spectra and the dark current curves of this detector are shown in Figs. 35 and 36. The peak responsivities $R = 0.39$ and 0.31 A/W for $V_b = \pm 8$ V as well as the dark current curve are quite similar to the $L_b = 500$-Å QWIP (with similar cutoff wavelength). Furthermore, in order to accurately determine whether there are any performance improvements in increasing L_b, the noise current and the photoconductive gains were measured (Fig. 37). The shape of this curve and the saturation value of the photoconductive gain ($g = 0.28$) are also quite similar to that of $L_h = 500$-Å QWIPs.

B. p-DOPED QWIPs

Levine et al. (59) experimentally demonstrated the first QWIP that uses hole intersubband absorption in the GaAs valence band. The strong mixing between the light and heavy holes (60–64) (at $k \neq 0$) allows the desirable normal incidence illumination geometry to be used.

The samples were grown on a (100) semi-insulating substrate, using gas

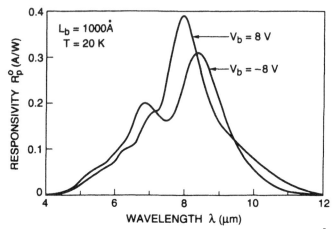

FIG. 35. Experimental responsivity spectrum for a thick-barrier $L_b = 1000$-Å QWIP, at $T = 20$ K, and $V_b = \pm 8$ V.

source MBE, and consisted of 50 periods of $L_w = 30$-Å (or $L_w = 40$-Å) quantum wells (doped $N_D = 4 \times 10^{18}$ cm^{-3} with Be) separated by $L_b = 300$-Å barriers of $Al_{0.3}Ga_{0.7}As$, and capped by $N_D = 4 \times 10^{18}$ cm^{-3} contact layers. In order to measure the responsivity $R(\lambda)$, these p-QWIPs were processed into 200-μm-diameter mesas. One detector had a 45° angle

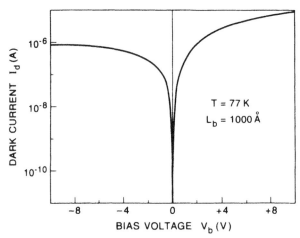

FIG. 36. Dark current–voltage curves at $T = 77$ K, for a thick-barrier $L_b = 1000$-Å QWIP.

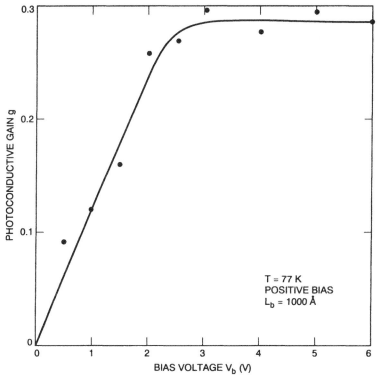

FIG. 37.   Measured photoconductive gain g (points) and smooth curve drawn as a guide, at $T = 77$ K for a thick-barrier $L_b = 1000$-Å QWIP.

of incidence facet polished on it (similar to n-QWIP geometry), and the other had the back of the wafer polished for normal incidence illumination through the substrate (as shown in the inset of Fig. 38).

The unpolarized responsivity spectra (measured to be the same at $T = 10$ and 80 K) of the $L_w = 40$-Å QWIP (corrected for 28% reflection from the GaAs surface) are compared for the two geometries in Fig. 38. Note that the two spectra are essentially identical (peak wavelength $\lambda_p = 7.2$ μm and long wavelength cutoff $\lambda_c = 7.9$ μm) and that, in fact, the normal incidence responsivity is larger than that of 45° illumination, consistent with both polarizations contributing to the photoresponse. The peak unpolarized responsivities (for the $L_w = 40$ Å sample at $\lambda_p = 7.2$ μm and $V_b = +4$ V) are $R_p = 39$ and 35 mA/W for normal and 45° incidence, respectively, which are approximately an order of magnitude

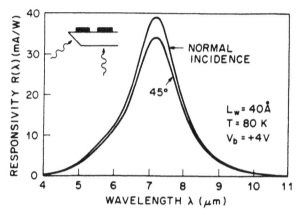

FIG. 38. Comparison between the normal incidence and 45° responsivity spectra measured at $T = 80$ K, $V_b = +4$ V for the 40-Å well sample. The illumination geometry is shown in the inset (59).

smaller than the responsivities for n-QWIPs discussed in Section IV.A. As a further test of polarization behavior, Levine *et al.* (59) found that by using the 45° geometry, s-polarized light had twice the photoresponse of p-polarization. This is again in strong contrast with n-QWIPs for which the s-polarized photoresponse is forbidden by symmetry. A comparison between the $L_w = 30$- and 40-Å detectors (Fig. 39) shows that, as expected, the narrower well QWIP has a broader spectral response ($\lambda_c = 8.6$ μm) due to the excited state being pushed further up into the continuum and thereby broadening the absorption (65). The 30-Å well detector also has a slightly longer peak wavelength ($\lambda_p = 7.4$ μm) consistent with the ground state being pushed up even further than the excited state. Similar line-shape effects (65) have been seen in n-QWIPs.

The current–voltage curves were measured at $T = 77$ and 4.2 K as shown in Fig. 40. Note that the 30-Å well QWIP (for which $\lambda_c = 8.6$ μm, i.e., 144 meV) has approximately an order of magnitude higher dark current, I_d, than the $L_w = 40$ Å detector (for which $\lambda_c = 7.9$ μm, i.e., 157 meV). This is an excellent agreement with the expected thermionic emission current $I_d \propto e^{-\Delta E/kT}$. That is, taking $\Delta E = (144-157)$ meV = -13 meV and $kT = 6.65$ meV (at $T = 77$ K) yields a calculated increase of approximately $e^2 = 10$. It is also worth noting that, for these p-QWIPs, the positive bias direction (defined as mesa top positive) is the low dark current direction, whereas for n-QWIPs the negative bias direction (defined as mesa top negative) is also the low I_d direction. That is, in both

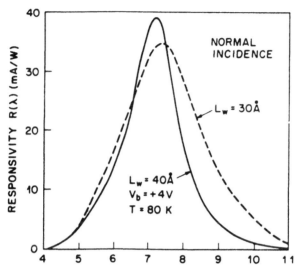

FIG. 39. Comparison between the normal incidence responsitivity spectra for the 30- and 40-Å quantum well detectors, at $T = 80$ K and $V_b = +4$ V (*59*).

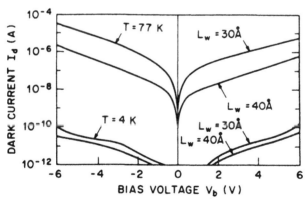

FIG. 40. Dark current vs. bias voltage for the 30- and 40-Å well sample at $T = 77$ and 4.2 K (*59*).

cases the lowest dark current (and also lowest photocurrent) direction corresponds to carriers flowing into the substrate.

In order to determine the peak detectivity, D_λ^*, and photoconductive gain g (48), we measured the detector noise, i_n, using a low noise spectrum analyzer. For $|V_b| < 3$ V the noise of the $L_w = 40$-Å QWIP was so low that we could not measure it. However, for $V_b = +4$ V and $T = 77$ K we measured $i_n = 4 \times 10^{-14}$ A. We can now determine D_λ^* by using Eq. (27) and $R_p = 39$ mA/W, which yields $D_\lambda^* = 1.7 \times 10^{10}$ cm $\sqrt{\text{Hz}}$/W. Similarly, $D_\lambda^* = 3.5 \times 10^9$ cm $\sqrt{\text{Hz}}$/W at $V_b = +4$ V for $L_w = 30$ Å (using the measured values $R_p = 35$ mA/W and $i_n = 1.8 \times 10^{-13}$ A at $T = 77$ K). By substituting the measured noise and dark current values at $V_b = +4$ V and $T = 77$ K to Eq. (17), we can obtain the experimental photoconductive gains $g = 3.4 \times 10^{-2}$ and $g = 2.4 \times 10^{-2}$ for $L_w = 30$ and 40 Å, respectively. Now Eq. (18) allows us to determine that hot-hole mean free path $L = 510$ and 360 Å for the two samples (i.e., photoexcited carrier travels only one or two periods before being recaptured). It should be noted that both g and L are over an order of magnitude smaller than the corresponding values for n-QWIPs (where photoconductive gains of $g \approx 1$ have been obtained (14). Since $L = v_s \tau_r$ (where τ_r is the well recapture time (30)), we can see that the lower g is due to a lower velocity associated with the higher hole effective mass, and a shorter lifetime due to increased scattering between the light and heavy hole bands. Having now obtained g, we can relate it to the peak responsivity Eq. (16), and thus using the measured values for R_p and g directly determine the low-temperature quantum efficiency. This yields double pass values of $\eta = 17$ and 28% for the $L_w = 30$- and 40-Å QWIPs, respectively. Thus, in spite of the larger effective mass of the holes compared to that of electrons, the different symmetry (normal incidence) of the intersubband absorption, and the much smaller photoconductive gain and mean free path, the quantum efficiency (and hence escape probability) for bound-to-continuum n- and p-QWIPs are similar.

V. Asymmetrical GaAs / $Al_x Ga_{1-x}$ As QWIPs

For typical QWIP structures with symmetrical rectangular wells the electric field shift of intersubband absorption is relatively small, since the linear shift term is forbidden by symmetry. This was demonstrated by Harwit and Harris (66) who obtained a shift of less than 2 meV for a field of 36 kV/cm across symmetrical QWIP structure. Large Stark shifts can be obtained by designing QWIP structures with asymmetrical quantum wells such as a stepped well. Martinet *et al.* (*67*) have demonstrated a linear Stark shift of ~ 10 meV for an electric field shift of ~ 15 kV/cm, whereas Mii *et al.* (*68*) measure a shift of 8 meV at 18 kV/cm. These shifts are an order of magnitude larger than that for the symmetrical rectangular well at the same electrical field. In addition to absorption spectral changes due to asymmetrical wells or barriers in the structure, there will be much stronger bias-dependent behavior in escape probability, photoconductive gain and, hence, in the responsivity spectrum.

A. *n*-DOPED QWIPS

1. Asymmetrical Barriers. By introducing QWIP structure with asymmetrical barriers, Levine *et al.* (*53*) demonstrated an electrically tunable photodetector without sacrificing the responsivity. A 50-period graded-barrier structure (shown in Fig. 41) was grown (*53*) using an electron-beam MBE technique with 300-Å-thick $Al_xGa_{1-x}As$ barriers, which were graded from $x = 0.24$ (substrate side) to $x = 0.28$ and alternated with 40-Å GaAs wells doped $n = 2 \times 10^{18}$ cm^{-3}. This barrier Al compositional grading corresponds to an energy grading of 32 meV (i.e., ~ 0.1 meV/Å). A similar control sample was grown in the same run but having uniform rectangular barriers of $x = 0.26$. The responsivity spectrum of the conventional control sample showed essentially no change with bias, while that of the graded

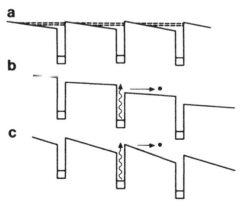

FIG. 41. Schematic band diagram of a graded-barrier QWIP at (a) zero bias, showing two bound excited states, (b) forward bias, and (c) reverse bias (*129*).

barrier QWIP depended strongly on bias voltage (53). As can be readily seen from Fig. 42, both the peak wavelength (which shifts from $\lambda_p = 8.2$ μm at $V_b = +5$ V to $\lambda_p = 9.1$ μm at $V_b = -5$ V) as well as the spectral width (which doubles from $\Delta\lambda/\lambda = 20$ to 43%) are strongly field dependent. The reason for this large shift and increase in spectral width is that at zero bias the excited state is bound (indicated as dotted lines in Fig. 41a) and, thus, the intersubband absorption is a bound-to-bound transition. For large forward bias, the excited state is now in continuum leading to a broader absorption. Since, in addition, at forward bias the photoexcited electron can escape over a low barrier, it has a longer cutoff wavelength than at reverse bias where it must escape over the high barrier (Fig. 41). The peak responsivity in Fig. 42, $R_p = 0.5$ A/W is the same for both bias polarities $V_b = \pm4$ V, and is quite similar to that of the conventional control sample (53) $R_p = 0.6$ A/W at $V_b = +3$ V.

Lacoe et al. (69) demonstrated a large difference in spectral width and cutoff wavelength by using a graded barrier QWIP with much stronger 81-meV grading in the barrier. They also found that the responsivity spectral line width is much narrower ($\Delta\lambda < 1$ μm) and the cutoff wavelength much shorter ($\lambda_c = 7.6$ μm) in the negative-bias direction where the QWIP operates on a bound-to-bound transition, compared to the

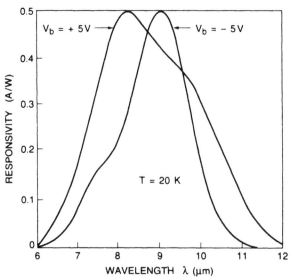

FIG. 42. Responsivity spectra of the graded-barrier QWIP showing the strong dependence on bias voltage (129).

broad ($\Delta \lambda$ = 6 μm) longer-wavelength cutoff (λ_c = 11.4 μm) positive-bias bound-to-continuum transition.

2. *Asymmetrical Quantum Wells.* Recently, Levine, Gunapala, and Hong have discussed (70) LWIR photoinduced charge polarization and storage in graded asymmetrical quantum wells. The structure designed for this experiment (shown in Fig. 43) was grown using MBE. The wide asymmetrical quantum well consists of a 40-Å GaAs section (doped n = 1 $\times$ 10^{18} cm^{-3}) and an undoped 1000-Å Al$_x$Ga$_{1-x}$As section linearly graded from x = 0.28 to x = 0.15 by varying the oven temperatures. This asymmetrical well is alternated with 500-Å barriers of undoped Al$_{0.35}$Ga$_{0.65}$As. Twenty periods of this structure are then sandwiched by GaAs contact layers (0.5-μm top- and 1-μm bottom-doped n = 1 $\times$ 10^{18} cm^{-3}). The

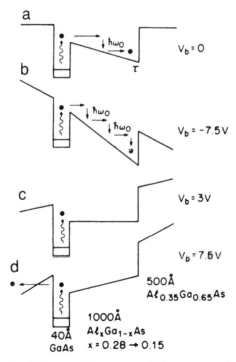

FIG. 43.   Conduction–band diagram of the asymmetrical quantum-well structure. Photoexcitation is followed by charge transfer, the loss of energy via emission of phonons ($\hbar \omega$), and the trapping and storage τ of charge in the triangular graded region. (a) is for V_b = 0, (b) is V_b = -7.5 V, (c) is V_b = $+3$ V, and (d) is V_b = $+7.5$ V (70).

40-Å doped section is designed to contain only one bound state, with the excited state a few meV above the $x = 0.28$ corner, and thus ~ 50 meV below the top of the $Al_{0.35}Ga_{0.65}As$ barrier. This excited state, which is bound with respect to the high $Al_{0.35}Ga_{0.65}As$ barriers is in a quasicontinuum state with respect to the 40-Å quantum-well section. Thus an electron, which is resonantly photoexcited out of the doped 40-Å well section via intersubband absorption, will be attracted by the quasielectric field of the graded gap section where it can relax (via phonon emission) into the wide 1000-Å trough and remain stored for a time t. It can then relax back to the 40-Å well via tunneling, thermionic emission, dielectric relaxation, or current flow through the external circuit.

The electron transfer out of the narrow 40-Å well side and storage on the wide graded well side will produce a polarization P, which can be expressed as $P = ne\Lambda p$, where n is the number of excited charges per cm^3 stored, $\Lambda = 1000$ Å is the transfer distance, and p is the transfer probability. n can be related to the optical power P_0, angle of incidence ϕ, and storage time τ using $n = (\alpha/h\nu)(P_0/A)\tau$, where α is the absorption coefficient, A is the area, and ν is the infrared frequency. The signal voltage V produced by the infrared illumination is related to the induced polarization by $P = eV/l$, where l is the total thickness of all 20 periods. By combining the above relations, we can write the voltage responsivity (70)

$$R_\nu = (e/h\nu)(\alpha\Lambda p)(\tau/C)\cos\phi, \tag{37}$$

where $C = \varepsilon A/l$ is the device capacitance. As shown in the above equation, the responsivity is proportional to the storage time τ of the optically transferred charge. After calculating the tunneling time (through the 1000-Å barrier), thermionic emission time (at $T < 80$ K), and the intrinsic dielectric relaxation time, Levine et al. found (70) that, for load resistors of interest ($R_L \leq 10$ MΩ), the main charge relaxation time is dominated by the external circuit, i.e., $\tau = R_L C$. This long intrinsic storage time is consistent with previous work on memory devices (71). After substituting this, Eq. (37) yields

$$R_\nu = (e/h\nu)(\alpha\Lambda p)R_L\cos\phi \tag{38}$$

The absorption coefficient α was measured and the spectrum is shown in Fig. 44. It is extremely broad with a half height extending from $\lambda = 6$ to 10 μm. This is about twice as wide as the usual case of a 40-Å well with simple rectangular barriers of $Al_{0.28}Ga_{0.72}As$ on both sides (see Section

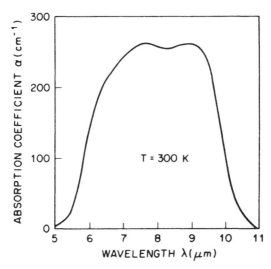

FIG. 44. Measured absorption spectrum at $T = 300$ K (*70*).

IV). This width and double resonance is most likely due to the lower symmetry in the well and an increase in the number of allowed final states.

The responsivity spectrum was measured using a 45° polished facet for the incident infrared light ($\phi = 45°$) and the result is given in Fig. 45 for zero bias ($V_b = 0$). The peak responsivity at $\lambda = 6.8$ μm is $R_v = 30$ V/W. As the bias is made more negative (Fig. 43b), R_v increases (as shown in Fig. 46) due to the increasing probability p of charge transfer and reaches $R_v = 1200$ V/W at $V_b = -7.5$ V. For positive bias, the responsivity initially decreases reaching $R_v = 0$ at $V_b = +3$ V (see inset in Fig. 46) when the external bias balances out the built-in graded-gap field and the trough disappears (*70*) (Fig. 43c). Increasing the positive bias further allows the photoexcited electrons to tunnel out of the high-barrier side causing the responsivity to again increase reaching $R_v = 1000$ V/W at $V_b = +7.5$ V (shown in Fig. 43d). The shape of the responsivity spectrum, however, remains approximately constant as a function of bias voltage.

After substituting measured values for responsivity in Eq. (38), p can be determined. At $V_b = -7.5$ V calculated $p = 90\%$, i.e., as expected, at high bias the charge transfer probability approaches unity. As the temperature is raised from $T = 30$ to 90 K, R and hence p remain highly constant. This is in fact to be expected from a large electric field and hence rapid escape from the 40-Å well compared with the slower recapture time back into the well. The $V_b = -7.5$ V voltage responsivity $R_v = 1200$ V/W

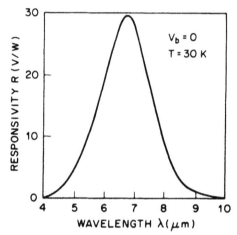

FIG. 45. Photovoltaic responsitivity R_v spectrum measured at zero bias at $T = 30$ K for the asymmetrical quantum-well QWIP (*70*).

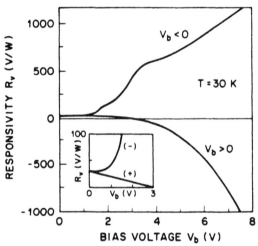

FIG. 46. Positive- and negative-bias dependence of the responsivity measured at $T = 30$ K for the asymmetrical quantum-well QWIP. The inset is an expanded scale for low bias (*70*).

(measured with $R_L = 100$ kΩ) corresponds to a current responsivity $R_A = 12$ mA/W, which is only about 1% of that for symmetrical, flat-barrier photoconductors, because the photoexcited electron is transferred a distance of $L = 1000$ Å instead of being transported to the hot-electron mean free path $L > 3$ μm. In this context, it is worth noting that although the photoexcited electron has sufficient energy to surmount the 500-Å barrier and escape into the next period (Fig. 43b), the ballistic mean free path is much shorter than L so that energy has been lost (via phonon emission) by the time the electron reaches this barrier and it is thus trapped in the trough.

At zero bias, the calculated [from Eq. (38)] escape probability is quite small, $p = 2\%$. This arises from the fact that after photoexcitation the electron wave function is still highly localized in the region of the 40-Å well, as expected from the large measured oscillator strength (i.e., wave function overlap) between the excited state and highly localized ground state. This localization in the 40-Å well region means that the photoexcited electron has a high probability of emitting a phonon and being recaptured into the original ground state without charge transfer into the graded-gap region. In fact, as the temperature is increased from $T = 30$ to 90 K, the $V_b = 0$ values of R and p decrease by more than an order of magnitude from the values given above, which is consistent with an increase in the well recapture rate.

B. p-DOPED QWIPS

Because of the importance of the light- and heavy-hole mixing in allowing normal incidence absorption in p-QWIPs, and the fact that this mixing vanishes at $k = 0$, it is necessary to heavily dope the quantum well so that the large Fermi wave vector permits $k \neq 0$ mixing. Thus, it is interesting to study an asymmetrical p-doped quantum well because such a structure strongly affects the valence-band mixing and could potentially increase the normal-incidence absorption without increasing the doping. The asymmetrical quantum well, which was investigated by Bandara, Levine and Leibenguth (72) consisted of a 30-Å GaAs section and a 20-Å $Al_{0.1}Ga_{0.9}As$ section both p-doped ($N_D = 5 \times 10^{18}$ cm^{-3}) with Be. The barriers were 300 Å of undoped $Al_{0.3}Ga_{0.7}As$ and 50 periods of this structure were capped with the lower-doped GaAs layer $N_D = 1 \times 10^{18}$ cm^{-3} (to avoid unnecessary free-carrier absorption in the contacts).

The measured normal-incidence responsivity spectra are shown in Fig. 47 for both positive and negative bias $V_b = \pm 6$ V (the band structure for

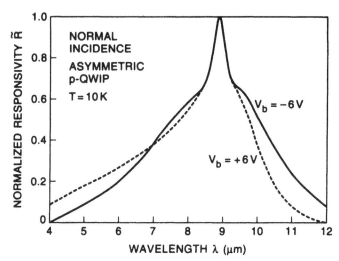

FIG. 47. Normalized responsivity spectra (measured at normal incidence) for an asymmetric
p-QWIP (72).

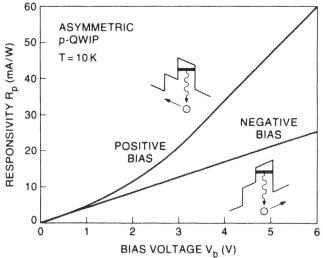

FIG. 48. Peak responsivity ($\lambda_p = 8.8~\mu$m) vs. bias voltage for positive and negative bias for
an asymmetric *p*-QWIP. The inserts show the valance band structure for positive and
negative bias and thus define the polarity (72).

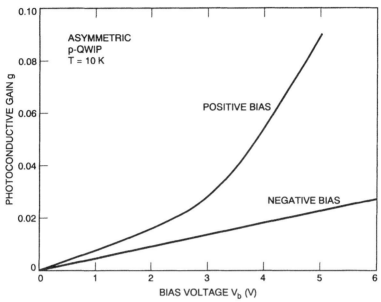

FIG. 49. Photoconductive gain for an asymmetric *p*-QWIP as a function of positive and negative bias (*72*).

positive and negative bias are indicated in the inserts in Fig. 48). Note that the narrow central peak ($\Delta\lambda/\lambda \sim 4\%$) in the responsivity spectrum is strikingly different from that of the symmetrical *p*-QWIPs, suggesting that it is due to a strong resonance mixing between excited light- and heavy-hole states. The bias dependence of the responsivity (*72*) is also substantially different from that of the symmetrical *p*-QWIPs. As can be seen in Fig. 48, at $V_b = 6$ V the positive bias responsivity is 2.5 times larger than that for negative bias, whereas for the symmetrical *p*-QWIPs the responsivity is nearly independent of the bias polarity. The origin of this polarity sensitivity is the asymmetrical quantum well capture probability p_c. This can be seen from the measured (*72*) photoconductive gain shown in Fig. 49. For example, $g = 0.023$ and $g = 0.09$ at $V_b = \pm 5$ V corresponding (using $g = 1/p_c N_w$) to $p_c = 87\%$ at $V_b = +5$ V and $p_c = 22\%$ at $V_b = -5$ V. This suggests that the asymmetrical quantum-well structure discussed above does not significantly improve the valence-band mixing.

VI. Single QWIPs

Although there has been extensive research on multiquantum-well infrared photodetectors, which typically contain many ($\sim$ 50) periods, there has been relatively limited experimental work (73–75) done on QWIPs containing only a single quantum well. Bandara *et al.* have performed a complete series of experiments (76–78) on single quantum-well structures with an *n*-type doped well, undoped well, and *p*-type doped well. These doped single-well detectors are, in fact, particularly interesting since they have exceptionally high photoconductive gain compared to MQW detectors. In addition, their simple band structures allow accurate calculations of the bias voltage dependence of the potential profiles of each of the two barriers, band bending effects in the contacts, as well as charge accumulation (or depletion) in the quantum well.

A. DIRECTLY DOPED WELL *n*-TYPE

We have grown a single well QWIP via MBE consisting of a $L_w = 40$-Å GaAs quantum well (doped $N_D = 1 \times 10^{18}$ cm^{-3} with Si) surrounded by two $L_e = L_c = 500$-Å (where L_e and L_c are emitter and collector barriers respectively) undoped Al$_{0.27}$Ga$_{0.73}$As barriers, and sandwiched between a 0.5-μm-thick top and 1-μm-thick bottom doped ($N_D = 1 \times 10^{18}$ cm^{-3}) GaAs contact layer. The sample was processed into 200-μm-diameter mesas (area $A = 3.1 \times 10^{-4}$ cm^2) using wet chemical etching. Under an applied bias voltage of V_b across the entire structure (Fig. 50) a current $I_e(V_e)$ will be injected, via field emission, from the emitter contact by the voltage drop V_e across the emitter barrier. In addition, a current $I_w(V_c)$ will also be generated from the quantum well composed of both a field emission component $I_w^{fe}(V_c)$ and an optically excited component $I_w^{opt}(V_c)$ by the voltage drop V_c across the collector barrier. These currents and voltages can be related (74, 75) using

$$I_c(V_c) = I_w^{fe}(V_c) + I_w^{opt}(V_c) \tag{39}$$

$$V_b = V_e + V_c + V_w \tag{40}$$

$$e(N_D - n)L_w A = C_e V_e - C_c V_c = 2C_w V_w, \tag{41}$$

where n is the 3-dimensional free-carrier density in the well, V_w is the voltage drop across the well, and the capacitance of the emitter (C_e) and collector (C_c) are given by $C_e = C_c = \varepsilon A / L_b$, and C_w is the capacitance of the quantum well $C_w = \varepsilon A / L_w$. The expressions for 3-dimensional emitter current I_e, and 2-dimensional quantum-well field emission current

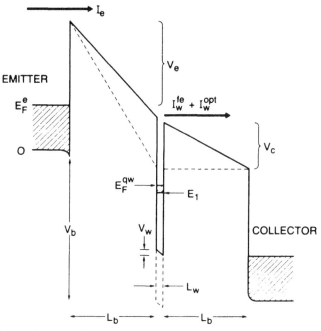

FIG. 50. Schematic conduction band diagram of an n-type doped single QWIP. The solid line is for $T = 12$ K in the dark; the dashed line is under illumination (76).

$I_w^{fc}(V_c)$, are given by (14, 74, 75)

$$I_c(V_c) = \left[\frac{em^*kTA}{2\pi^2\hbar^3}\right]\int_0^\infty T_c(E, V_c)\ln\{1 + e^{(E_F - E)/kT}\}\,dE \qquad (42)$$

and

$$I_w^{fc}(V_c) = \left[\frac{em^*kTA}{\pi\hbar^2 l}\right]v(V_c)\int_{E_1}^\infty T_c(E, V_c)\{1 + e^{(E - E_1 - E_F^{qw})/kT}\}^{-1}\,dE, \qquad (43)$$

where $l = (L_b + L_w)$ and E_F^c and E_F^{qw} are the Fermi energies in the emitter (relative to the bottom of the conduction band) and quantum well (relative to the ground state energy E_1), respectively. The energy-dependent transmission coefficients of the emitter and collector barriers are $T_c(E, V_c)$ and $T_c(E, V_c)$, which were calculated numerically by matching boundary conditions (across the entire structure for T_c, and across the collector barrier for T_c using matrix techniques). The collector velocity is given by (14)

$$v(V_c) = \mu F_c / \sqrt{1 + (\mu F_c / v_s)^2} \tag{44}$$

At zero bias, these equations require the equality of the Fermi levels $E_F^c = E_1 + E_F^{qw}$ and thus determine the $V_b = 0$ band alignment, which has $V_e > 0$ and $V_c < 0$. As V_b is increased, the potential drop on both barriers increases. Using Eqs. (39)–(44) the potential V_e and V_c as well as the currents can be calculated. These theoretical dark currents at $T = 12$ K and $T = 77$ K (dashed curves) are compared with experiments (solid curves) in Fig. 51. The excellent agreement in the dark currents (over

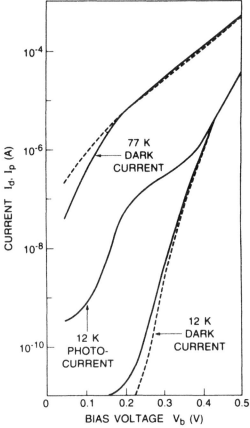

FIG. 51.   Dark current I_d, and photocurrent I_p vs. bias voltage V_b at $T = 12$ and 77 K for the single-quantum-well QWIP. The solid curves are experiment; the dashed lines are theory (76).

several orders of magnitude) are obtained by fitting (76) μ and v_s in Eq. (44). Also shown in Fig. 51 is the large increase in current with the $T = 300$-K background illumination through the dewar window (with the QWIP at $T = 12$ K).

The responsivity spectrum has been measured using a 45° facet polished on the substrate and spectral shape as well as the peak responsivity $R_p = 0.325$ A/W are similar (76) to bound-to-continuum 50-period QWIPs. This can be understood from $R \propto g$ and the increase of the quantum efficiency η with number of quantum wells (N), together with the decrease of the gain g ($g \propto N^{-1}$). Thus, large values of responsivity similar to 50-well QWIP are expected in single-well QWIPs. The spectral shape of this responsivity spectrum (76) was essentially unchanged for different biases, but the peak value R_p (measured at $\lambda = \lambda_p$) has the unusual behavior shown in Fig. 52. Note the large voltage offset of $V_b = 0.1$ V before the photoresponse increases, the maximum in R_p as a function of bias, and the large response at lower temperature (with maximum values of $R_p = 0.325$ A/W at $T = 12$ K and 0.15 A/W at $T = 77$ K). All of these features are significantly different from those of the usual 50-period bound-to-continuum QWIPs, for which R_p increases nearly linearly with V_b from $V_b = 0$, and then saturates with V_b becoming essentially constant at high bias. Furthermore, the 50-period QWIP re-

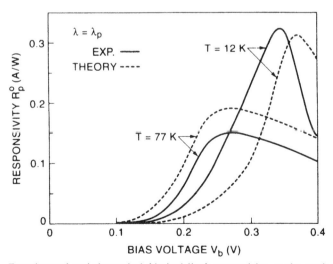

FIG. 52. Experimental and theoretical (dashed line) responsivity, at the peak wavelength λ_p, as a function of bias V_b for $T = 12$ and 77 K, for n-type doped single QWIP (76).

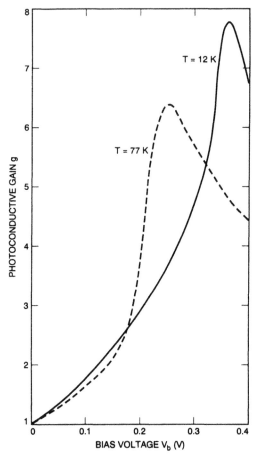

FIG. 53. Experimental photoconductive gain g as a function of bias V_b for $T = 12$ and 77 K, for n-type doped single QWIP (76).

sponsivity is only weakly dependent on temperature in this range of $T = 12-77$ K.

In order to understand this behavior, Bandara et al. also measured (76) the current noise i_n and determined the photoconductive gain at $T = 77$ K from the dark current and at $T = 12$ K from the photocurrent produced from a black-body source focused through the dewar window (the dark current noise was too small to measure directly at $T = 12$ K) (see Fig. 53). Note the striking maxima in Fig. 53, for the gain vs. bias curves at both temperatures. The initial increase of g is due to the usual increase in velocity with field, i.e., $g(V_b) = v(V_b)\tau_L/l = \mu\tau_L V_b/l^2$, where τ_L is the hot-electron lifetime. However, at high bias where the gain for the usual 50-period QWIP saturates, g decreases strongly for these single-quantum well detectors. This is a result of high field tunneling of electrons from the emitter contact directly into the quantum well and thus "short circuits" the hot electron transport process. This tunneling process is more important in these single QWIPs than in the MQW detectors, since most of the bias voltage drop occurs across the emitter tunnel barrier.

In order to further understand this device, we have plotted in Fig. 54 the 3-dimensional free electron density, n, in the quantum well calculated from Eqs. (39)–(44), at $T = 12$ K (for both illuminated and dark conditions) and at $T = 77$ K (in the dark). Note that in the dark, the free-carrier density decreases with increasing bias. This can be understood from the following considerations. From Eqs. (42) and (43) we can see, since $E_F^{qw} + E_1 > E_F^e$, that for the same voltage drop across the barriers (i.e., $V_c = V_c = V$), $I_w^{fe}(V) \geq I_c(V_c)$. Thus, as discussed by Rosencher et al. (74) the emitter contact will not be able to supply a sufficient current to replenish that generated from the quantum well. Therefore, to satisfy the dark current requirement $I_w^{fe}(V) = I_c(V_c)$ necessitates that $V_c > V_c$. In order to achieve this voltage inequality, Eq. (42) shows that $(N_D - n)$ must increase and hence n decreases. A similar photoinduced drop in n also occurs at $T = 12$ K. That is, in order to extract the large required field emission current from the emitter to balance the large photocurrent generated by the quantum well, V_c must increase, and hence according to Eq. (41), n must dramatically decrease (74, 77). (The strong potential shift in the barrier under illumination is indicated in Fig. 50). This photoinduced carrier depletion in the well also explains the rise in the 12-K photocurrent free-carrier density curve in Fig. 54. As V_b increases, the photocurrent becomes a smaller and smaller fraction of the total current and thus the photoinduced depletion becomes less and less, eventually becoming negligible for $V_b > 0.4$ V where the dark and photocurrent curves become essentially identical.

Combining all of these effects, the responsivity can be calculated (76) as a function of V_b (see dashed curves in Fig. 52). Note that all of the experimental features are now well explained. The bias offset of $V_b = 0.1$ V is a result of $V_c < 0$ for low V_b and hence the escape probability of photoexcited electrons is zero for $V_b < 0.1$ V. The increase in the maximum responsivity at $T = 12$ K over that for 77 K is due to several factors: the increase in the gain g, the large electron density n in the well (at $V_b > 0.4$ V and $T = 12$ K) resulting in a large absorption quantum efficiency. Finally, the broadening of the $T = 77$-K responsivity vs. bias

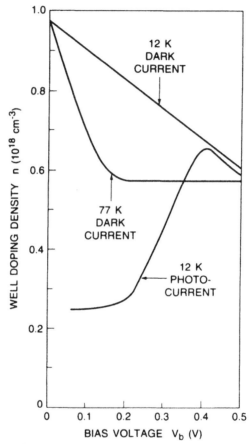

FIG. 54.   Electron density n in the quantum well as a function of bias V_b for $T = 12$ K (both illuminated and dark conditoins) and for $T = 77$ K, for n-type doped single QWIP (76).

curve can be understood due to the increase of n with V_b (Fig. 54). The reason that 50 quantum-well QWIPs do not show these interesting effects is that, except for a few quantum wells near the contacts, their periodic and symmetrical structure ensures that the input and output barriers have identical potential drops (i.e., $V_e = V_c$ in the present notation) and hence from Eq. (41) that $n = N_D$.

B. Undoped Tunneling Filled Well

The optical response of the undoped single quantum-well detector is strongly dependent on the bias voltage because the well is filled by tunneling through a thin emitter barrier. The structure used by Bandara *et al.* (77), shown schematically in Fig. 55, consists of a 1-μm GaAs contact layer (doped $N_D = 1.3 \times 10^{18}$ cm^{-3}) grown on a semi-insulating substrate followed by a $L_e = 155$-Å Al$_{0.27}$Ga$_{0.73}$As emitter barrier, a $L_w = 42$-Å undoped GaAs quantum well, a $L_c = 500$-Å Al$_{0.27}$Ga$_{0.73}$As collector barrier and a top 0.5-μm (doped $N_D = 1.3 \times 10^{18}$ cm^{-3}) GaAs contact layer.

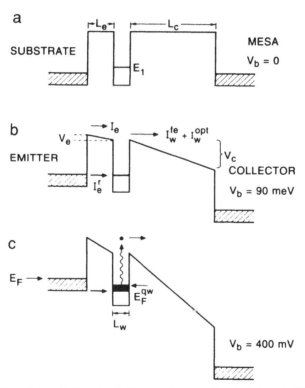

FIG. 55. Schematic conduction–band diagram of the undoped single-well QWIP. (a) Zero bias $V_b - 0$; (b) $V_b - 90$ meV corresponding to resonance between the emitter Fermi energy and bound-state energy in the well E1; (c) $V_b = 400$ meV corresponding to resonance between the bottom of the emitter conduction band and E$_1$. Also shown is the absorption of an infrared photon and photoexcitation of and electron out of the well (77).

The bias dependence of the responsivity for forward and reverse bias is shown in Fig. 56. As expected (73), the photoresponse for forward bias (where electrons from the emitter contact can rapidly tunnel through the thin L_e barrier and populate the quantum well) is nearly an order of magnitude larger than for reverse bias (where the injected electrons must tunnel through the thick L_c barrier and thus the electron density in the well is low). Two features of this responsivity vs. bias curve are noteworthy, namely the finite bias voltage required to get nonzero responsivity (i.e., the responsivity does not start to increase from $V_b = 0$ as expected for a bound-to-continuum QWIP) and the sharp drop at high bias (73). In order to understand this behavior, Bandara et al. have done a self-consistent theoretical calculation (77) of the voltage drops across the individual barriers, the electron density of the well populated by emitter resonant tunneling through the thin barrier, and the resulting dark current. The equations relating these quantities are similar to Eqs. (39)–(41) except that both resonant and nonresonant tunneling contributions to emitter current must be included (77). As indicated in Fig. 55c, the origin of the striking drop in n is that at this bias the quantum-well level E_1 drops below the conduction band edge of the emitter and thus the emitter resonant tunneling contribution is suppressed. For higher biases, the nonresonant tunneling contribution becomes dominant. The results of the calculation

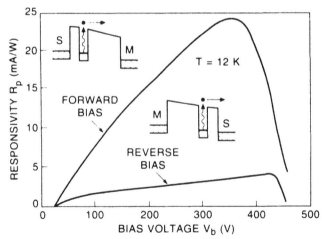

FIG. 56.   Measured responsivity vs. bias voltage V_b at $T = 77$ K for both forward ($V_b > 0$) and reverse ($V_b < 0$) bias for undoped single QWIP. The inserts show the conduction–band structure under bias where S is the substrate side and M is the mesa top (77).

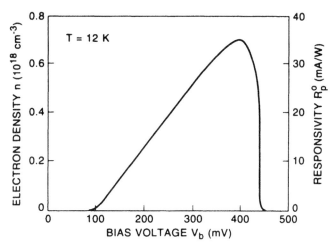

FIG. 57. Theoretical responsivity and electron density n in the quantum well at $T = 12$ K as a function of bias voltage V_b (77).

(77) for the electron density in the well and responsivity are shown in Fig. 57. The similarity to Fig. 56 is obvious, explaining the unusual dependence of the responsivity on bias. Note that the responsivity does not increase immediately at low bias but requires a significant bias offset. This is because the quantum well is essentially empty at low bias since $E_1 > E_F^c$, thus requiring a substantial bias voltage V_b to lower the ground level and allow filling of the well (Fig. 55b). The other feature of the data that is reproduced in the theory is the drop in responsivity for $V_b > 400$ mV, as a result of E_1 dropping below the conduction band edge at high bias thus cutting off the resonant tunneling (Fig. 55c).

For the n-doped single QWIP, gain is so large (maximum $g = 7.4$) since $g = \tau_L/\tau_T$, where τ_L is the photoconductive carrier lifetime, and τ_T is the transient time across the structure (i.e., $\tau_T = l/v$). Thus, for this single well, QWIP l and τ_T are more than an order of magnitude smaller, and thus g is more than an order of magnitude larger than the usual 50-period n-doped multiple-well QWIPs. However, this scaling (48, 79) $g \propto l^{-1}$ in well-doped QWIPs results from the lifetime τ_L being independent of the transient time τ_T. This is not true for the undoped single QWIP for which the quantum well is filled by tunneling through the emitter barrier. That is, after the photoexcited electron escapes from the well, it is immediately refilled from the emitter contact and thus $\tau_L = \tau_T$ and $g = 1$. Instead of making many passes through the single quantum well before being recaptured (as for doped QWIPs), the photoelectron is recaptured by the well in only a single pass (*49*). To check the expectation that $g = 1$, the current noise i_n was measured and determined to be the gain from $i_n = \sqrt{2eI_D g \Delta f}$, where Δf is the noise bandwidth. Note the factor of 2 in the noise expression rather than the usual factor of 4. This is because the usual photoconductor $\tau_L \neq \tau_T$ and the hot carrier "recirculates" through the quantum well. Thus, in addition to the noise due to the photogeneration process there is also the noise due to the recombination (i.e., recapture) process. However, when the well is immediately refilled by emitter tunneling, there is no uncertainty in the recapture process and no excess noise is produced. This is analogous to p–n junction photodetectors, where, because of the non-ohmic contacts, the photocarriers are not rejected. The results of our noise measurements taken using photocurrent at 77 K are shown in Fig. 58. Note that as expected $g = 1$, and is within experimental error.

C. p-DOPED WELL

In this subsection we discuss the properties of a single-well p-doped QWIP, which were found by Bandara *et al.* (*78*) to be dramatically different from those of the single n-QWIP. The structure, schematically

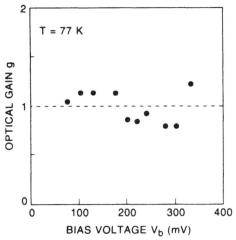

FIG. 58.   Measured photoconductive gain g at $T = 77$ K as a function of bias voltage V_b (*77*).

shown in Fig. 59a, was grown on a semi-insulating GaAs substrate via gas source MBE, and consisted of a single p-doped ($N_D = 2 \times 10^{18}$ cm^{-3} with Be) $L_w = 40$-Å quantum well surrounded by two $L_b = 300$-Å undoped Al$_{0.3}$Ga$_{0.7}$As barriers. The GaAs contact layers (top 0.5 μm and bottom 1 μm) were also p-doped to the same carrier density.

The responsivity was measured (78) at normal incidence (55) since the strong mixing between the light and heavy holes at $k \neq 0$ allows this advantageous geometry. The resulting spectrum is shown in Fig. 60 for a bias of $V_b = -0.4$ V (i.e., mesa top negative) and was found to be independent of temperature from $T = 10$ to 80 K to within experimental error ($< 10\%$). The peak position $\lambda_p = 7.2$ μm as well as the absolute magnitude of the responsivity are, as expected, quite similar to those of a previously discussed (55) 50-period p-QWIP having the same values of $L_w = 40$-Å, $L_b = 300$-Å, and Al$_{0.3}$Ga$_{0.7}$As barrier composition. Note that the single p-QWIP responsivity increases approximately linearly with bias at low voltage, demonstrating the bound-to-continuum nature of the

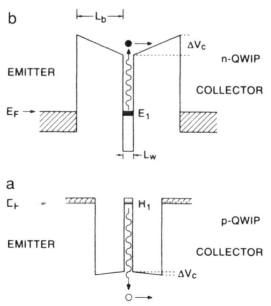

FIG. 59. (a) Schematic valence-band diagram of a single p-doped QWIP. The ground-state heavy-hole energy level is indicated as H_1. (b) Conduction–band diagram of a single n-doped QWIP. The ground-state electron level is indicated as E_1. The Fermi levels are E_F, and the barrier bending is ΔV_c (78).

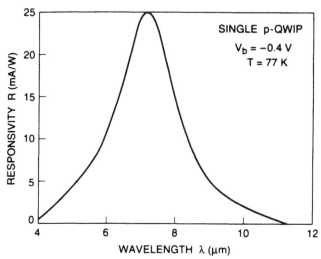

FIG. 60.   Responsivity spectrum measured at $V_b = -0.4$ V and $T = 77$ K for a single *p*-QWIP (*78*).

intersubband transition, where as the *n*-type single QWIP requires a large bias offset (i.e., the responsivity is essentially zero for $V_b < 0.1$ V). In addition, for the *p*-single QWIP there is no saturation of the responsivity with increasing bias even for very large voltages of $V_b = -0.5$ V (i.e., voltage drop per barrier of -0.25 V corresponding to 12.5 V for a 50-period QWIP), which is also strikingly different from that of an *n*-single QWIP.

The responsivity bias offset for a *p*-single QWIP is less than for *n*-single QWIPs (see Fig. 61) because as indicated in Fig. 59, the *n*-single QWIP has a much stronger band bending of the barriers than the *p*-QWIP. This is a result of the larger-hole effective mass relative to the electron mass and the consequent small ground-state hole energy $H_1 = 25$ meV (in Fig. 59a) with respect to the ground-state electron energy $E_1 = 88$ meV (in Fig. 59b). That is, the alignment of the Fermi level in the doped quantum well with the Fermi levels in the emitter and collector contact layers requires a large drop in the E_1 electron level (and thus a substantial barrier band bending), whereas for H_1 this adjustment is much less (*78*). The barrier band bending for the single *n*-QWIP is $\Delta V_c = 50$ meV, whereas it is only $\Delta V_c = 15$ meV for the single *p*-QWIP discussed above. Furthermore, the responsivity peak at $\lambda_p = 7.2$ μm (= 172 meV) is sufficiently high in the continuum ($\Delta E = 22$ meV) above the $V_b = 150$ meV barrier height that $\Delta E > \Delta V_c$. In strong contrast, the *n*-QWIP band

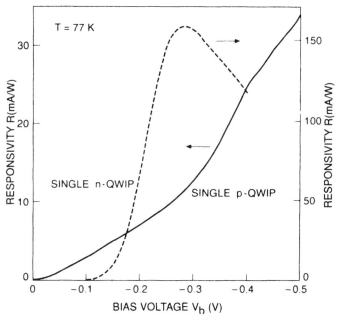

FIG. 61. Responsivity vs. bias voltage measured at $\lambda_p = 7.2$ μm and $T = 77$ K. The solid curve is for the single-p QWIP (left-hand-side scale) and the dashed curve is for the single n-QWIP (right-hand scale) (78).

bending of $\Delta V_c = 50$ meV is so large that the photoexcited electron ($\Delta E = 13$ meV) remains bound (i.e., $\Delta E < \Delta V_c$) and thus cannot escape without a large applied bias voltage to eliminate ΔV_c. That is, for the p-QWIP, escape probability of photoexcited carriers p_c is large, even at low bias, and thus remains constant as the bias is increased, whereas, for the n-QWIP, p_c is very small at low V_b and thus requires a substantial bias to overcome the band bending ΔV_c and allows the photoexcited electrons to escape.

The other important difference shown in Fig. 61 is the continuing increase of the single p-QWIP responsivity with bias, while the n-QWIP reaches a maximum after which it strongly decreases with increasing bias. This n-QWIP maximum was found to be due to the maximum in the gain, as shown in the Fig. 53. Thus, the current noise of this single p-QWIP at 77 K was measured and the calculated gain shown in Fig. 62. Note that the gain increases monotonically with increasing bias, thereby explaining the similar increase of the responsivity. This is again a result of the large effective hole mass. For both n- and p-QWIPs the initial increase in g is due to the decrease in photoexcited carrier transient time across the

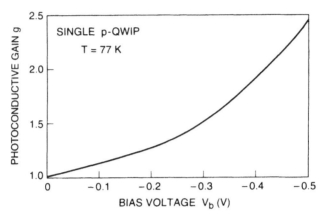

FIG. 62.   Photoconductive gain vs. bias voltage measured at $T = 77$ K for a single p-QWIP (78).

quantum well. However, at high bias the single n-QWIP gain decreases due to the direct tunneling of the low effective mass electrons from the emitter contact, which effectively "short circuits" the hot carrier transport process. For the high-mass holes, this direct tunneling process is inhibited and thus the gain continues to increase as shown in Fig. 62.

VII. Superlattice Miniband QWIPs

The superlattice miniband detector uses the concept of infrared photoexcitation between minibands (ground state and first excited state) and transport of these photoexcited electrons along the excited state miniband. When the carrier de Broglie wavelength becomes comparable to the barrier thickness of the superlattice, the wave functions of the individual wells tends to overlap due to tunneling, and an energy miniband is formed. The miniband occurs when the bias voltage across one period of the superlattice becomes smaller than the miniband width (80).

A. BOUND-TO-BOUND MINIBAND DETECTORS

Experimental work on IR detectors involving the miniband concept was initially carried out by Kastalsky et al. (81) at BellCore. The spectral response of this GaAs/AlGaAs detector was in the range 3.6–6.3 μm and indicated that low-noise IR detection was feasible without the use of external bias. O *et al.* (*82*) reported experimental observations and related theoretical analysis for this type of detector with absorption peak in the LWIR spectral range (8–12 μm). Both of these detectors consist of a bound-to-bound miniband transition (i.e., two minibands below the top of the barrier) and a graded barrier between the superlattice and the collector layer as a blocking barrier for ground-state miniband tunneling dark current.

In order to further reduce ground-state miniband tunneling dark current, Bandara *et al.* used a square step barrier (*83*) at the end of the superlattice. This structure, illustrated in Fig. 63, was grown by MBE and consists of 50 periods of 90-Å GaAs quantum wells and 45-Å $Al_{0.21}Ga_{0.79}As$ barriers. A 600-Å $Al_{0.15}Ga_{0.85}As$ blocking layer was designed so that it has two minibands below the top of the barrier, with the top of the step-blocking barrier being lower than the bottom of the first excited-state miniband, but higher than the top of the ground-state miniband. Detectors were fabricated by etching 100×100-μm mesas, and IR measurements were carried out on the structure with a substrate face polished to allow backside illumination at a 45° incident angle to the MQW active region.

Figure 64 shows the spectral photoresponse measured at 20 K with a 240-mV bias voltage across the detector and at 60 K with a 200-mV bias voltage. The experimental response band of this detector is in the extended LWIR range with peak response at 14.5 μm, which is the longest operating peak wavelength yet achieved in a GaAs/AlGaAs superlattice *miniband* detector. The theoretical spectral response band, shown in Fig. 65, was calculated using the resulting wave functions of the Kronig–Penney model in the standard dipole approximation. Details of this calculation are given in Refs. (*82, 83*). Figure 66 shows the bias voltage

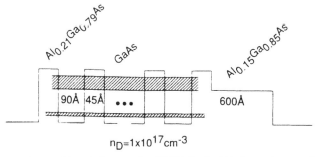

FIG. 63. Parameters and band diagram for LWIR GaAs/Al_xGa_{1-x}As superlattice miniband detector with $\lambda_c \sim 15$ μm (*83*).

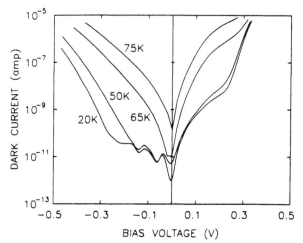

FIG. 64.   Dark current–voltage curves at different temperatures (*83*).

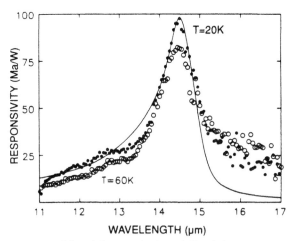

FIG. 65.   Spectral responsivity of the superlattice miniband detector measured at 20 K with 240-mV bias and 60 K with 200-mV bias. Since the light intensity of the globar source at longer wavelengths becomes smaller, the photocurrent signal of the detector becomes noisier. The solid line is based on the theoretical calculation (*83*).

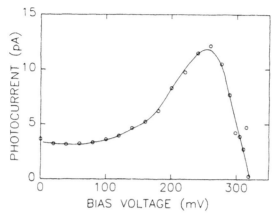

FIG. 66.   Bias voltage dependence of the photocurrent at the wavelength 14.5 μm at 20 K
(*83*).

dependence of the photocurrent at the wavelength 14.5 μm at 20 K. The
rapid falloff in the photocurrent at higher bias voltage values can probably
be attributed to progressive decoupling of the miniband as well as the
rapid decrease in the impedance of the detector.

The peak responsivity of this detector at 20 and 60 K are 97 and
86 mA/W for unpolarized light. Based on these values and noise measurements, the estimated detectivity at 20 and 60 K are 1.5×10^9 and
9×10^8 cm $\sqrt{Hz}$/W, respectively. Although this detector operates under a
modest bias and power conditions, the exhibited detectivity is relatively
lower than for usual QWIPs. This is mainly due to lower quantum
efficiency. Although there is enough absorption between minibands, only
photoexcited electrons in a few quantum wells near the collector contact
contribute to the photocurrent.

B. Bound-to-Continuum Miniband Detectors

It is anticipated that placing the excited-state miniband in the continuum levels would improve the transportation of the photoexcited electrons, i.e., responsivity of the detector. This is just as in the case of
the wide-barrier detectors discussed previously; a detector based on photoexcitation from a single miniband below the top of the barriers to one
above the top of the barriers is expected to show a higher performance.

Gunapala *et al.* (*84*) proposed and demonstrated this type of bound-to-continuum miniband photoconductor based on a GaAs/Al$_x$Ga$_{1-x}$As superlattice operating in the 5–9-μm spectral range. Their structure shows more than an order of magnitude improvement in electron transport and detector performance, compared with previous bound-to-bound state miniband detectors.

Two structures studied by Gunapala *et al.* (*84*) consisted of 100 periods of GaAs quantum wells of either L_b = 30- or 45-Å barriers of Al$_{0.28}$Ga$_{0.72}$As, and L_w = 40-Å GaAs wells (doped N_D = 1 × 10^{18} cm^{-3}) sandwiched between doped GaAs contact layers. The normalized room-temperature absorption spectra of these samples as well as the miniband structure are shown in Fig. 67. Note that the peak position $\lambda_p \sim 9$ μm is relatively independent of barrier thickness as expected, since λ_p is determined mostly by the well width. The absolute values for the peak absorption coefficients are α = 3100 and α = 1800 cm^{-1} for the L_b = 30- and 45-Å structures, respectively. The structure with narrower barrier (L_b = 30-Å) has a higher peak absorption coefficient as well as a broader spectrum resulting in significantly larger integrated absorption strength. Also, the dark current of this structure (L_b = 30-Å) is much larger than that for the other (L_b = 45-Å) structure.

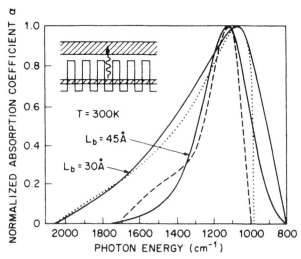

FIG. 67. Normalized room-temperature absorption coefficient spectra for the L_b = 30- and 45-Å bound-to-continuum miniband structures. The solid lines are experiment; the dashed and dotted lines are theory. The insert shows the schematic miniband structure (*84*).

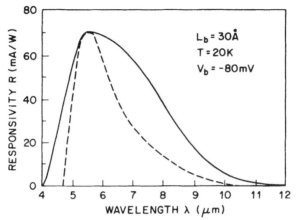

FIG. 68. Responsivity spectrum at $T = 20$ K and $V_b = -80$ mV for the $L_b = 30$-Å detector. The solid line is experiment; the dashed line is theory (84).

The measured responsivity values at 20 K (corrected for reflection losses) for both samples shown in Fig. 68 and 69 are one to two orders of magnitude larger than bound-to-bound miniband results (82). Although the absorption spectra (shown in Fig. 67) for both samples peak at ~ 9 μm, the responsivity spectra are quite different, peaking at 5.4 and 7.3 μm. This is due to the shift in the miniband group velocity curve as already

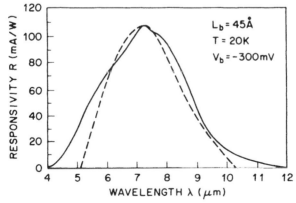

FIG. 69. Responsivity spectrum at $T = 20$ K and $V_b = -300$ mV for the $L_b = 45$-Å detector. The solid line is experiment; the dashed line is theory (84).

discussed (84). The dotted curves in Figs. 68 and 69 are the responsivities calculated (84) from miniband theory. Miniband optical absorption and group velocity were obtained by applying the Kronig–Penney Model for calculating allowed and forbidden energy bands. The reason that the spectral peaks in the responsivity curves do not coincide with the absorption peaks is that the group velocity peaks near the center of the band and vanishes at the edges, whereas the absorption curves peak near the low-energy edge of the band.

Detectivities at the peak wavelength for the above miniband detectors were calculated using measured responsivities and dark currents. For the $L_b = 30$-Å structure, the result is $D^* = 2.5 \times 10^9$ and 5.4×10^{11} cm $\sqrt{\text{Hz}}$/W for $T = 77$ and 4 K at a -80-mV bias. For the $L_b = 45$-Å structure, $D^* = 2.0 \times 10^9$ and 2.0×10^{10} cm $\sqrt{\text{Hz}}$/W for $T = 77$ and 4 K at a -300 mV bias were obtained. These values are significantly larger than the previous bound-to-bound miniband results. Although the responsivity is improved by placing the excited state in the continuum, it also increases the thermionic dark current because of the lower barrier height. This fact is more critical for longer-wavelength IR detectors because the photoexcitation energy becomes even smaller, i.e., the detector operating temperature will be limited.

C. BOUND-TO-MINIBAND DETECTORS

Yu et al. (85, 86) proposed and demonstrated a miniband transport QWIP, which contains two bound states with higher energy level being resonance with the ground state miniband in the superlattice barrier (see Fig. 70a). In this approach, infrared radiation is absorbed in the doped quantum wells, exciting an electron into the miniband and transporting it in the miniband until it is collected or recaptured into another quantum well. Thus the operation of this miniband QWIP is analogous to that of a weakly coupled MQW bound-to-continuum QWIP. In this device structure, the continuum states above the barriers are replaced by the miniband of the superlattice barriers. These miniband QWIPs show lower photoconductive gain than bound-to-continuum QWIPs because the photoexcited electron transport occurs in the miniband where electrons have to transport through many thin heterobarriers resulting in a lower mobility. The bandwidth of the absorption spectrum is controlled by the position of the miniband relative to the barrier threshold as well as the width of the miniband which is exponentially dependent on the thickness of the super-

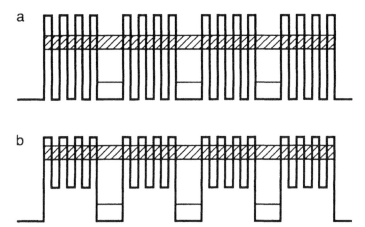

FIG. 70. Band diagrams for (a) bound-to-miniband and (b) step bound-to-miniband QWIP structures (*129*).

lattice barriers. Faska *et al.* (*87*) adopted this bound-to-miniband approach and demonstrated excellent IR images from a 256 × 256 focal plane array camera. These bound-to-miniband QWIPs have been demonstrated using a GaAs/Al_xGa_{1-x}As material system. In order to further improve performance (by decreasing dark current) of these miniband QWIPs, Yu *et al.* (*88*) proposed a step bound-to-miniband QWIP, which is shown in Fig. 70b. This improved structure consists of GaAs/Al_xGa_{1-x}As superlattice barriers and $In_{0.07}Ga_{0.93}$As strained quantum wells which are deeper than superlattice barrier wells, as shown in Fig. 70b.

VIII. Indirect Band-Gap QWIPs

Infrared detectors operating in the MWIR spectral range ($\lambda \sim 3$–5 μm) are also of interest due to the atmospheric window in this range. However, the short-wavelength limit in the GaAs/Al_xGa_{1-x}As materials system imposed by keeping the Al_xGa_{1-x}As barriers direct is $\lambda = 5.6$ μm. That is, if the Al concentration x is increased beyond $x = 0.4$ the indirect X valley becomes the lowest band gap. This has been thought to be highly undesirable since Γ–X scattering together with GaAs X-barrier trapping can result in inefficient carrier collection and thus poor responsivity. However, in view of the technology advantages in the more mature GaAs/Al_xGa_{1-x}As material system, it would be highly desirable to design detectors using indirect $Al_xGa_{1-x}As$. Levine, Gunapala, and Kopf (89) successfully demonstrated the first bound-to-continuum GaAs/ $Al_xGa_{1-x}As$ indirect band-gap QWIP operating at $\lambda_p = 4.2\ \mu m$.

A single period of the structure (89) consists of $L_w = 30$-Å of GaAs (doped $n = 1 \times 10^{18}\ cm^{-3}$) and 500-Å of undoped $Al_{0.55}Ga_{0.45}As$. This is repeated 50 times and is sandwiched between a 0.5-μm top and 1-μm bottom contact layer (also doped $n = 1 \times 10^{18}\ cm^{-3}$). The conduction-band diagram under bias for the structure is shown in Fig. 71. The solid line in Fig. 71 is the Γ-valley band edge and the dotted line is that of the X valley. The Γ-barrier height of $E_b^\Gamma = 452$ meV results in a calculated single bound Γ state at an energy $E_0^\Gamma = 165$ meV, and an optical absorption peak (from E_0^Γ to the excited continuum state E_1^Γ) at an energy of $\Delta E_{01}^\Gamma = 289$ meV (i.e., 2 meV above E_b^Γ) corresponding to a calculated absorption peak at $\lambda_p = 4.3\ \mu m$.

The operation of the detector (89) is schematically indicated by the numbers in Fig. 71. A photon excites an electron from E_0^Γ to E_1^Γ (No. 1), where it will escape out of the well and begin to transport above the Γ barrier (No. 2). Up to this point, the photoexcitation process is exactly the same as that for a direct-gap barrier detector. However, for the indirect barrier (90) the X-valley minimum lies $E_b^{\Gamma X} = 109$ meV below E_b^Γ. Since

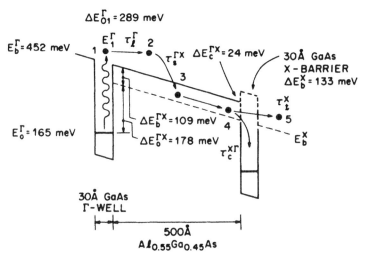

FIG. 71.   Schematic conduction-band diagram of indirect $Al_{0.55}Ga_{0.45}As$ barrier QWIP. The solid lines are the direct Γ-valley band edge and the dashed lines are the indirect X-band edge (89).

the scattering time (91) from the Γ valley to the X valley, $\tau_s^{\Gamma X} \sim 0.1$ ps, is much shorter than the Γ-valley hot-electron lifetime (14) $\tau_1^\Gamma \sim 40$ ps, the photoexcited electron will be rapidly scattered into the X valley (No. 3), where it will be transported by the field toward the $L_w = 30$ Å GaAs region (No. 4). Although the GaAs region is a quantum well for Γ electrons, it is a barrier of height (89) $\Delta E_b^X = 133$ meV for the X electrons. Therefore, in order for these electrons to be efficiently collected as photocurrent the GaAs X-barrier tunneling time, $\tau_{L_w}^X$ (No. 5), must be much shorter than the recapture time $\tau_c^{X\Gamma}$ back into the Γ-valley GaAs well.

In order for this X-valley electron to be captured into the Γ well it must first scatter back to the Γ valley. This scattering time (89) $\tau_s^{X\Gamma}$ is more than an order of magnitude larger than the reverse process $\tau_s^{\Gamma X}$ due to the higher density of states in the X valley compared with the Γ valley. In addition, the capture time is further increased by the well barrier duty cycle (L_w/L_p). After scattering into the Γ valley, the well recapture process requires an additional period of time of order τ_1^Γ, so that we approximately have $\tau_c^{X\Gamma} \sim (\tau_s^{X\Gamma} + \tau_1^\Gamma) \geq 40$ ps. Efficient photocarrier transport and collection therefore requires $\tau_t^X \ll \tau_c^{X\Gamma}$. In summary, the IR photon first excites an electron out of the GaAs Γ-valley quantum well. This hot Γ electron gets rapidly scattered into the $Al_{0.55}Ga_{0.45}As$ X valley where it must tunnel through the GaAs X barrier before being recaptured into the GaAs-Γ well.

It remains now to calculate this tunneling time from $\tau_{L_w}^X = L_w/vT$, where the velocity $v = \hbar k/m^* = \pi\hbar/L_w m^*$ and where the transmission factor is given approximately by

$$T = \exp\left[-(2t/\hbar)\sqrt{2m_X^*(\Delta E_b^X - E)}\right], \qquad (45)$$

where E is average hot-electron energy (we take $E \sim \hbar\omega_0 = 36$ meV to be a photon energy above the bottom of the X-band edge), and m_X^* is the appropriate effective mass in the X-valley. If transverse momentum is conserved, the longitudinal X-valley mass $m_X^* = 1.1$ should be used ($92, 93$).

This leads to a tunneling time for the longitudinal mass electrons of $\tau_t^X = 680$ ps $\gg \tau_c^{X\Gamma}$. Thus, most of the photoexcited electrons would get recaptured back into the Γ wells and not be collected as photocurrent. Fortunately, tunneling experiments have shown that momentum is not conserved (e.g., due to interface roughness, alloy, and phonon scattering) and therefore the transverse mass $m_X^* = 0.2$ should be used ($92, 93$). This leads to a very rapid tunneling time from Eq. (45) of $\tau_{L_w}^X = 0.4$ ps $\ll \tau_c^{X\Gamma}$.

Thus, the GaAs X barriers are virtually transparent, leading to excellent hot-electron transport and efficient collection comparable to the usual Γ direct gap transport. This analysis shows that the GaAs regions should be thin ($L_w \sim 30$ Å) for efficient tunneling, which is also desirable in order to have only one bound state in the Γ well for efficient photoabsorption. These considerations led to the structure described in Fig. 71.

The IR absorption coefficient was measured (89) at room temperature using a 45° multipass waveguide and, as shown in Fig. 72, has a peak value of $\alpha = 185$ cm^{-1} at $\lambda_p = 4.4$ μm with an absorption half height at 3.8 and 4.8 μm (i.e., full width half maximum of $\Delta\nu = 550$ cm^{-1}). At low temperature, the half width narrows and the peak absorption increases (21, 22) by a factor of 1.3 so that $\alpha = 240$ cm^{-1} at $T = 77$ K corresponding to an unpolarized quantum efficiency of $\eta = (1 - e^{-2\alpha l})/2 = 6.0\%$.

Detectors were fabricated and responsivity R was measured (80) (Fig. 73) to be independent of temperature from $T = 20$ to 100 K. The peak value is $R_p = 0.61$ A/W at $\lambda_p = 4.2$ μm and a bias of $V_b = 4$ V, and

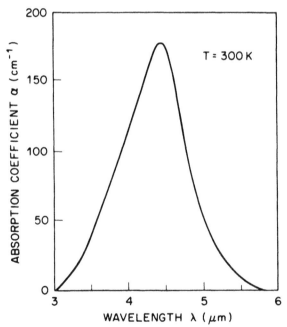

FIG. 72. Absorption coefficient spectrum at $T = 300$ K for the indirect Al$_{0.55}$Ga$_{0.45}$As barrier QWIP (89).

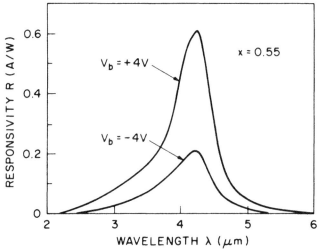

FIG. 73.   Responsivity spectrum at $T = 80$ K and $V_b = \pm 4$ V for the indirect $Al_{0.55}Ga_{0.45}As$ barrier QWIP (89).

$R_p = 0.20$ A/W at $V_b = -4$ V. The long wavelength cutoff $\lambda_c = 4.5$ μm (i.e., responsivity half height) and short-wavelength cutoff $\lambda_c = 3.8$ μm lead to a half width of $\Delta \nu = 410$ cm^{-1} which is, as expected, significantly narrower than the room-temperature absorption width (21, 22).

The bias dependence of R at $\lambda_p = 4.2$ μm has also been measured (89) and found to have significant zero-bias photovoltaic effect with $R_p = 0.05$ A/W at $V_b = 0$. This yields the photoconductive gain $g = 3.0$ and 1.0 for $V_b = 4$ and -4 V, respectively, and (using the superlattice length $l = 2.65$ μm) the corresponding values of the hot-electron mean free path $L = 8.0$ and 2.65 μm. These large values for g and L are comparable to (or even larger than) the best results in the usual Γ direct-gap devices (14, 48) and thus confirm our calculations of excellent transport and efficient carrier collection expected in this indirect barrier QWIP. That is, this large mean free path is due to the difficulty of an X electron being captured by a Γ well. The peak detectivity D_λ^* at $V_b = -1$ V and $T = 77$ K is $D_\lambda^* = 4 \times 10^{10}$ cm $\sqrt{Hz}$/W. However, an even larger value $D_\lambda^* = 1.1 \times 10^{12}$ cm $\sqrt{Hz}$/W is obtained at zero bias [determined by the Johnson noise generated by QWIP differential resistance (89)]. This large D^* is comparable to that obtained using the direct-gap material $In_{0.53}Ga_{0.47}As/In_{0.52}Al_{0.48}As$ which is discussed in Section IX.A.1.

IX. QWIPs with Other Materials Systems

A. n-DOPED QWIPs

1. $In_{0.53}Ga_{0.47}As$ / $In_{0.52}Al_{0.48}As$. In order to shift the intersubband absorption resonance into the higher energy spectral region ($\lambda = 3$–5 μm), Levine *et al.* (*94*) have investigated lattice-matched quantum-well superlattices of $In_{0.53}Ga_{0.47}As/In_{0.52}Al_{0.48}As$ grown using MBE on an InP substrate and reported intersubband absorption in this heterosystem. This direct-gap heterostructure has a conduction band discontinuity of 550 meV, which is significantly higher than that of the direct gap $GaAs/Al_xGa_{1-x}As$ system, therefore allowing for shorter-wavelength operation.

A 50-period multiquantum-well superlattice consisting of $In_{0.53}Ga_{0.47}As$ wells (doped $N_D = 1 \times 10^{18}$ cm^{-3}) having a width $L_w = 50$-Å, and 150-Å barriers of $In_{0.52}Al_{0.48}As$ was grown on an InP substrate. The experimental absorption peak is at $\lambda = 4.4$ μm (as shown in Fig. 74) in good agreement with the theoretical estimation of the energy separation of the bound states [see Ref. (*94*) for details]. In order to achieve higher performances in the MWIR range, Hasnain *et al.* designed a MQW structure (*95*) of the same materials system involving bound-to-continuum intersubband absorption. This structure consisting of 50 periods of 30-Å $In_{0.53}Ga_{0.47}As$ wells (doped $N_D = 2 \times 10^{18}$ cm^{-3}) and 300-Å $In_{0.52}Al_{0.48}As$ barriers was grown by MBE on an InP substrate. The absorption spectrum shown in Fig. 75 is peaked at 279 meV ($\lambda = 4.44$ μm) with a full width at half maximum of 93 meV. Although the peak absorption of this bound-to-continuum detector is 4.2 times lower than that of a bound-to-bound detector (*94*), the line width is five times greater. Thus, it has comparable (20% higher) absorption strength covering the full 3–5 μm MWIR band. The responsivity spectrum and bias dependence of this detector are shown in Figs. 76 and 77. The responsivity increases super linearly with bias up to 5 V and gradually saturates at a maximum value of 25 mA/W.

The noise measured in these MQW detectors at 500 Hz corresponds to the shot noise of the dark current resulting a peak detectivity at 77 K of $D^* = 1.5 \times 10^{12}$ cm $\sqrt{Hz}$ /W with a background limited (for a 180° field of view) $D_B^* = 2.3 \times 10^{10}$ cm $\sqrt{Hz}$ /W at 120 K and lower temperatures. These values are comparable to those demonstrated with the Pt–Si devices (*96*) presently used in the MWIR band.

2. $In_{0.53}Ga_{0.47}As$ / InP. An InGaAs/InP materials system has been used extensively for optical communication devices and therefore has a highly

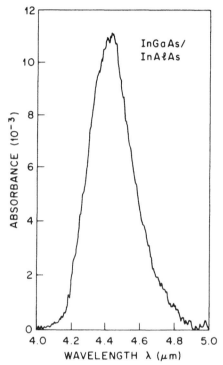

FIG. 74. Measured absorption at Brewster's angle for a 50-period MQW superlattice of $In_{0.53}Ga_{0.47}As/In_{0.52}Al_{0.48}As$ having a well width $L_w = 50$-Å and a barrier thickness of 100 Å (*129*).

developed growth and processing technology. Since the quality of the barriers is extremely important for optimum QWIP performance, and InP is binary whereas $Al_xGa_{1-x}As$ is a ternary alloy, Gunapala *et al.* investigated the hot-electron transport and performance of detectors fabricated from these two materials. Two structures were grown by metal-organic-molecular-beam epitaxy (MOMBE) with arsine and phosphine as group-V sources, trimethylindium and trimethylgallium as group-III sources, and elemental Sn as *n*-type dopant sources (*97, 98*). The first structure consisted of 20 periods of $L_w = 60$-Å $In_{0.53}Ga_{0.47}As$ quantum wells that were lattice matched to 500-Å InP barriers, whereas a second sample contained 50 periods of $L_w = 50$-Å $In_{0.53}Ga_{0.47}As$ wells separated by 500-Å InP barriers. These MQWs were doped $N_D = 5 \times 10^{17}$ cm^{-3}, and had top and bottom 0.4-μm contact layers of $N_D = 1 \times 10^{18}$ cm^{-3} doped $In_{0.53}Ga_{0.47}As$. The intersubband absorption was measured on a 45° multi-

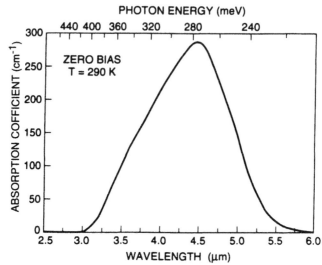

FIG. 75. Room-temperature intersubband absorption spectrum of bound-to-continuum $In_{0.53}Ga_{0.47}As/In_{0.52}Al_{0.48}As$ QWIP (*129*).

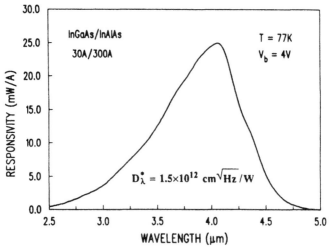

FIG. 76. Measured responsivity spectrum for an $In_{0.53}Ga_{0.47}As/In_{0.52}Al_{0.48}As$ QWIP at 77 K (*129*).

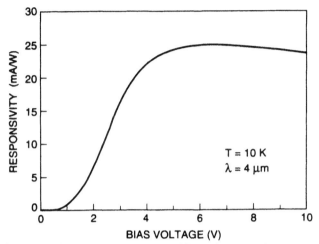

FIG. 77. Bias-voltage dependence of the responsitivity of an $In_{0.53}Ga_{0.47}As/In_{0.52}Al_{0.48}As$ QWIP measured at $\lambda_p = 4\ \mu m$ and $T = 10$ K *(129)*.

pass waveguide and the result shown in Fig. 78 for the 60-Å well sample. The peak ($\lambda_p = 8.1\ \mu m$) room-temperature absorption coefficient $\alpha = 950\ cm^{-1}$ is expected to increase by a factor 1.3 at $T = 77$ K resulting in a low-temperature quantum efficiency of $\eta = 12\%$ (*97*). For the 50-Å well QWIP, the corresponding value is $\eta = 11\%$. These values are quite comparable to those of GaAs samples, when the lower doping level of $N_D = 5 \times 10^{17}\ cm^{-3}$ in the wells is taken into account.

The measured responsivity spectrum for the 50-Å well QWIP is shown in Fig. 79 and the spectrum was substantially broader than that of a 60-Å well QWIP (*97*). Gunapala *et al.* have calculated the energy levels in the wells, using parameters $E_g(InP) = 1.424$ eV, $E_g(InGaAs) = 0.812$ eV, $m^*(InP) = 0.079$, $m^*(InGaAs) - 0.041$, $\Delta E_c = 242$ meV, and a non-parabolicity parameter $\gamma = 9.3 \times 10^{-15}\ cm^2$. They found that for a 60-Å quantum well, the excited level is just barely bound and for slightly different parameters (e.g., $L_w = 55$-Å) this excited state is just barely in the continuum, which explains the difference between the two spectrums (*97*).

The bias dependence of the responsivities (which was essentially independent of temperature $T = 10-80$ K) was measured and the result is shown for the 60-Å sample in Fig. 80 (the behavior of the 50-Å QWIP was similar at comparable electric fields). Note the extremely large values of

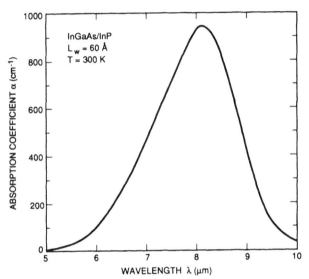

FIG. 78.   Room-temperature absorption coefficient for $In_{0.53}Ga_{0.47}As/InP$ QWIP (97).

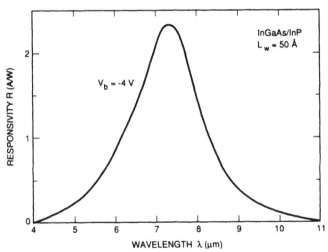

FIG. 79.   Responsivity spectrum for an $In_{0.53}Ga_{0.47}As/InP$ QWIP at $T = 20$ K and $V_b = -4$ V (97).

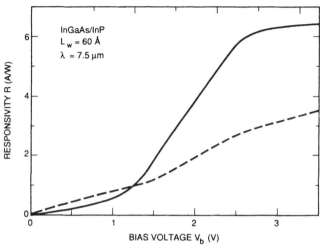

FIG. 80. Responsivity bias dependence for an $In_{0.53}Ga_{0.47}As/InP$ QWIP at $T = 20$ K and $\lambda_p = 7.5$ μm. The solid curve is forward bias, and the dashed curve is for reverse bias (*97*).

the responsivity, reaching $R = 6.5$ A/W at $V_b = +3.5$ V and $R = 3.5$ A/W at $V_b = -3.5$ V. This responsivity is five times larger than that of similar $GaAs/Al_xGa_{1-x}As$ QWIPs, which demonstrates the excellent transport in this materials system. These large responsivity values yield very large values of photoconductive gain $g = 9.0$ for $V_b = +3.5$ V and $g = 4.8$ for $V_b = -3.5$ V (*97*). The corresponding value of the hot-electron mean free path is $L = 10$ μm at $V_b = +3.5$ V, which is five times larger than that for similar $GaAs/Al_xGa_{1-x}As$ QWIPs. This excellent transport may be associated with the high-quality binary InP barriers, and higher mobility of InP compared with $Al_xGa_{1-x}As$. The calculated (*97*) peak detectivity $D_\lambda^* = 9 \times 10^{10}$ cm $\sqrt{Hz}$ /W based on noise current and measured responsivity at $V_b = +12$ V and $T = 77$ K of 60 Å QWIP compares favorably with $GaAs/Al_xGa_{1-x}As$ QWIPs operating at this wavelength.

3. InGaAsP / InP. In the previous subsection, we discussed lattice-matched InGaAs/InP QWIPs operating in the wavelength range 6–9 μm with excellent transport properties, which are attributed to the presence of binary InP barriers in the MQW structure. By using the lattice-matched InGaAsP/InP system, we can lower the well depth while keeping the InP barrier fixed and extend the IR response out to longer wavelengths.

For these experiments, Gunapala *et al.* investigated three structures grown in a VG–V80H system modified for metal-organic-molecular-beam epitaxy on semi-insulating (100) InP wafers (*99, 100*). The first structure consisted of 20 periods of L_w = 54-Å 1.55-μm band-gap InGaAsP quantum wells, lattice matched to 500-Å InP barriers (i.e., a superlattice length of 1.17-μm). A second structure consisted of 20 periods of L_w = 54-Å 1.3-μm band-gap InGaAsP quantum wells, lattice matched to 500-Å InP barriers, and a third 1.3-μm band-gap sample was similar, with L_w = 63 Å. The quantum wells were doped with N_D = 5 × 10^{17} cm^{-3} and sandwiched between 0.4-μm In$_{0.53}$Ga$_{0.47}$As contact layers of N_D = 1 × 10^{18} cm^{-3}.

The responsivity spectra of all three samples were measured at a bias voltage V_b = − 1 V at 77 K using a 200-μm-diameter mesa by illuminating through a 45° polished facet (see Fig. 81). The responsivity spectrum of the 1.55-μm InGaAsP sample is peaked at a wavelength of λ_p = 8.6 μm with long-wavelength λ_c = 9.4 μm. The two 1.3-μm quaternary band gap responsivity spectra were peaked at wavelengths of λ_p = 11.0 μm (L_w = 54 Å) and λ_p = 12.1 μm (L_w = 63 Å), with λ_c = 12.5 and 13.2 μm, respectively. This difference in spectral width for L_w = 54- and 63-Å QWIP detectors is in excellent agreement with theoretical analysis of the quantum well absorption, since for a 54-Å quantum well the excited state is high in the continuum, whereas for a 63-Å quantum well the excited

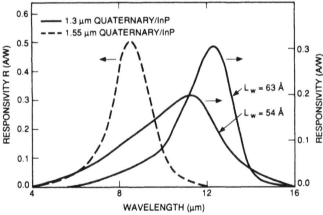

FIG. 81.   Responsivity spectra measured at T = 77 K and V_b = − 1 V, for a 1.55-μm bandgap InGaAsP quantum-well QWIP (dashed line), and for the 1.3-μm bandgap InGaAsP QWIPs (solid lines) (*99*).

state is just barely bound (99). The cutoff wavelengths of these two 1.3-μm band gap detectors were shifted to longer wavelengths with respect to the 1.55-μm band gap detector due to the reduced quantum well depth.

The bias dependence of the responsivities was measured and results are shown for the 1.55-μm band gap sample in Fig. 82, for both positive and negative bias. It is worth noting that the large values of responsivity reached 2.8 A/W at $V_b = -2.4$ V, and 1.2 A/W at $V_b = +2.5$ V. This value of 2.8 A/W for the responsivity is approximately twice as large as the responsivity of similar GaAs/Al$_x$Ga$_{1-x}$As QWIPs, again clearly demonstrating the excellent hot-electron transport in this materials system. It should be noted also, however, that this large maximum responsivity is still two times smaller than that of InGaAs/InP QWIPs previously discussed. This reduction of responsivity can be attributed to an increase in scattering and hence reduction in photoconductive gain of the quaternary GaInAsP material. The decrease of the responsivity at high bias, as shown in Fig. 82, is due to the decreasing photoconductive gain with increasing electric field.

The peak detectivities D_λ^* can be calculated using the above optical parameters and measured noise current (99). At an operating bias voltage

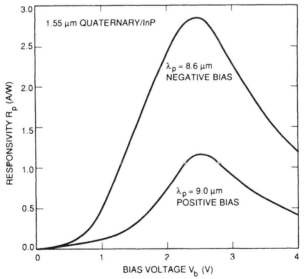

FIG. 82. Positive and negative bias dependence of peak responsivity (measured at λ_p) for 1.55-μm band-gap InGaAsP quantum-well QWIP obtained at $T = 77$ K (99).

of $V_b = -1$ V and $T = 77$ K, the optical parameters for the 1.55-μm quaternary sample are peak responsivity $R_p = 0.51$ A/W, quantum efficiency $\eta = 8.7\%$, photoconductive gain $g = 0.85$, dark current $I_d = 5.6$ mA, noise current $i_n = 1.7$ pA, and thus $D_\lambda^* = 5.3 \times 10^9$ cm $\sqrt{\text{Hz}}$/W. This detectivity is comparable to GaAs/Al$_x$Ga$_{1-x}$As detectors operating at similar wavelengths. The detectivity of the third sample at $T = 77$ K and $V_b = -1$ V, is $D_\lambda^* = 2.2 \times 10^8$ cm $\sqrt{\text{Hz}}$/W with cutoff wavelength $\lambda_c = 13.2$ μm. When the noise current was calculated at $T = 50$ K for this sample (at bias voltage $V_b = -1$ V) the detectivity increased to $D_\lambda^* = 1.3 \times 10^9$ cm $\sqrt{\text{Hz}}$/W (99). The peak responsivities (corrected for a 45° angle of incidence) of the first, second, and third sets of detectors were 0.51, 0.2, and 0.3 A/W, respectively, at a bias voltage $V_b = -1$ V.

4. *GaAs / Ga$_{0.5}$In$_{0.5}$P.* In a GaAs/Ga$_{0.5}$In$_{0.5}$P materials systems, Ga$_{0.5}$In$_{0.5}$P acts as the barrier material for the transport of electrons with an effective mass similar to that of Al$_{0.3}$Ga$_{0.7}$As. Therefore, Gunapala *et al.* investigated (101) this lattice-matched GaAs/Ga$_{0.5}$In$_{0.5}$P MQW structure grown on GaAs substrate as an alternative system to GaAs/Al$_x$Ga$_{1-x}$As for LWIR detection. In addition, these IR photoconductive measurements have allowed us to accurately determine the GaAs/Ga$_{0.5}$In$_{0.5}$P band offset.

The QWIP structure consisting of ten periods of 40-Å GaAs quantum wells (doped $N_D = 2 \times 10^{18}$ cm^{-3}) and 300-Å of undoped Ga$_{0.5}$In$_{0.5}$P barriers was grown by atmospheric pressure metal-organic-vapor-phase epitaxy (MOVPE) at a substrate temperature of 675°C on a semi-insulating undoped GaAs substrate. Figure 83 shows the absorption spectrum of the QWIP measured at room temperature using a 45° multipass waveguide configuration. The absorption spectrum is peaked at 8 μm (155 meV) with a full width at half maximum of 82 meV; i.e., $\Delta\lambda/\lambda = 53\%$. Figure 84 shows that the measured responsivity spectrum is also peaked at 8 μm, where the room-temperature absorption peaked, but the width decreases from $\Delta\lambda/\lambda = 53$ to 21%. This small spectral width at 20 K indicates that there is not much broadening from inhomogeneities in the quantum well width or other material parameters, and thus that the width at room temperature is due to the phonon broadening. The measured peak responsivity is 0.34 A/W corresponding to a photoconductive gain of g = 0.86.

In order to determine the lattice-matched GaAs/Ga$_{0.5}$In$_{0.5}$P conduction band offset ΔE_c, we calculated the position and bandwidth of the low-temperature absorption spectrum and adjusted the quantum well depth (i.e., ΔE_c) to fit the experiment. The theory includes both non-

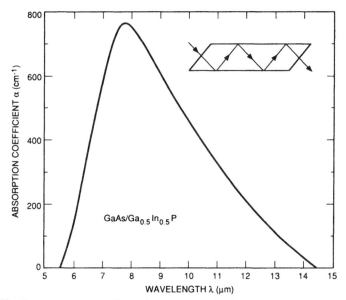

FIG. 83. Room-temperature absorption spectrum of a GaAs/Ga$_{0.5}$In$_{0.5}$P QWIP (*101*).

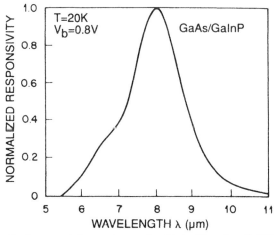

FIG. 84. Normalized responsivity spectrum of a GaAs/Ga$_{0.5}$In$_{0.5}$P QWIP at $T = 20$ K (*101*).

parabolicity as well as the exchange interaction. The exchange effect is significant and lowers the bound-state energy level (26, 102) by 19 meV for a doping density of $N_D = 2 \times 10^{18}$ cm^{-3}. From these results, the conduction band offset ΔE_c was determined to be 221 ± 15 meV. From the known band gaps of GaAs and $Ga_{0.5}In_{0.5}P$, the band gap difference $\Delta E_g = 483$ meV was obtained, and thus from $\Delta E_c + \Delta E_v = \Delta E_g = 483$ meV, we determine that $\Delta E_v = 262 \pm 15$ meV.

5. *GaAs / Al$_{0.5}$In$_{0.5}$P.* The GaAs/(Al$_x$Ga$_{1-x}$)$_{0.5}$In$_{0.5}$P quaternary has a direct Γ-valley bandgap from $x = 0$ to 0.7, which then becomes an indirect X-valley conduction band from $x = 0.7$ to $x = 1$ (103). The limiting composition GaAs/Al$_{0.5}$In$_{0.5}$P heterobarrier (although having an indirect gap) has a very large Γ-valley conduction band discontinuity of $\Delta E_c \sim$ 0.5 eV, (103) and thus is an interesting system for short-wavelength QWIPs. A multiquantum-well structure was grown (104) on a GaAs substrate via gas-source MBE, and consisted of 20 periods of 30-Å doped ($N_D = 1 \times 10^{18}$ cm^{-3}) quantum wells of GaAs and 500-Å barriers of undoped Al$_{0.5}$In$_{0.5}$P, surrounded by GaAs contact layers (doped $N_D = 2 \times 10^{18}$ cm^{-3}). The responsivity measured (104) on a 45° polished QWIP (shown in Fig. 85) had a narrow spectral shape ($\Delta\lambda/\lambda = 12\%$ indicating a nearly resonant bound-to-bound transition) and was peaked at $\lambda_p = 3.25$

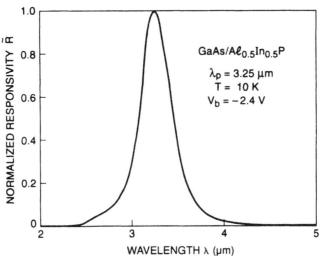

FIG. 85. Normalized responsivity of a GaAs/Al$_{0.5}$In$_{0.5}$P QWIP having a very-short-wavelength peak at $\lambda_p = 3.25$ μm (104).

μm. The fact that this $\lambda \approx 3$ μm QWIP is grown lattice matched to a GaAs substrate means that it can be integrated with a long-wavelength GaAs/Al$_x$Ga$_{1-x}$As QWIP ($\lambda = 6$–20 μm), grown on the same wafer, allowing for the fabrication of monolithic multiwavelength detectors.

6. In$_{0.15}$Ga$_{0.85}$As / GaAs. For all of the GaAs-based QWIPs demonstrated thus far, GaAs is the low bandgap *well* material and the barriers are *lattice matched* Al$_x$Ga$_{1-x}$As, Ga$_{0.5}$In$_{0.5}$P or Al$_{0.5}$In$_{0.5}$P. However, it is interesting to consider GaAs as the *barrier* material since the transport in binary GaAs is expected to be superior to that of a ternary alloy, as was previously found to be the case in the In$_{0.53}$Ga$_{0.47}$As/InP binary barrier structures (97, 99). To achieve this, Gunapala *et al.* (105) have used the lower bandgap nonlattice matched alloy In$_x$Ga$_{1-x}$As as well material together with GaAs barriers. Band-edge discontinuities and critical thicknesses of quantum-well structures of this materials system have been studied (106, 107). It has been demonstrated (108, 109) that strain-layer heterostructures can be grown for lower In concentrations ($x < 0.15$), which results in lower barrier heights. Therefore, this heterobarrier system is very suitable for very-long-wavelength ($\lambda > 14$ μm) QWIPs.

The samples were grown on semi-insulating GaAs substrates by MBE. The first structure shown in Fig. 86a consisted of 5 sets of 80-Å In$_{0.15}$Ga$_{0.85}$As quantum wells doped $N_D = 5 \times 10^{17}$ cm^{-3} separated by 500-Å barriers of undoped GaAs, with the top and bottom contacts being $N_D = 1 \times 10^{18}$ cm^{-3} doped GaAs. It should be noted that, unlike all the other III–V QWIPs demonstrated thus far, in this structure the heavily doped contacts are made using the *high* bandgap (i.e., GaAs) semiconductor. This is quite different from the GaAs/Al$_x$Ga$_{1-x}$As heterosystem in which the GaAs is the low-bandgap quantum well and contact material. This reversal for the In$_{0.15}$Ga$_{0.85}$As/GaAs structure is necessary because a thick contact layer of the strained nonlattice matched In$_{0.15}$Ga$_{0.85}$As material would have too many defects and threading dislocations. It is worth noting that the band structure of these QWIPs is quite similar to the band structure of Si$_{1-x}$Ge$_x$/Si QWIPs (110). The design shown in Fig. 86a overcomes this problem by making use of the strong band bending between the heavily doped GaAs contact layers and the first and last In$_{0.15}$Ga$_{0.85}$As quantum wells. This results in a large tunneling current (schematically indicated by the double arrow in Fig. 86a), which essentially "short circuits" the first and last wells, thus effectively contacting the low bandgap material. The active QWIP structure therefore consists of the central *three* quantum wells.

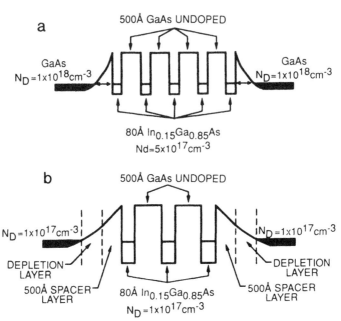

FIG. 86. (a) Conduction–band diagram of the first $In_{0.15}Ga_{0.85}As/GaAs$ QWIP structure.
(b) Conduction–band diagram of the second $In_{0.15}Ga_{0.85}As/GaAs$ QWIP structure. Unlike
the first structure, the second structure has two additional undoped GaAs spacer layers
between the quantum wells and the top and bottom contact layers (112).

The second structure, schematically shown in Fig. 86b, is slightly different from the first structure, which consisted of 3 sets of 85-Å $In_{0.15}Ga_{0.85}As$
quantum wells doped $N_D = 1 \times 10^{17}$ cm^{-3} separated by *two* 500-Å barriers of undoped GaAs, with the top and bottom contacts being $N_D = 1 \times$
10^{17} cm^{-3} doped GaAs. Also, this structure has two additional undoped
GaAs spacer layers between the quantum wells and the top and bottom
contact layers. As a result of these undoped spacer layers and the lower
contact doping, the *tunneling injection* current from contacts to the quantum wells is expected to be smaller in this structure in comparison to the
first structure.

The dark current–voltage curves for both samples were measured as a
function of temperature from $T = 30$–60 K as shown in Fig. 87. As
expected, Fig. 87 clearly shows that the dark current of the second
structure is many orders of magnitude smaller than the dark current of the
first structure for temperatures up to 60 K (dark current of the first
structure at 60 K is out of range of Fig. 87), which indicates that the
undoped spacer layers and the lower contact doping significantly reduced

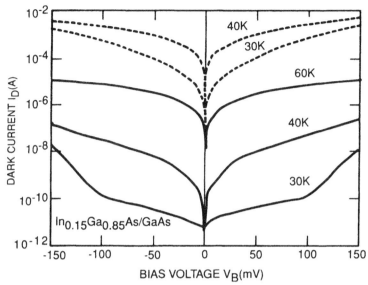

FIG. 87. Dark current vs. bias voltage at different temperatures for the first and the second device structure. This clearly indicates the reduction of dark current as a result of spacer layers, which reduces the current injection into the photosensitive MQW region (*112*).

the tunneling injection current to the quantum wells. Note the reduced asymmetry in the dark current I_d of this device structure. This can be attributed to the lower well doping (higher well doping will increase the Si dopant migration into the growth direction (*111*) and hence higher asymmetry in the band structure).

In order to further investigate this materials system Gunapala *et al.* (*112*) have used three samples designed to give a very wide variation in the QWIP absorption and transport properties. Sample A was designed to have an intersubband IR absorption transition occurring between a single localized *bound* state in the quantum well and a delocalized state in the *continuum* (*14, 35*). Sample C was designed with a wider well width, $L_w = 70$-Å, yielding two *bound* states in the well. Therefore, the intersubband transition is from the *bound* ground state to the *bound* excited state (*10, 40*) and requires electric field-assisted tunneling for the photoexcited carrier to escape into the continuum (*10, 40*). Due to the low effective mass of the electrons in the GaAs barriers, the electric field required for the field-assisted tunneling is expected to be smaller in these structures in comparison to the usual $GaAs/Al_xGa_{1-x}As$ QWIP structures. Sample B was designed such that the second bound level is resonant with the conduction band of the GaAs barrier. Thus, the intersubband transition is from the *bound* ground state to the *quasibound* excited state, which is intermediate between a strongly bound excited state and a weakly bound continuum state. (See the inserts in Fig. 88 for a schematic conduction band diagram of all three types of QWIP structures.)

The dark current–voltage curves for all three samples were measured as a function of temperature from $T = 20$–50 K; Fig. 88 shows the dark current–voltage curves of all three samples at two different temperatures ($T = 30$ and 50 K). Note that the dark current shown in Fig. 88 decreases with increasing quantum well width L_w. This is due to the fact that the effective barrier height of the ground-state electrons ($E_E = \Delta E_C - E_0 - E_F$) increases with the increasing quantum well width L_w, as shown in the inset of Fig. 88. This increase in effective barrier height reduces both tunneling and thermionic emission of ground-state electrons into the continuum transport states, resulting in a lower dark current.

Responsivity spectrums of the samples are shown in Fig. 89. The responsivities of samples A, B, and C peak at 12.3, 16.0 and 16.7 μm, respectively. The absolute peak responsivities (R_p) of the detectors were measured using a calibrated black-body source and the peak responsivities of samples A, B and C are 293, 510, and 790 mA/W, respectively, at bias $V_b = 300$ mV. As can readily be seen in Fig. 89, the responsivity spectra of the *bound-to-continuum* QWIP (sample A) is much broader than the *bound-to-bound* (sample C) or *bound-to-quasibound* (sample B) QWIPs. Correspondingly, the magnitude of the peak absolute responsivity (R_p) is

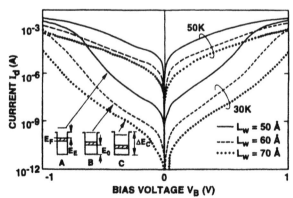

FIG. 88. Dark current–voltage curves of all three samples at temperatures $T = 30$ and 50 K. The inset shows the schematic conduction–band diagram for sample A (*bound-to-continuum*), sample B (*bound-to-quasibound*), and sample C (*bound-to-bound*) (*112*).

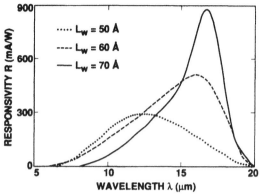

FIG. 89. Responsivity spectrums of samples A, B, and C at temperature $T = 50$ K. The peak responsivities are $R_p = 293, 510$, and 790 mA/W (at $V_B = 300$ mV) for samples A, B, and C, respectively (*112*).

significantly lower than that of the *bound-to-bound* or *bound-to-quasibound* QWIPs, due to the reduction of absorption coefficient α. This reduction in the absorption coefficient is a result of the conservation of oscillator strength (*42*). These peak wavelengths and the spectral widths are in good agreement with theoretical estimates (see Table V) of *bound-to-continuum* and *bound-to-bound* intersubband transition based on the 55% conduction band offset $(\Delta E_C/E_g)$ of GaAs/In$_{0.2}$Ga$_{0.8}$As materials system (i.e., $\Delta E_V/E_g = 45\%$). The measured absolute responsivities of all three samples increase nearly linearly with the bias, reaching $R_p = 0.85$, 1.1, and 2.2 A/W at $V_b = 1.3$ V for the samples A, B, and C, respectively. The higher responsivity of the sample C (*113, 114*) (*bound-to-bound*) in comparison to the samples A and B (*bound-to-continuum* and *bound-to-quasi-*

TABLE V
EXPERIMENTAL DETECTIVITIES AND EXPERIMENTAL AND THEORETICAL SPECTRAL DATA
OF ALL THREE SAMPLES AT TEMPERATURE $T = 50$ K (*112*)

Samples	Peak detectivity ($\times 10^8$ cm$\sqrt{\text{Hz}}$/W) Bias $V_B = 0.3$ V	Experimental peak wavelength (μm)	Theoretical peak wavelength (μm)	Experimental $\Delta\lambda/\lambda$ (%)	Theoretical $\Delta\lambda/\lambda$ (%)
A	3	12.3	12.5	60	61
B	10	16.0	16.3	41	34
C	30	16.7	16.8	18	22

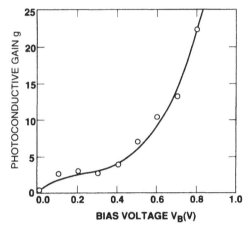

FIG. 90. Photoconductive gain vs. bias voltage for sample-C device structures at temperature $T = 40$ K (*112*).

bound) is attributed to the higher absorption coefficient of sample C relative to the other samples and the higher tunneling probability associated with the lower effective mass in the GaAs barriers (in comparison to the higher effective mass in the $Al_xGa_{1-x}As$ barriers of *bound-to-bound* $GaAs/Al_xGa_{1-x}As$ QWIPs).

The current noise i_n was measured using a spectrum analyzer and experimentally determined the photoconductive gain (*49*) g using $g = i_n^2/4eI_d\Delta f + 1/2N$, where Δf is the measurement band width and N is the number of quantum wells. As shown in Fig. 90, photoconductive gain of sample C reached 22.5 at $V_b = 800$ mV, which is very large compared to the photoconductive gains of usual $Al_xGa_{1-x}As/GaAs$ QWIPs. Since the gain of QWIP is proportional to the number of quantum wells N, the better comparison would be the well-capture probability p_c, which is directly related to the gain (*49*) by $g = 1/Np_c$. The calculated well capture probabilities are 20% at very low bias ($V_b = 10$ mV) and 0.4% at high-bias voltage ($V_b = 800$ mV) which indicates the excellent hot-electron transport in these device structures. This may be a result of the high-mobility binary GaAs barriers. The peak detectivity D_λ^* can now be calculated from $D_\lambda^* = R\sqrt{A\Delta f}/i_n$, where A is the area of the detector and $A = 3.14 \times 10^{-4}$ cm^2. Table V shows the D_λ^* values of all three samples at temperature $T = 50$ K at a bias of $V_B = 300$ mV. As shown in the Table VI, the detectivity values of these detectors increase with decreasing temperature.

The large responsivity and detectivity D^* values are comparable to those achieved with the usual lattice-matched $GaAs/Al_xGa_{1-x}As$ materials system (52, 115). The high photoconductive gains and the small carrier-capture probabilities demonstrate the excellent carrier transport of the GaAs barriers and the potential of this heterobarrier system for very long-wavelength ($\lambda > 14$ μm) QWIPs. By comparing the theoretically calculated peak wavelengths and spectral widths, we have determined the band offsets ΔE_C and ΔE_V for the nonlattice matched GaAs/In$_x$Ga$_{1-x}$As heterobarrier.

B. p-DOPED QWIPS

In Sections IV.B and VI.C, we have discussed the GaAs/Al$_x$Ga$_{1-x}$As QWIPs, which are based on hole intersubband absorption in p-doped GaAs quantum wells. Due to the complex GaAs valence-band interactions at nonzero wave vector, the IR absorption at normal incidence is allowed. Also in Section IX.A.2 and 3 we have indicated that, due to the high-quality InP barriers, n-doped lattice-matched In$_{0.53}$Ga$_{0.47}$As/InP ternary and InGaAsP/InP quaternary QWIPs have even larger responsivities than GaAs/Al$_x$Ga$_{1-x}$As n-QWIPs. Therefore, Gunapala et $al.$ have investigated the p-QWIPs in a lattice-matched In$_{0.53}$Ga$_{0.47}$As/InP materials system (116).

Since most of the GaAs/Al$_x$Ga$_{1-x}$As band-gap discontinuity is in the conduction band (i.e., $\Delta E_c/\Delta E_g = 65\%$), whereas in In$_{0.53}Ga_{0.47}$As/InP most of the band-gap difference is in the valence band (i.e., $\Delta E_v/\Delta E_g = 60\%$), the In$_{0.53}Ga_{0.47}$As/InP p-QWIP intersubband absorption occurs at a much shorter wavelength. The GaAs p-QWIPs discussed previously (Sections IV.B and VI.C), operated with a peak wavelength of $\lambda_p = 7.2$–7.4 μm and a cutoff wavelength of $\lambda_c = 7.9$–8.6 μm. In strong contrast, the In$_{0.53}$Ga$_{0.47}$As/InP p-QWIPs discussed here have a responsivity peaked at $\lambda_p = 2.7$ μm. This is in fact the shortest wavelength QWIP ever reported.

TABLE VI

EXPERIMENTAL DETECTIVITIES OF SAMPLE C AT VARIOUS TEMPERATURES (112)

Temperature (K)	Peak detectivity D_λ^* (cm$\sqrt{\text{Hz}}$/W)
10	1.3×10^{13}
20	5.4×10^{12}
30	2.9×10^{11}
40	1.8×10^{10}
50	3.0×10^{9}

The devices discussed below were grown using MOMBE and consisted of 20 periods of 25-Å quantum wells of $In_{0.53}Ga_{0.47}As$ (doped $N_D = 2 \times 10^{18}$ cm^{-3} with Be) and 500-Å slightly p-doped barriers of InP. This slight p-doping is necessary to compensate for the n-background ($N_D \approx 4 \times 10^{16}$) of the InP barriers. Otherwise, n-InP and p-InGaAs result in a series of $p-n$ junctions, which would significantly reduce the available carriers in the quantum well, and thus lower the IR absorption. In addition, the $p-n$ junctions would increase the series resistance and thus impede the transport. These MQWs were sandwiched between $In_{0.53}Ga_{0.47}As$ contact layers (0.4-μm top and bottom) having the same doping as the wells.

The responsivity spectra (shown in Fig. 91 for $V_b = +3$ V and also for zero bias), demonstrate the shortest-wavelength QWIP ever reported with a responsivity peak at $\lambda_p = 2.7$ μm and a cutoff wavelength of $\lambda_c = 3.0$ μm (the long-wavelength side where the responsivity drops to half of its peak value). The peak unpolarized responsivity, at normal incidence, is $R_p = 29$ mA/W at $V_b = +3$ V (corresponding to a double optical pass through the QWIP). This value is essentially unchanged from $T = 20$ to 80 K. The unpolarized responsivity measured using the 45° polished substrate-control detector was nearly identical, with $R_p = 31$ mA/W for the same conditions. This is in good agreement with the $GaAs/Al_xGa_{1-x}As$ p-QWIPs discussed in the Section IV.B, where both

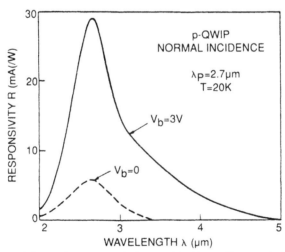

FIG. 91.   Responsivity spectra measured at normal incidence (for unpolarized radiation) at $T = 20$ K, $V_b = +3$ V, and $V_b = 0$ (116).

the magnitude and spectral shape of the normal-incidence and 45° responses were also very similar. It is worth noting that there is a zero-bias responsivity of R_p = 6 mA/W (shown in Fig. 91), which also peaked at λ_p = 2.7 μm. This is due to the asymmetry in the InGaAsP/InP and InP/InGaAsP heterobarriers on the two sides of the quantum well, giving the photoexcited carriers a preferred direction. The relatively low values of these responsivities (compared with n-QWIPs) is a result of the large mass and low transport velocity of the holes, as well as impurity scattering in the doped InP barriers. It is also worth noting the long low-energy tail on the responsivity spectrum of Fig. 91. This is due to the band bending (discussed above) and the resulting tunneling of the low-energy photoexcited carriers out of the well. The peak detectivity for this unoptimized device was $D_\lambda^* = 3 \times 10^{10}$ cm $\sqrt{\text{Hz}}$ /W at T = 77 K.

X. Light Coupling Methods

QWIPs do not absorb radiation incident normal to the surface since the light polarization must have an electric field component normal to the superlattice (growth direction) to be absorbed by the confined carriers. As shown in Fig. 92, when the incoming light contains no polarization component along the growth direction, the matrix element of the interaction vanishes (i.e., $\vec{\varepsilon} \cdot \vec{p}_z$ = 0 where $\vec{\varepsilon}$ is the polarization and $\vec{p}_z$ is the momentum along z direction). As a consequence, these detectors have to be illuminated through a 45° polished facet (10) (see Fig. 93a). Clearly, this illumination scheme limits the configuration of detectors to linear arrays and single elements. For imaging, it is necessary to be able to couple light uniformly to two-dimensional arrays of these detectors. Goosen et al. (117) and Hasnain et al. (118) have demonstrated efficient light coupling to QWIPs using linear gratings, which removed the light-coupling limitations and made two-dimensional QWIP imaging arrays feasible (see Fig. 93b). These line gratings were made of metal on top of each detector (117) or crystallographically groove etched through a cap layer on top of the quantum well structure. These gratings deflect the incoming light away from the direction normal to the surface, enabling intersubband absorption. The normal-incidence light-coupling efficiency of line gratings is comparable to the light-coupling efficiency of a 45° polished facet illumination scheme and yields a quantum efficiency (the fraction of the incident photons effective in producing the emitted electrons) of about 10–20% (for QWIPs having 50 periods and $N_D = 1 \times 10^{18}$ cm^{-3}). This relatively low quantum efficiency is due to the poor light-coupling efficiency and the fact that only one polarization of the light is absorbed. In order to

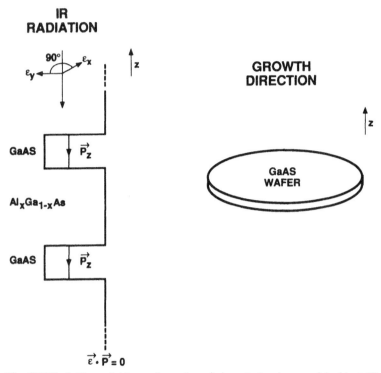

FIG. 92. QWIPs (with no light coupling scheme) do not absorb normal incident IR light since there is no light polarization component along the quantum-well direction (growth direction) (*116*).

maintain high absorption and thus high quantum efficiency, the quantum-well doping density must be kept very high, leading to a higher dark current. In order to further increase the quantum efficiency without increasing the dark current, Anderson *et al.* (*119, 121*) and Sarusi *et al.* (*122*) developed the cross grating (or two-dimensional grating) for QWIPs operating at the 8–10-μm spectral range. In this case, the periodicity of the grating is repeated in both directions, leading to the absorption of both polarizations of incident IR light. The addition of optical cavity results in two optical passes through the MQW structure before the light is diffracted out through the substrate, as shown in Fig. 93c.

Many more passes of IR light and significantly higher absorption can be achieved with a randomly roughened reflecting surface, as shown in Fig. 94a. Sarusi *et al.* (*123*) have shown experimentally that by careful

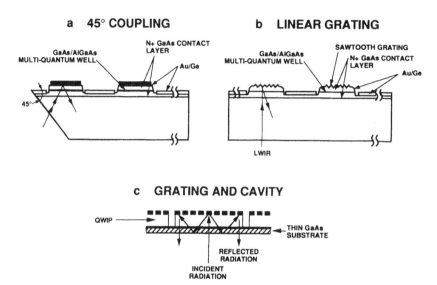

FIG. 93. Different light-coupling mechanisms used in QWIPs. (a) 45° polished facet, (b) linear or two-dimensional gratings on each detector, and (c) gratings with optical cavity.

design of surface texture randomization, efficient light trapping can be obtained. They demonstrated nearly an order of magnitude enhancement in responsivity compared to 45° illumination geometry. The random structure on top of the detector prevents the light from being diffracted normally backward after the second bounce as happens in the case of cross-grating (see Fig. 93c). After each bounce, light is scattered at a different random angle and the only chance for light to escape out of the detector is when it is reflected toward the surface within the critical angle of the normal. For the GaAs/air interface, this angle is about 17°, defining a very narrow escape cone for the trapped light. Some considerations have been taken into account when designing such random scatters to reduce the probability of light being diffracted into the escape cone. The reflector was designed with three levels of scattering surfaces located at quarter wavelength separations, as shown in Figs. 94b and 94d. As shown in Fig. 94b and 94c, these scattering centers were arranged in cells to prevent clustering of the scattering centers, having the same dimension as the light wavelength in GaAs. The combined area of the top unetched level and the bottom level ($\lambda/2$ deep) are $U^2/2$ (where U^2 is the area of a unit cell as shown in Fig. 94b). The area of the intermediate level ($\lambda/4$ deep) is also $U^2/2$. These reflecting areas and depths were chosen such that the

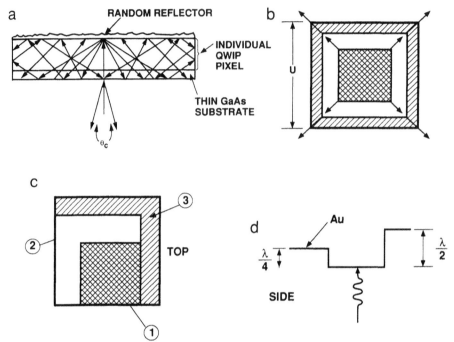

FIG. 94. (a) Schematic side view of a thin QWIP pixel with a random reflector. Ideally all the radiation is trapped except for a small fraction which escape through the escape cone (defined by critical angle Θ_c), (b) top view of the unit cell of the scattering surface (arrows indicate the 16 random possibilities), (c) top view of the one of the 16 possibilities, (d) side view of the unit cell (*129*).

normally reflected light intensities from two adjacent surfaces are equal and 180° out of phase, thus maximizing the destructive interference at normal reflection (and hence lowering the light leakage through the escape cone). These scattering centers were organized randomly inside the cell as shown in Fig. 94c. This random structure was fabricated on the detectors by using standard photolithography and selective dry etching. The advantage of the photolithographic process over a completely random process is the ability to accurately control the feature size and to preserve the pixel-to-pixel uniformity, which is a prerequisite for high-sensitivity imaging focal-plane arrays. It is clearly evident from the experiments that maximum responsivity is obtained when the unit cell size is equal to the wavelength of the QWIP maximum response. When this condition is met, light scattering becomes very efficient. If the unit cell is larger than the wavelength of the IR radiation (in GaAs), the number of scattering centers on the detector surface will decrease and the light is scattered less efficiently. On the other hand, if the unit cell size is smaller than the wavelength (in GaAs), the scattering surface becomes effectively smoother, and, as a consequence, scattering efficiency again decreases. Naturally, thinning down the substrate enables more bounces of light and therefore higher responsivity. One of the main differences between the effect of the cross grating and the random reflector is the shape of the responsivity curve; unlike the cross grating, the random reflector has little impact on the bandwidth of the response curve since the scattering efficiency of the random reflector is significantly less wavelength dependent than for the regular grating. Therefore, for the QWIPs with random reflectors, the integrated responsivity is enhanced by nearly the same amount as the peak responsivity.

XI. Imaging Arrays

A. EFFECT OF NONUNIFORMITY

The general figure of merit that describes the performance of a large imaging array is the noise equivalent temperature difference (NEΔT). NEΔT is the minimum temperature difference across the target that would produce a signal-to-noise ratio of unity and it is given by (43, 124)

$$\text{NE}\Delta\text{T} = \frac{\sqrt{A \Delta f}}{D_\text{B}^*(dP_\text{B}/dT)}, \qquad (46)$$

where D_B^* is the black-body detectivity, defined by Eq. (30), and (dP_B/dT) is the change in the incident integrated black-body power in the spectral range of detector, with temperature. The integrated black-body power P_B in the spectral range from λ_1 to λ_2 can be written as

$$P_\text{B} = A \sin^2\left(\frac{\theta}{2}\right) \cos \phi \int_{\lambda_1}^{\lambda_2} W(\lambda) \, d\lambda, \qquad (47)$$

where θ, ϕ, and $W(\lambda)$ are the optical field of view, angle of incidence, and black-body spectral density, respectively, and are defined by Eqs. (13) and (14). Before discussing the array results, it is also important to understand the limitations on the focal-plane-array imaging performance due to pixel nonuniformities (125). This point has been discussed in detail by Shepherd et al. and others (96, 126–128) for the case of PtSi infrared focal plane arrays, which have low response but very high uniformity. The general figure of merit to describe the performance of a large imaging array is the NEΔT, including the spatial noise, which has been derived by Shepherd (96), and given by

$$\text{NE}\Delta T = \frac{N_n}{dN_b/dT_b},\tag{48}$$

where T_b is the background temperature and N_n is the total number of noise electrons per pixel, given by

$$N_n^2 = N_t^2 + N_b^2 + u^2 N_b^2.\tag{49}$$

The photoresponse-independent temporal noise (electrons) is N_t, the shot noise (electrons) from the background radiation is N_b, a residual nonuniformity after correction by the electronics is u. The temperature derivative of the background flux can be written to a good approximation as

$$\frac{dN_b}{dT_b} = \frac{hcN_b}{k\bar{\lambda}T_b^2},\tag{50}$$

where $\bar{\lambda} = (\lambda_1 + \lambda_2)/2$ is the average wavelength of the spectral band between λ_1 and λ_2. When temporal noise dominates, NEΔT reduces to Eq. (46). In the case where residual nonuniformity dominates, Eqs. (48) and (50) reduce to

$$\text{NE}\Delta T = \frac{u\bar{\lambda}T_b^2}{1.44}.\tag{51}$$

The unit of constant is cm K, $\bar{\lambda}$ is in cm, and T_b is in K. Thus, in this spatial noise limited operation NEΔT α u and higher uniformity means higher imaging performance. Levine (129) has shown as an example, taking $T_b = 300$ K, $\bar{\lambda} = 10$ μm, and $u = 0.1\%$ leads to NEΔT $= 63$ mK, while an order of magnitude uniformity improvement (i.e., $u = 0.01\%$) gives NEΔT $= 6.3$ mK. By using the full expression, Eq. (49) Levine (129) has calculated NEΔT as a function of D^* as shown in Fig. 95. It is important to note that when $D^* \geq 10^{10}$ cm $\sqrt{\text{Hz}}$/W, the performance is uniformity limited and thus essentially independent of the detectivity; i.e., D^* is not the relevant figure of merit (130).

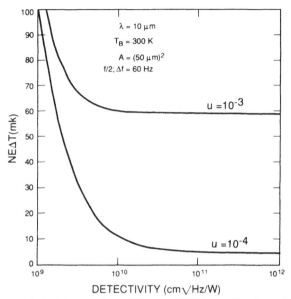

FIG. 95. Noise equivalent temperature difference NEΔT as a function of detectivity D^*. The effects of nonuniformity are included for $u = 10^{-3}$ and 10^{-4}. Note that for $D^* > 10^{10}$ cm $\sqrt{\text{Hz}}$ /W detectivity is not the relevant figure of merit for focal plane array (*129*).

B. ARRAY PERFORMANCE

1. LWIR Arrays. IR imaging systems that work at the 8–12 μm (LWIR) band have many applications, including night vision, navigation, flight control, and early warning systems. Several research groups have demonstrated (*8, 87, 131–139*) the excellent performance of QWIP arrays. For example, Faska *et al.* (*87*) have obtained very good images using a bound-to-miniband QWIP. The array we discuss here consisted of a bound-to-contiuum QWIP containing 50 periods of L_w = 40-Å of doped ($N_D = 1 \times 10^{18}$ cm^{-3}) GaAs, and L_b = 500-Å of undoped Al$_{0.27}$Ga$_{0.73}$As. This QWIP had a peak responsivity of R_p = 210 mA/W at λ_p = 8.6 μm and a cutoff wavelength of λ_c = 9.6 μm (see Fig. 96). The peak absorption coefficient was 560 cm^{-1} at T = 77 K corresponding to a quantum efficiency of η = 13% (for a 45° double pass). Device structures were grown on 2-inch GaAs wafers and processed (*131, 140, 135, 137–139*) into 50-μm pitch 128 × 128 arrays. Linear gratings (fabricated by reactive-ion etching) were used to couple the normal incident light into the focal plane array detectors. These gratings are covered with Au (i.e., used for ohmic contact and reflection), and In bumps were evaporated on top for hybridization to the Si multiplexer. A single QWIP array was chosen and bonded to a Si multiplexer. Detectors of the focal plane array were back-illuminated through the flat polished substrate, which is identical to that used in HgCdTe or InSb imaging systems.

As described by Refs. (*131, 135, 137*) this QWIP focal plane array gave excellent images with 99% of the pixels working, demonstrating the high yield of this new GaAs focal-plane-array technology. This high yield is a result of the excellent GaAs MBE growth uniformity and the mature GaAs processing technology. Figure 96 clearly shows the excellent uncorrected photocurrent uniformity of the 16,384 pixels with a standard deviation of only $\sigma = 3.6\%$. The uniformity after two point correction was $u = 0.02\%$.

Video images (*131, 135, 137–140*) were taken at a frame rate of 109 Hz with f/2 optics at temperatures high as $T = 65$ K, using a multiplexer having charge capacity of 4×10^7 electrons. A closeup image of a face is shown in Fig. 97, where the warm air emanating from the nostrils can be seen. The measured noise equivalent temperature difference of NEΔT = 10 mK, as well as the very large dynamic range of 85 dB, and the negligible 1/f noise (*8*) (noise corner frequency < 0.05 Hz) clearly show the promise of this imaging technology.

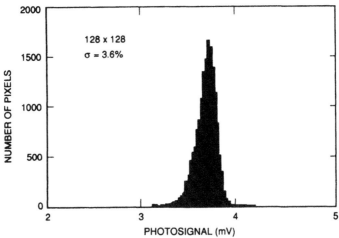

FIG. 96.   Photosignal histogram of the 16,384 pixels of the 128 × 128 array showing a high uniform uncorrected standard deviation of only $\sigma = 3.6\%$ (*129*).

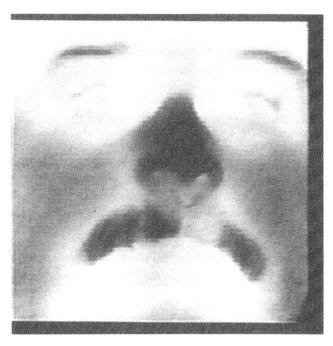

FIG. 97. One frame from a QWIP video image of a man's face NEΔT = 10 mK. The warm air from his nostrils can be clearly seen (*129*).

2. *VLWIR Arrays.* The absorption lines of many gas molecules, such as ozone, water, carbon monoxide, carbon dioxide, and nitrous oxide occur in the wavelength region from 3 to 18 μm. Thus, IR imaging systems that operate in the very-long-wavelength IR (VLWIR) region (12–18 μm) are also required in many space applications, such as monitoring the global atmospheric temperature profiles, relative humidity profiles, cloud characteristics, and the distribution of minor constituents in the atmosphere, which are being planned for the NASA's Earth Observing System (EOS). These space applications have placed stringent requirements on the performance of the IR detectors and arrays, including high detectivity, low dark current, low $1/f$ noise, uniformity, radiation hardness, and lower power dissipation. Also, this spectral region is rich in information vital to the understanding of the composition, structure, and energy balance of molecular clouds and star forming regions of our galaxy. Thus, there is a

great interest in IR imaging systems operating both inside and outside the atmospheric windows up to 18 μm.

QWIPs of GaAs/Al_xGa_{1-x}As are thus an attractive alternative which can overcome most of the difficulties mentioned above for this spectral region. By carefully designing the quantum-well structure as well as the light coupling (as discussed in Section X) to the detector, it is possible to optimize the material in order to achieve an optical response in the desired spectral range, to determine the spectral response shape, and to reduce the parasitic dark current and, therefore, to increase the detector impedance. Generally, in order to tailor the QWIP spectral response to the VLWIR spectral region, the barrier height should be lowered and the well width increased relative to the shorter cutoff wavelength QWIPs. See Ref. (*141*) for a detailed analysis of design and performance optimization of VLWIR QWIPs.

The VLWIR focal-plane arrays we have discussed here in detail consisted of bound-to-quasicontinuum QWIPs. The advantage of the bound-to-quasicontinuum QWIP over the bound-to-continuum QWIP is that in the case of bound-to-quasicontinuum QWIP the energy barrier for the thermionic emission is the same as it is for the photoionization as shown in Fig. 98. In the case of a bound-to-continuum QWIP, the energy barrier for the thermionic emission is 10–15 meV less than the photoionization energy. Thus, the dark current of bound-to-quasibound QWIPs is reduced by an order of magnitude (i.e., $I_d \propto e^{-\Delta E/kT} \approx e^{-2}$ for $T = 55$ K) as shown in Fig. 98. Samples were grown using MBE and their well widths L_w vary from 65- to 75-Å, whereas barrier widths are approximately constant at $L_b = 600$-Å. These QWIPs consisted of 50 periods of doped ($N_D = 2 \times 10^{17}$ cm^{-3}) GaAs quantum wells, and undoped Al_xGa_{1-x}As barriers. Very low doping densities were used to minimize the parasitic dark current. The Al molar fraction in the Al_xGa_{1-x}As barriers varies from $x = 0.15$ to 0.17 (corresponding to cutoff wavelengths of 14.9 to 15.7 μm). These QWIPs had peak wavelengths from 14 to 15.2 μm as shown in Fig. 99. The peak quantum efficiency was 3% (lower quantum efficiency is due to the lower well doping density) for a 45° double pass.

Four device structures were grown on 3-inch GaAs wafers and each wafer processed into 35 128 × 128 focal-plane arrays as shown in Fig. 100. An expanded corner of the a focal plane array is shown in Fig. 101. The pitch of the focal-plane array is 50 μm and the actual pixel size is 38 × 38 μm^2. Two-level random reflectors used to improve the light coupling, can be seen on top of each pixel. These random reflectors, which were etched to a depth of quarter peak wavelength in GaAs using reactive-ion etching, had a square profile. These reflectors are covered with Au/Ge

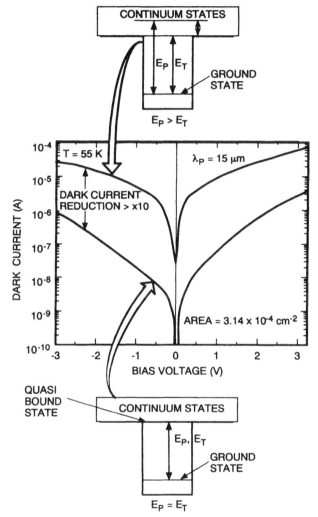

FIG. 98. Comparison of dark currents of bound-to-continuum and bound-to-quasicontinuum VLWIR QWIPs as a function of bias voltage at temperature $T = 55$ K.

and Au (for ohmic contact and reflection), and In bumps are evaporated on top for Si multiplexer hybridization. A single QWIP focal-plane array was chosen from sample number 7060 (cutoff wavelength of this sample is 14.9 μm) and bonded to a Si multiplexer. The focal-plane array was back-illuminated through the flat thinned substrate. Dark current of a

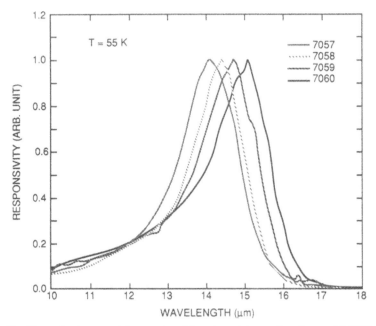

FIG. 99. Normalized responsivity spectra of four bound-to-quasicontinuum VLWIR QWIP focal-plane-array samples at temperatures T = 55 K.

38×38 μm^2 pixel is shown in Fig. 102 at various temperatures and Fig. 103 shows the peak responsivity as a function of bias voltage. This initial array gave excellent images with 99.9% of the pixels working, demonstrating the high yield of GaAs technology. As mentioned earlier, this high yield is due to the excellent GaAs growth uniformity and the mature GaAs processing technology. The uniformity after two point correction was u = 0.03%.

Video images were taken at various frame rates varying from 50 to 200 Hz with f/2.6 KRS-5 optics at temperatures high as T = 45 K, using a multiplexer having a charge capacity of 4×10^7 electrons. However, the total charge capacity was not available during the operation because the charge storage capacitor was partly filled to provide the high operating bias voltage required by the detectors (i.e., V_b = -3 V). Figure 104 shows an image of a truck with NEΔT = 30 mK. The warm tires, motor, and the

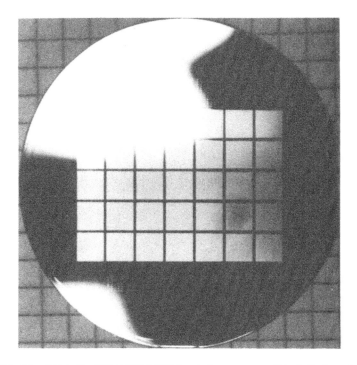

FIG. 100.   Thirty five 128 × 128 QWIP focal-plane arrays on 3-inch GaAs wafer.

driver's cabin can clearly be seen. It should be noted that these initial unoptimized focal-plane-array results are far from optimum. The QWIP device structures were not optimized; the gratings were also not optimized for the maximum light coupling efficiency; no microlenses were used; no antireflection coatings were used; no substrate thinning was used (in the case of VLWIR imaging the hybrid was thinned to 25 μm; however, it was not sufficient to improve the light coupling efficiency to a small pixel); and finally the multiplexer used was a multiplexer developed for photovoltaic InSb focal plane arrays, which is certainly not optimized to supply the proper bias and impedance levels required by photoconductive QWIPs. Implementation of these improvements should significantly enhance the QWIP focal-plane-array operating temperatures (i.e., 77 K for 10 μm and 55 K for 15 μm).

FIG. 101.   Two-level random reflectors on pixels (38×38 μm^2) of QWIP focal-plane array. The random reflectors increase the light coupling efficiency by factor of eight when the substrate thins down to < 1 μm.

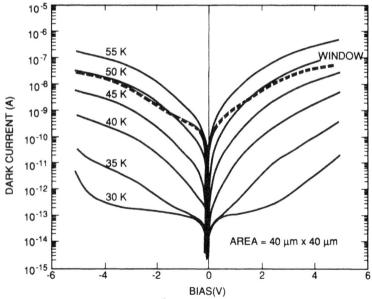

FIG. 102.   Dark current of 40×40 μm^2 QWIP pixel ($\lambda_c = 14.9$ μm) as a function of bias voltage at various temperatures.

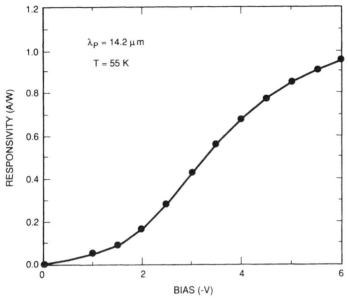

FIG. 103. Bias dependent peak responsitivity R_p measured at temperature $T = 55$ K for sample 7060.

XII. Summary

The intersubband absorption in MQWs, bound-to-bound state QWIPs, bound-to-continuum state QWIPs, asymmetrical QWIPs, single quantum-well QWIPs, indirect bandgap QWIPs, QWIPs with other materials systems, light coupling methods, and QWIP imaging arrays are reviewed. Our discussion of QWIPs has been necessarily brief and the literature references cited are not all inclusive, but represent a selection of key articles from a historic and technical point of view. As discussed in Section IV.A.1 the major breakthrough in this technology is the invention of a *bound-to-continuum* QWIP, which dramatically enhanced the signal-to-noise ratio of these devices, and has made QWIP detectivities comparable to the state-of-art HgCdTe detectors in the wavelength regime of 8–12 μm. Exceptionally rapid progress has been made in the understanding of intersubband absorption and carrier transport in the QWIP device structure and practical demonstration of large, very sensitive, two-dimensional QWIP imaging focal plane arrays since they were first demonstrated only several years ago. This remarkable progress was possibly due to the highly mature

FIG. 104. One frame from a 15-μm QWIP video image of a truck with NEΔT = 30 mK. The warm tires, motor, and the driver's cabin can be seen clearly.

III–V growth and processing technologies. As discussed in Section IV.A.8, an important advantage of QWIPs over the state-of-art HgCdTe is that as the temperature is reduced D^* increases dramatically reaching $D^*_\lambda = 10^{13}$ cm $\sqrt{\text{Hz}}$ /W at $T = 35$ K and even larger values at lower temperatures. Even though a detailed discussion of cooling methods is outside of the scope of this chapter, it should be noted that advances in cooling technologies have kept pace with the detector developments. Efficient, light-weight mechanical and sorption coolers have been developed that can achieve temperatures down to 10 K. Therefore, when QWIP technology utilizes the advanced cooling technologies, it can easily meet the stringent spectroscopic and imaging requirements of ground-based and space-based applications in the LWIR and VLWIR spectral bands. As can be seen from Fig. 105, rapid progress has been made in the performance (detectivity) of long-wavelength QWIPs, starting with bound-to-bound QWIPs, which had relatively poor sensitivity, and culminating in high performance bound-to-quasicontinuum QWIPs with random reflectors. In conclusion,

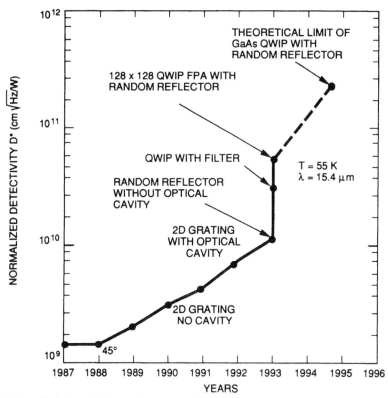

FIG. 105. Evolution of the performance of very long-wavelength QWIP. All the data is normalized to wavelength $\lambda = 15.4 \mu m$ at temperature $T = 55$ K.

very sensitive, low cost, large (1024 × 1024) LWIR and VLWIR QWIP focal-plane arrays can be expected in the near future.

Acknowledgments

We are grateful to C. A. Kukkonen, V. Sarohia, S. K. Khanna, K. M. Koliwad, B. A. Wilson, and P. J. Grunthaner of the Jet Propulsion Laboratory for encouragement and support during the preparation of this chapter. The authors would like to give their special thanks to B. F. Levine of AT&T Bell Laboratories for his support and guidance received over the years. In addition, the authors would like to express their thanks to M. T. Asom, C.

G. Bethea, N. Chand, A. Y. Cho, R. Hamm, G. Hasnain, W. S. Hobson, T. Hoelter, M. Hong, S. P. Hui, J. de Jong, R. F. Kopf, T. N. Krabach, J. M. Kuo, R. E. Leibenguth, T. L. Lin, R. A. Logan, J. K. Liu, M. B. Panish, J. S. Park, S. S. Pei, L. Pfeiffer, D. Ritter, G. Sarusi, C. A. Shott, D. L. Sivco, K. West, and A. Zussman who worked closely with us for stimulating technical discussions, device processing, and crystal growth. The VLWIR QWIP research described in this chapter was performed partly by the Center for Space Microelectronics Technology, Jet Propulsion Laboratory, California Institute of Technology, and was jointly sponsored by the Ballistic Missile Defense Organization/Innovative Science and Technology Office, and the National Aeronautics and Space Administration, Office of Advanced Concepts and Technology.

References

1. L. Esaki and H. Sakaki, *IBM Tech. Disc. Bull.* **20**, 2456 (1977).
2. J. S. Smith, L. C. Chiu, S. Margalit, A. Yariv and A. Y. Cho, *J. Vac. Sci. Technol. B* **1**, 376 (1983).
3. D. D. Coon and R. P. G. Karunasiri, *Appl. Phys. Lett.* **45**, 649 (1984).
4. L. C. West and S. J. Eglash, *Appl. Phys. Lett.* **46**, 1156 (1985).
5. B. F. Levine, K. K. Choi, C. G. Bethea, J. Walker and R. J. Malik, *Appl. Phys. Lett.* **50**, 1092 (1987).
6. B. F. Levine, C. G. Bethea, G. Hasnain, J. Walker and R. J. Malik, *Appl. Phys. Lett.* **53**, 296 (1988).
7. C. G. Bethea, B. F. Levine, M. T. Asom, R. E. Leibenguth, J. W. Stayt, K. G. Glogovsky, R. A. Morgan, J. Blackwell and W. Parrish, *IEEE Trans. Electron. Devices* **40**, 1957 (Nov. 1993).
8. L. J. Kozlowski, G. M. Williams, G. J. Sullivan, C. W. Farley, R. J. Andersson, J. Chen, D. T. Cheung, W. E. Tennant and R. E. DeWames, *IEEE Trans. Electron Devices* **ED-38**, 1124 (1991).
9. W. A. Beck, T. S. Faska, J. W. Little, J. Albritton and M. Sensiper, "Second International Symposium on 2–20 μm Wavelength Infrared Detectors and Arrays: Physics and Applications, Miami Beach, Florida, October 10–12, 1994."
10. B. F. Levine, R. J. Malik, J. Walker, K. K. Choi, C. G. Bethea, D. A. Kleinman and J. M. Vandenberg, *Appl. Phys. Lett.* **50**, 273 (1987).
11. J. Y. Andersson and G. Landgren, *J. Appl. Phys.* **64**, 4123 (1988).
12. M. J. Kane, M. T. Emeny, N. Apsley, C. R. Whitehouse and D. Lee, *Semicond. Sci. Technol.* **3**, 722 (1988).
13. M. J. Kane, M. T. Emeny, N. Apsley and C. R. Whitehouse, *Electron. Lett.* **25**, 230 (1989).
14. B. F. Levine, C. G. Bethea, G. Hasnain, V. O. Shen, E. Pelve, R. R. Abbott and S. J. Hsieh, *Appl. Phys. Lett.* **56**, 851 (1990).
15. D. D. Coon and K. M. S. V. Bandara, *in* "Physics of Thin Films," Vol. 15 (M. H. Francombe and J. L. Vossen, eds.), Academic Press, New York, 1991.
16. K. K. Choi, *J. Appl. Phys.* **73**, 5230 (1993).
17. M. Ramsteiner, J. D. Ralston, P. Koidl, B. Dischler, H. Biebl, J. Wagner and H. Ennen, *J. Appl. Phys.* **67**, 3900 (1990).
18. P. von Allmen, M. Berz, F. K. Reinhart and G. Harbeke, *Superlattices and Microstructures* **5**, 259 (1989).

19. B. C. Covington, C. C. Lee, B. H. Hu, H. F. Taylor and D. C. Streit, *Appl. Phys. Lett.* **54**, 2145 (1989).
20. H. Asai and Y. Kawamura, *Appl. Phys. Lett.* **56**, 1427 (1990).
21. G. Hasnain, B. F. Levine, C. G. Bethea, R. R. Abbott and S. J. Hsieh, *J. Appl. Phys.* **67**, 4361 (1990).
22. M. O. Manasreh, F. Szmulowicz, D. W. Fischer, K. R. Evans and C. E. Stutz, *Phys. Rev.* B **43**, 9996 (1991).
23. M. O. Manasreh, F. Szmulowicz, T. Vaughan, K. R. Evans and C. E. Stutz, *Appl. Phys. Lett.* **57**, 1790 (1990).
24. J. P. Loehr and M. O. Manasreh, *in* "Semiconductor Quantum Wells and Superlattices for Long-Wavelength Infrared Detectors," (M. O. Manasreh, ed.), Artech House, Norwood, 1992, p. 19.
25. K. K. Choi, M. Taysing-Lara, P. G. Newman, W. Chang and G. J. Iafrate, *Appl. Phys. Lett.* **61**, 1781 (1992).
26. K. M. S. V. Bandara, D. D. Coon, O. Byungsung, Y. F. Lin and M. H. Francombe, *Appl. Phys. Lett.* **53**, 1931 (1988).
27. T. Ando, A. B. Fowler and F. Stern, *Rev. Mod. Phys.* **54**, 437 (1982).
28. R. G. Wheeler and H. S. Goldberg, *IEEE Trans. Electron Devices* **ED-22**, 1001 (1975).
29. K. K. Choi, B. F. Levine, N. Jarosik, J. Walker and R. J. Malik, *Appl. Phys. Lett.* **50**, 1814 (1987).
30. B. F. Levine, A. Zussman, S. D. Gunapala, M. T. Asom, J. M. Kuo and W. S. Hobson, *J. Appl. Phys.* **72**, 4429 (1992).
31. B. F. Levine, *in* "Proceedings of the NATO Advanced Research Workshop on Intersubband Transitions in Quantum Wells, Cargese, France, Sept. 9–14, 1991" (E. Rosencher, B. Vinter and B. F. Levine, eds.), Plenum, New York, 1992.
32. J. Y. Andersson and L. Lundqvist, *Appl. Phys. Lett.* **59**, 857 (1991).
33. B. K. Janousek, M. J. Daugherty, W. L. Bloss, M. L. Rosenbluth, M. J. O'Loughlin, H. Kanter, F. J. De Luccia and L. E. Perry, *J. Appl. Phys.* **67**, 7608 (1990).
34. S. R. Andrews and B. A. Miller, *J. Appl. Phys.* **70**, 993 (1991).
35. A. G. Steele, H. C. Liu, M. Buchanan and Z. R. Wasilewski, *Appl. Phys. Lett.* **59**, 3625 (1991).
36. M. J. Kane, S. Millidge, M. T. Emeny, D. Lee, D. R. P. Guy and C. R. Whitehouse, *in* "Proceedings of the NATO Advanced Research Workshop on Intersubband Transitions in Quantum Wells, Cargese, France, Sept. 9–14, 1991" (E. Rosencher, B. Vinter and B. F. Levine, eds.), Plenum, New York, 1992.
37. C. S. Wu, C. P. Wen, R. N. Sato, M. Hu, C. W. Tu, J. Zhang, L. D. Flesner, L. Pham and P. S. Nayer, *IEEE Trans. Electron Devices* **39**, 234 (1992).
38. B. F. Levine, C. G. Bethea, K. K. Choi, J. Walker and R. J. Malik, *J. Appl. Phys.* **64**, 1591 (1987).
39. B. F. Levine, C. G. Bethea, K. K. Choi, J. Walker and R. J. Malik, *Appl. Phys. Lett.* **53**, 231 (1987).
40. K. K. Choi, B. F. Levine, C. G. Bethea, J. Walker and R. J. Malik, *Phys. Rev. Lett.* **59**, 2459 (1987).
41. M. S. Kiledjian, J. N. Schulman and K. L. Wang, *Phys. Rev.* B **44**, 5616 (1991).
42. S. D. Gunapala, B. F. Levine, L. Pfeiffer and K. West, *J. Appl. Phys.* **69**, 6517 (1990).
43. A. Zussman, B. F. Levine, J. M. Kuo and J. de Jong, *J. Appl. Phys.* **70**, 5101 (1991).
44. E. Pelve, F. Beltram, C. G. Bethea, B. F. Levine, V. O. Shen, S. J. Hsieh and R. R. Abbott, *J. Appl. Phys.* **66**, 5656 (1989).

45. M. A. Kinch and A. Yariv, *Appl. Phys. Lett.* **55**, 2093 (1989).
46. N. Vodjdani, B. Vinter, V. Berger, E. Bockenhoff and E. Costard, *Appl. Phys. Lett.* **59**, 555 (1991).
47. B. F. Levine, K. K. Choi, C. G. Bethea, J. Walker and R. J. Malik, *Appl. Phys. Lett.* **51**, 934 (1987).
48. G. Hasnain, B. F. Levine, S. Gunapala and N. Chand, *Appl. Phys. Lett.* **57**, 608 (1990).
49. W. A. Beck, *Appl. Phys. Lett.* **63**, 3589 (1993).
50. K. K. Choi, B. F. Levine, R. J. Malik, J. Walker and C. G. Bethea, *Phys. Rev.* B **35**, 4172 (1987).
51. H. C. Liu, *Appl. Phys. Lett.* **60**, 1507 (1992).
52. B. F. Levine, A. Zussman, J. M. Kuo and J. de Jong, *J. Appl. Phys.* **71**, 5130 (1992).
53. B. F. Levine, C. G. Bethea, V. O. Shen and R. J. Malik, *Appl. Phys. Lett.* **57**, 383 (1990).
54. S. D. Gunapala, B. F. Levine, D. Ritter, R. A. Hamm and M. B. Panish, *Appl. Phys. Lett.* **58**, 2024 (1991).
55. B. F. Levine, S. D. Gunapala, J. M. Kuo, S. S. Pei and S. Hui, *Appl. Phys. Lett.* **59**, 1864 (1991).
56. The factor of $1/\lambda$ in the fractional bandwidth arises from the factor ν in the oscillator strength.
57. B. F. Levine, *Appl. Phys. Lett.* **56**, 2354 (1990).
58. Y. Namirovsky and D. Rosenfeld, *J. Appl. Phys.* **63**, 2435 (1988).
59. B. F. Levine, S. D. Gunapala, J. M. Kuo, S. S. Pei and S. Hui, *Appl. Phys. Lett.* **59**, 1864 (1991).
60. L. C. Chiu, J. S. Smith, S. Margalit, A. Yariv and A. Y. Cho, *Infrared Phys.* **23**, 93 (1983).
61. A. Pinczuk, D. Heiman, R. Sooryakumar, A. C. Gossard and W. Wiegmann, *Surf. Sci.* **170**, 573 (1986).
62. R. P. G. Karunasiri, J. S. Park, Y. J. Mii and K. L. Wang, *Appl. Phys. Lett.* **57**, 2585 (1990).
63. Y.-C. Chang and R. B. James, *Phys. Rev.* B **39**, 12672 (1989).
64. A. D. Wieck, E. Batke, D. Heitman, and J. P. Kotthaus, *Phys. Rev.* B **30**, 4653 (1984).
65. B. F. Levine, G. Hasnain, C. G. Bethea and N. Chand, *Appl. Phys. Lett.* **54**, 2704 (1989).
66. A. Harwit and J. S. Harris, Jr., *Appl. Phys. Lett.* **50**, 685 (1987).
67. E. Martinet, F. Luc, E. Rosencher, Ph. Bois, E. Costard, S. Delaitre and E. Bockenhoff, *in* "Proceedings of the NATO Advanced Research Workshop on Intersubband Transitions in Quantum Wells, Sept. 9–14, 1991" Cargese, France, (E. Rosencher, B. Vinter and B. Levine, eds.), Plenum, New York, 1992, p. 299.
68. Y. J. Mii, R. P. G. Karunasiri, K. L. Wang, M. Chen and P. F. Yuh, *Appl. Phys. Lett.* **55**, 2417 (1989).
69. R. C. Lacoe, M. J. O'Loughlin, D. A. Gutierrez, W. L. Bloss, R. C. Cole, P. A. Dafesh and M. Isaac, *Proc. SPIE* (1992).
70. B. F. Levine, S. D. Gunapala and M. Hong, *Appl. Phys. Lett.* **59**, 1969 (1991).
71. F. Beltram, F. Capasso, J. F. Walker and R. J. Malik, *Appl. Phys. Lett.* **53**, 376 (1988).
72. K. M. S. V. Bandara, B. F. Levine and R. E. Leibenguth, unpublished manuscript.
73. H. C. Liu, G. C. Aers, M. Buchanan, Z. R. Wasilewski and D. Landheer, *Appl. Phys. Lett.* **70**, 935 (1991).
74. E. Rosencher, F. Luc, Ph. Bois, and S. deLaitre, *Appl. Phys. Lett.* **61**, 468 (1992).
75. H. C. Liu, M. Buchanan, G. C. Aers and Z. R. Wasilewski, *Semicond. Sci. Technol.* **6**, C124 (1991).
76. K. M. S. V. Bandara, B. F. Levine, R. E. Leibenguth and M. T. Asom, *J. Appl. Phys.* **74**, 1826 (1993).

77. K. M. S. V. Bandara, B. F. Levine and M. T. Asom, *J. Appl. Phys.* **74**, 346 (1993).
78. K. M. S. V. Bandara, B. F. Levine and J. M. Kuo, *Phys. Rev. B* **48**, 7999 (1993).
79. A. G. Steele, H. C. Liu, M. Buchanan, and Z. R. Wasilewski, *J. Appl. Phys.* **72**, 1062 (1992).
80. F. Capasso, K. Mohammed and A. Y. Cho, *IEEE J. Quantum Electron.* **22**, 1853 (1986).
81. A. Kastalsky, T. Duffield, S. J. Allen and J. Harbison, *Appl. Phys. Lett.* **52**, 1320 (1988).
82. O. Byungsung, J.-W. Choe, M. H. Francombe, K. M. S. V. Bandara, D. D. Coon, Y. F. Lin and W. J. Takei, *Appl. Phys. Lett.* **57**, 503 (1990).
83. K. M. S. V. Bandara, J.-W. Choe, M. H. Francombe, A. G. U. Perera and Y. F. Lin, *Appl. Phys. Lett.* **60**, 3022 (1992).
84. S. D. Gunapala, B. F. Levine and N. Chand, *J. Appl. Phys.* **70**, 305 (1991).
85. L. S. Yu and S. S. Li, *Appl. Phys. Lett.* **59**, 1332 (1991).
86. L. S. Yu, S. S. Li and P. Ho, *Appl. Phys. Lett.* **59**, 2712 (1991).
87. T. S. Faska, J. W. Little, W. A. Beck, K. J. Ritter, A. C. Goldberg and R. LeBlanc, *in* "Innovative Long Wavelength Infrared Detector Workshop, Pasadena, CA, April 7–9," 1992,
88. L. S. Yu, Y. H. Wang, S. S. Li and P. Ho, *Appl. Phys. Lett.* **60**, 992 (1992).
89. B. F. Levine, S. D. Gunapala and R. F. Kopf, *Appl. Phys. Lett.* **58**, 1551 (1991).
90. B. A. Wilson, *IEEE J. Quantum Electron.* **QE-24**, 1763 (1988).
91. P. C. Becker, H. L. Fragnito, C. H. Brito Cruz, J. Shah, R. L. Fork, J. E. Cunningham, J. E. Henry and C. V. Shank, *Appl. Phys. Lett.* **53**, 2089 (1988).
92. P. M. Solomon, S. L. Wright and C. Lanza, *Superlattices and Microstructures* **2**, 521 (1986).
93. C. S. Kyono, V. P. Kesan, D. P. Neikirk, C. M. Maziar and B. G. Streetman, *Appl. Phys. Lett.* **54**, 2234 (1991).
94. B. F. Levine, A. Y. Cho, J. Walker, R. J. Malik, D. A. Kleinman and D. L. Sivco, *Appl. Phys. Lett.* **52**, 1481 (1988).
95. G. Hasnain, B. F. Levine, D. L. Sivco and A. Y. Cho, *Appl. Phys. Lett.* **56**, 770 (1990).
96. F. D. Shepherd, *in* "Infrared Detectors and Arrays," *Proc. SPIE* **930**, 2 (1988).
97. S. D. Gunapala, B. F. Levine, D. Ritter, R. Hamm and M. B. Panish, *Appl. Phys. Lett.* **58**, 2024 (1991).
98. D. Ritter, R. A. Hamm, M. B. Panish, J. M. Vandenberg, D. Gershoni, S. D. Gunapala and B. F. Levine, *Appl. Phys. Lett.* **59**, 552 (1991).
99. S. D. Gunapala, B. F. Levine, D. Ritter, R. A. Hamm, and M. B. Panish, *Appl. Phys. Lett.* **60**, 636 (1992).
100. S. D. Gunapala, B. F. Levine, D. Ritter, R. A. Hamm and M. B. Panish, *SPIE* **11**, 1541 (1991).
101. S. D. Gunapala, B. F. Levine, R. A. Logan, T. Tanbun-Ek and D. A. Humphrey, *Appl. Phys. Lett.* **57**, 1802 (1990).
102. J. W. Choe, B. O, K. M. S. V. Bandara and D. D. Coon, *Appl. Phys. Lett.* **56**, 1679 (1990).
103. M. O. Watanabe and Y. Ohba, *Appl. Phys. Lett.* **50**, 906 (1987).
104. B. F. Levine, K. M. S. V. Bandara and J. M. Kuo, unpublished manuscript.
105. S. D. Gunapala, K. M. S. V. Bandara, B. F. Levine, G. Sarusi, D. L. Sivco and A. Y. Cho, *Appl. Phys. Lett.* **64**, 2288 (1994).
106. T. G. Andersson, Z. G. Chen, V. D. Kulakovskii, A. Uddin and J. T. Vallin, *Phys. Rev. B* **37**, 4032 (1988).
107. J. Y. Yao, T. G. Andersson and G. L. Dunlop, *J. Appl. Phys.* **69**, 2224 (1991).

108. X. Zhou, P. K. Bhattacharya, G. Hugo, S. C. Hong and E. Gulari, *Appl. Phys. Lett.* **54**, 855 (1989).

109. B. Elman, E. S. Koteles, P. Melman, C. Jagannath, J. Lee and D. Dugger, *Appl. Phys. Lett.* **55**, 1659 (1989).

110. R. P. G. Karunasiri, J. S. Park and K. L. Wang, *Appl. Phys. Lett.* **56**, 2588 (1991).

111. Loren Pfeiffer, E. F. Schubert and K. W. West, *Appl. Phys. Lett.* **58**, 5101 (1991).

112. S. D. Gunapala, K. M. S. V. Bandara, B. F. Levine, G. Sarusi, J. S. Park, T. L. Lin, W. T. Pike and J. K. Liu, *Appl. Phys. Lett.* **64**, 3431 (1994).

113. The first excited state of the sample C is located 20 meV below the conduction band edge of the GaAs barrier. This state can become a quasibound state as a result of band bending due to Si dopant migration into the growth direction.

114. C. Y. Lee, M. Z. Tidrow, K. K. Choi, W. H. Chang and L. F. Eastman, unpublished manuscript.

115. The detectivity of lattice matched $GaAs/Al_xGa_{1-x}As$ QWIP with similar cutoff wavelength is about 7×10^9 cm $\sqrt{Hz}$ /W at temperature $T = 40$ K.

116. S. D. Gunapala, B. F. Levine, D. Ritter, R. Hamm and M. B. Panish, *J. Appl. Phys.* **71**, 2458 (1992).

117. K. W. Goosen and S. A. Lyon, *Appl. Phys. Lett.* **47**, 1257 (1985).

118. G. Hasnain, B. F. Levine, C. G. Bethea, R. A. Logan, J. Walker and R. J. Malik, *Appl. Phys. Lett.* **54**, 2515 (1989).

119. J. Y. Andersson, L. Lundqvist and Z. F. Paska, *Appl. Phys. Lett.* **58**, 2264 (1991).

120. J. Y. Andersson and L. Lundqvist, *Appl. Phys. Lett.* **59**, 857 (1991).

121. J. Y. Andersson, L. Lundqvist and Z. F. Paska, *J. Appl. Phys.* **71**, 3600 (1991).

122. G. Sarusi, B. F. Levine, S. J. Pearton, K. M. S. V. Bandara and R. E. Leibenguth, *J. Appl. Phys.* **76**, 4989 (1994).

123. G. Sarusi, B. F. Levine, S. J. Pearton, K. M. S. V. Bandara and R. E. Leibenguth, *Appl. Phys. Lett.* **64**, 960 (1994).

124. R. H. Kingston, "Detection of Optical and Infrared Radiation," Springer-Verlag, Berlin, 1978.

125. B. F. Levine, *Appl. Phys. Lett.* **56**, 2354 (1990).

126. N. Bluzer, *in* Ref. 96, p. 64.

127. D. A. Scribner, M. R. Kruer, K. Sarkady and J. C. Gridley, *in* "Infrared Detectors and Arrays," *Proc. SPIE* **930**, 56 (1988).

128. J. M. Mooney, F. D. Shepherd, W. S. Ewing, J. E. Murgia and J. Silverman, *Opt. Eng.* **28**, 1151 (1989).

129. B. F. Levine, *J. Appl. Phys.* **74**, R1 (1993).

130. I. Grave and A. Yariv, *in* "Proceedings of the NATO Advanced Research Workshop on Intersubband Transitions in Quantum Wells, Cargese, France, Sept. 9–14, 1991" (E. Rosencher, B. Vinter, and B. Levine, eds.), Plenum, New York, 1992, p. 15.

131. B. F. Levine, C. G. Bethea, K. G. Glogovsky, J. W. Stayt and R. E. Leibenguth, *Semicond. Sci. Technol.* **6**, C114 (1991).

132. "Proceedings of the NATO Advanced Research Workshop on Intersubband Transitions in Quantum Wells, Cargese, France, Sept. 9–14, 1991" (E. Rosencher, B. Vinter, and B. Levine, eds.), Plenum, New York, 1992,

133. "Quantum Well Intersubband Transition Physics and Devices," (H. C. Liu, B. F. Levine, and J. Y. Andersson, eds.), Plenum, New York, 1994,

134. L. J. Kozlowski, R. E. DeWames, G. M. Williams, S. A. Cabelli, K. Vural, D. T. Cheung, W. E. Tennant, C. G. Bethea, W. A. Gault, K. G. Glogovsky, B. F. Levine and J. W.

Stayt, Jr., *in* "Proceedings of the IRIS Specialty Group on Infrared Detectors, NIST, Boulder, CO, Aug. 13–16, 1991," p. I 29.

135. M. T. Asom, C. G. Bethea, M. W. Focht, T. R. Fullowan, W. A. Gault, K. G. Glogovsky, G. Guth, R. E. Leibenguth, B. F. Levine, G. Lievscu, L. C. Luther, J. W. Stayt, Jr., V. Swaminathan, Y. M. Wong and A. Zussman, *in* "Proceedings of the IRIS Specialty Group on Infrared Detectors, NIST, Boulder, CO, Aug. 13–16, 1991," p. I 13.

136. C. G. Bethea, B. F. Levine, V. O. Shen, R. R. Abbott and S. J. Hsieh, *IEEE Trans. Electron Devices* **ED-38**, 1118 (1991).

137. V. Swaminathan, J. W. Stayt, Jr., J. L. Zilko, K. D. C. Trapp, L. E. Smith, S. Nakahara, L. C. Luther, G. Livescu, B. F. Levine, R. E. Leibenguth, K. G. Glogovsky, W. A. Gault, M. W. Focht, C. Buiocchi and M. T. Asom, *in* "Proceedings of the IRIS Speciality Group on Infrared Detectors, Moffet Field, CA, Aug. 1992,"

138. C. G. Bethea and B. F. Levine, *in* "Proceedings of SPIE International Symposium on Optical Applied Science and Engineering, San Diego, CA, July 19–24, 1992,"

139. C. G. Bethea, B. F. Levine, M. T. Asom, R. E. Leibenguth, J. W. Stayt, K. G. Glogovsky, R. A. Morgan, J. Blackwell and W. Parish," *IEEE Trans. Electron Devices* **40**, 1957 (1993).

140. B. F. Levine, *in* "Proceedings of the NATO Advanced Research Workshop on Intersubband Transitions in Quantum Wells, Cargese, France, Sept. 9–14, 1991," (E. Rosencher, B. Vinter and B. Levine, eds.), Plenum, New York, 1992, p. 43.

141. G. Sarusi, S. D. Gunapala, J. S. Park and B. F. Levine, *J. Appl. Phys.* **76**, 6001 (1994).

Multiquantum-Well Structures for Hot-Electron Phototransistors

K. K. Choi

U. S. Army Research Laboratory, Physical Science Directorate, Fort Monmouth,
New Jersey

I. Introduction

In the area of long-wavelength infrared detection, the improvement of detector performance is usually focused on material quality rather than on detector structures. The research in infrared detectors is better characterized as a branch of material science. It is particularly true for HgCdTe detectors, since the material preparation process still needs to be improved. Further understanding of the material properties is critical in improving the resolution and the yield before HgCdTe detectors can be widely used in large-scale detector arrays. The situation is quite different in the case of the recently developed quantum-well infrared technology. A quantum-well infrared photodetector (QWIP) (*1, 2*) based on intersubband optical transitions (*3*) is usually made of a GaAs/AlGaAs multiple-quantum-well (MQW) structure. Since the GaAs material growth technology is relatively mature, detector performance is not limited so much by its material quality as by the optoelectronic properties intrinsic to the detector structure. In this case, it becomes very important to understand the physical phenomena that determine the basic properties of the detector.

In contrast to a bulk semiconductor, in which the material properties are fixed by nature, a MQW is a collection of different material layers whose parameters can be assigned at will. Consequently, its optoelectronic properties are not predetermined but vary with the device structure. Because there are no limits on the combinations of material layers, there are actually no definitive QWIP structures. Instead, the optimum structure depends on a number of external factors, such as the source spectrum, the temperature of operation, the readout electronics, the light-coupling scheme, and the material growth conditions. The task of the QWIP research is to find the best combination of material parameters for a given application. QWIPs with various sample parameters and material systems have been reviewed by Gunapala and Bandara in Chapter 3 of this volume.

The optoelectronic properties of QWIPs with different sets of structural parameters can be vastly different from each other. To understand these differences, it is useful to know the energy distribution of the conducting electrons in the structures under different experimental conditions. For this purpose, a quantum barrier placed next to a QWIP (*4*) to serve as an electron energy analyzer is extremely useful. It not only reveals the energy-level structure of the QWIP but also determines the energy of the carriers, with which a consistent transport picture in these structures can be established. A similar technique has also been used to study the electron energy relaxation in the base of a hot-electron transistor (*5, 6*). From these studies, the intrinsic optoelectronic properties of a QWIP as well as the extrinsic material properties can be much better understood. For example, using this technique, the dark current of thick-barrier QWIPs is shown to be determined by the impurity concentration inside the barriers at low temperatures, and is much higher than that predicted based on direct tunneling. The material quality, while sufficient for detector applications, is actually not perfectly under control.

It turns out that the same energy filter is not only valuable in the basic understanding of a QWIP, but also useful in improving detector performance. The new detector is known as an infrared hot-electron transistor (IHET) (7). An IHET serves a wide variety of functions. For example, it can be used to increase the detectivity of a QWIP (8), reduce the current level for higher-temperature operation (9), increase the detector impedance to lower the readout noise (10), increase the yield and relax the material quality requirements (11), amplify a photovoltage and a photopower (10), control the cutoff wavelength (12), select a specific detection wavelength (12), and reduce the generation–recombination noise and $1/f$ noise of a QWIP (13). With these improvements, an IHET is able to meet all the requirements for 10-μm thermal imaging at 77 K in the staring format.

In the following sections, we describe the characteristics and performance of different IHETs to illustrate some of their basic functions. The discussion will be confined to the GaAs/AlGaAs material system with n-type doping.

II. Intersubband Transitions in QWIP Structures

In this section, we summarize the optical properties of a QWIP with thick barriers (14). The band structure is shown in Fig. 1. Structures with thin barriers have been discussed by Helm et al. (15, 16) and Choi et al. (17, 18). The results represented here assume that the quantum wells are separated far apart such that the energy levels in each well are decoupled from the other wells and can be treated independently. We do not consider corrections from either the many-body effect or the effect related to the effective mass, as they tend to cancel each other (14). The discontinuity of the conduction band is assumed to be 60% of the bandgap discontinuity.

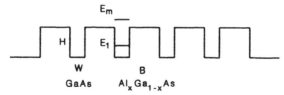

FIG. 1.   The energy level diagram of a MQW structure. E_1 is the ground state and E_m is one of the extended states above the barriers at which f is a maximum.

A. Absorption Wavelength

Let us consider an optical-beam incident at an angle θ_i onto a MQW material surface; the angle of reflection is θ. The electric vector is polarized in the plane of incidence. During an optical transition, an electron in the grond state E_1 absorbs a photon $\hbar\omega$ and is excited to an extended state E_2 above the barriers. The value of E_2 is determined by the incoming photon energy and is equal to $E_1 + \hbar\omega$. The absorption coefficient $\alpha(\hbar\omega)$ can be obtained by the Fermi Golden rule (17). If we denote the two-dimensional electron density in each well to be ρ_s, the length of one period L, the refractive index of the well n_w, and the effective mass of an electron in the well m^*, α can then be expressed as

$$\alpha(\hbar\omega) = \frac{\rho_s}{\mathscr{L}} \frac{\pi e^2 \hbar}{2 n_w \epsilon_0 m^* c} \sin^2 \theta \, f(E_2) g(E_2), \tag{1}$$

where $f(E_2)$ is the oscillator strength, $g(E_2)$ is the one-dimensional density of the final states, and $\mathscr{L} = L/(\cos\theta)$ is the optical path-length of each quantum well.

In order to obtain the position of the absorption peak λ_p, $g(E_2)$ is assumed to be relatively constant near the barrier height because of the scattering effect. The energy dependence of α then solely originates from f. The value of f as a function of the final state energy E_2 is first calculated. The value of E_2, denoted by E_m in this section, at which f is a maximum is then determined. The transition energy is equal to $E_m - E_1$, and finally, λ_p is equal to

$$\lambda_p = \frac{2\pi\hbar c}{E_m - E_1}. \tag{2}$$

In the following paragraphs, the results of the calculation are presented, which are based on the assumption that $m^* = 0.067\, m_0$, $H = 748x$, $n_b/n_w = (1 - 0.251x)^{1/2}$ and $m_b^*/m^* = 1 + 1.24x$, where H is the barrier height in meV, x is the aluminum molar ratio of the barriers, and n_w and n_b are the reflective indices of the well and the barrier, respectively.

Figure 2 shows the wavelength of the absorption peak λ_p as a function of well width W. Within the detector parameters shown, λ_p can be varied from 5 to over 25 μm. For a given x, there is a minimum value of λ_p, but there is no apparent upper limit as long as the quantum-well width is technologically feasible to produce. From these tuning curves, the peak wavelength for a given set of detector parameters can be obtained. Conversely, however, for a given x, there are usually two well widths which

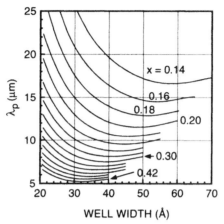

FIG. 2.   The position of the absorption peak λ_p for a given Al molar ratio x as a function of well width; x changes 0.02 in steps between each curve.

can satisfy a specific detection wavelength. The absorption lineshape of these two well widths, however, is not the same. As discussed below, the one with larger well width has a stronger absorption at λ_p but with a smaller linewidth.

Figure 3 shows the tuning curves as a function of the Al molar ratio x for different well widths. From this figure, λ_p is seen to be insensitive to

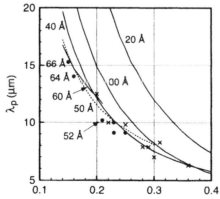

FIG. 3.   The position of the absorption peak λ_p for a given well width as a function of Al molar ratio. The figure also shows the existing experimental results for $d = 40$ Å unless indicated otherwise.

different well widths and is approximately fixed for a given x for W larger than 30 Å. In Fig. 3, we also show the experimental λ_p measured by Levine *et al.* (*19–21*) and, from our measurements, for well widths around 40 Å, except those indicated otherwise. Within the experimental error and theoretical uncertainty, the agreement is satisfactory.

In addition to the absorption wavelength, another parameter of interest is the absorption full width at half maximum, Γ. In determining Γ, one has to consider α over a wider range of energies; thus, it is necessary to include the energy dependence of the density of states. At large energies, the one-dimensional density of states $g_1(E_2)$ is expected to be applicable. Hence, one can obtain the approximate Γ by calculating the product of $f(E_2)$ and $g_1(E_2)$. The resultant Γ for three different well widths is shown in Fig. 4. It decreases as the Al molar ratio increases. It reaches a minimum of zero when $E_m = H$ if the scattering effect is not considered. Therefore, one cannot design a wide-band detector when E_m coincides with H. Figure 4 also shows the cutoff wavelength and the experimental results from our measurements. The agreement is satisfactory.

B. ABSORPTION STRENGTH

In this section, we estimate the magnitude of light absorption in a MQW structure. For this purpose, it is convenient to consider all optical transi-

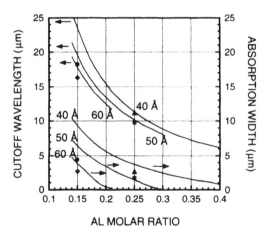

FIG. 4. The cutoff wavelength λ_c and the absorption width of a QWIP as a function of Al molar ratio x for different well widths. The figure also shows the experimental optical properties of four QWIPs with $W = 66$ Å (diamonds), 60 Å (circles), 50 Å (squares), and 44 Å (triangles).

tions to be between two specific quantum-well states, independent of whether the excited states are bound or unbound. Under this assumption, the oscillator strength f is a fixed number of order unity, and the integrated density of final states per well is one, so that the integrated absorption strength IA using Eq. (1) is

$$IA = \int_0^\infty \alpha(\hbar\omega)d(\hbar\omega) = \frac{\rho_s}{\mathscr{L}} \frac{\pi e^2 \hbar f}{2 n_w \epsilon_0 m^* c} \sin^2\theta. \tag{3}$$

Since IA is independent of the absorption lineshape, the peak absorption coefficient α_p is inversely proportional to the absorption width Γ in both bound-to-bound (B–B) and bound-to-extended (B–E) transitions. In the B–B case, the lineshape is Lorentzian (22), so that α_p is equal to

$$\alpha_p = \frac{\rho_s}{\mathscr{L}} \frac{e^2 \hbar f}{n_w \epsilon_0 m^* c} \frac{\sin^2\theta}{\Gamma}. \tag{4}$$

In the B–E case, the relation is more complicated. For simplicity, we will use Eq. (4) to estimate the peak absorption even in the B–E case. As the absorption peak moves away from the barrier height where the absorption lineshape resembles a Lorentzian (18), the estimation should become more accurate.

The peak absorption a_p is given by

$$\begin{aligned}
a_p &= 1 - \exp(-\alpha_p s N_w \mathscr{L}), \\
&= 1 - \exp\left(-\rho_s \frac{s N_w e^2 \hbar f}{n_w \epsilon_0 m^* c} \frac{\sin^2\theta}{\Gamma}\right), \\
&= 1 - \exp\left(-3.258 \times 10^{-3} \frac{N_d W f s N_w}{\Gamma} \sin^2\theta\right),
\end{aligned} \tag{5}$$

where s is the number of passes, N_w is the number of periods in a MQW structure, N_d is the doping density in 10^{18} cm^{-3}, W is in Å, and Γ is in meV. Note that a_p in a layered two-dimensional case is scaled as $\sin^2\theta$ instead of $\sin^2\theta/\cos\theta$ as usually stated (23). The additional factor of $1/\cos\theta$ in a three-dimensional isotropic case comes from the fact that with an inclined optical beam in a material slab, the number of atoms a photon interacts with varies as $1/\cos\theta$. In contrast, the number of quantum wells that a photon encounters in a MQW structure is independent on the angle θ, and, therefore, such a dependence is absent in the present structures. For typical parameter values $N_d = 0.5 \times 10^{18}$ cm^{-3}, $W = 50$ Å, $f = 1$, $s = 2$, $N_w = 30$, $\Gamma = 20$ meV, and $\theta = 45°$, the peak absorption will be $\approx 11.5\%$ for polarized light.

In a typical absorption measurement using a Fourier transform infrared spectrometer, Brewster's angle (73°) is used for the angle of incidence on an unpolished sample surface, so that θ is 17° and $s = 1$. The measured absorbance ABS_B, which is defined as $-\log_{10}$ transmission, is equal to

$$ABS_B = 1.210 \times 10^{-4} \frac{N_d W f N_w}{\Gamma}. \tag{6}$$

For the same material parameters given above, ABS_B is about 0.0045, which is consistent with experiments. In terms of ABS_B, the peak quantum efficiency η for unpolarized light, defined as $a_p/2$, is given by

$$\eta = \frac{1}{2}\left[1 - \exp(-26.92 s\, ABS_B \sin^2\theta)\right]. \tag{7}$$

III. Excitation Hot-Electron Spectroscopy

In this section, two hot-electron spectroscopies are described. They are performed by placing a single quantum barrier next to a QWIP to probe the hot-electron distribution injected from the structure. With these techniques, the electron transport mechanisms as well as the energy levels of the structure can be better understood. In Section III.A, we are interested in the electron transport caused by thermal excitation at different temperatures. The analytical techniques is referred as thermally stimulated hot-electron spectroscopy (TSHES) (24). Similarly, in Section III.B, photocurrent transport mechanisms will be studied in the optically stimulated hot-electron spectroscopy (OSHES) when intersubband optical excitation is used instead (4).

A. Thermally Stimulated Hot-Electron Spectroscopy

The structure of the entire device used in TSHES is shown in Fig. 5. It consists of a doped emitter contact layer, a QWIP structure under study, a doped base layer, a single barrier serving as an electron energy high-pass analyzer, and a doped collector layer. When a bias V_c is applied to the QWIP structure through the emitter and the base contacts, electron transport occurs within the QWIP. The contribution of conduction current by each energy state depends on the temperature T. Note that because of the layered structure of the material, one has to differentiate between the

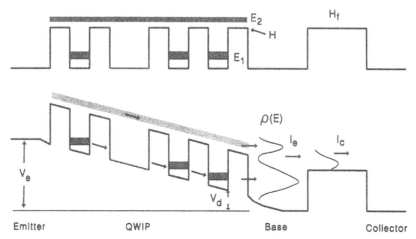

FIG. 5. The device structure used in TSHES. The upper figure shows the band structure of the device without bias. The lower figure shows the device under external emitter voltage V_e in a common base configuration. Usually, the collector voltage V_c is kept at 0 V. As V_e increases, V_d also increases, thereby shifting the electron distribution to higher energies and allowing more electrons to go through the analyzer.

energy component E, which is parallel to the superlattice axis, and the in-plane energy component E_{in}, which is perpendicular to the axis. At low temperatures, the conduction electrons from the dopant reside primarily in the lowest subband E_1, with E_{in} below E_F. The electrical conduction occurs only in this lowest subband through tunneling. As the temperature increases, the electrons gain energies both in the direction parallel to the superlattice axis in which higher subbands are populated, and in the direction perpendicular to the axis in which the average E_{in} will be increased. Although the number of the high-energy electrons may still be very small compared with those below E_F, the increase in the barrier transmission at higher energies greatly compensates for this factor and shifts the energy distribution of the injections to a higher energy even at a relatively low T. The hot electrons then travel through the base and are collected at the collector for those electrons with energy component E larger than the analyzer barrier height H_f. The electrons with less energy will then contribute to the base current.

In the first part of the measurement, the emitter-to-base voltage V_e is varied at each temperature with the collector-to-base voltage V_c fixed at zero volts. The signs of the applied voltages follow the convention of a common base configuration. The current transfer ratio α, defined as J_c/J_e, is then measured as a function of V_e, from which the derivative r ($= d\alpha/dV_e$) is obtained. J_e and J_c are the current densities of the emitter and the collector, respectively. In the second part of the measurement, α and $d\alpha/dV_c$ will be measured as a function of V_c at a fixed value of V_e.

In TSHES, $r(V_e)$ is directly proportional to the normalized electron-energy distribution $\rho(V_e, H_f)$ reaching the front boundary of the analyzer (4). Normally, when V_e increases, the hot-electron distribution shifts to higher energies E due to the increase of the depletion voltage drop V_d in the base as shown in Fig. 5. The upward energy shift leads to more electrons being collected at the collector. The incremental α and hence r are seen to be directly proportional to the energy distribution. Since a higher-energy electron is able to overcome the analyzer at a lower V_e, the value of r at a lower V_e reflects the electron distribution at a higher energy.

In this section, three QWIPs with relatively thick barriers are presented to exemplify this technique. They are labeled as devices A, B, and C. The device structural parameters are listed in Table I. With these device parameters, there is only one bound-state E_1 located in the well. The

TABLE I
DEVICE STRUCTURAL PARAMETERS

Device	N	W (Å)	N_d	B (Å)	x	W_b (Å)	B_f (Å)	x_f	N_{db}
A	50	52	1.5	300	0.23	500	2000	0.27	1.5
B	50	50	3.5	500	0.25	500	2000	0.29	3.5
C	30	50	1.2	500	0.25	500	2000	0.25	1.0
D	50	40	1.2	200	0.25	300	2000	0.25	1.2
E	30	50	0.5	500	0.25	500	2000	0.29	1.0
F	30	50	1.0	500	0.25	500	BF	—	1.0
G	30	66	0.5	500	0.16	500	BF	—	1.0
H	30	60	0.8	500	0.15	500	2000	0.17	1.0
I	50	65, 14	1.0	150, 40	0.25	—	—	—	—
J	50	72, 20	1.0	154, 39	0.31	—	—	—	—
K	30	72, 20	1.0	500, 39	0.31	—	—	—	—
L	30	50	1.2	160.8,	0.28,	500	BF	—	1.2
				166.7,	0.305,				
				172.9	0.33				

Note. N, number of quantum-well periods in the multiple quantum-well structure; W, well width; N_d, doping density in the well in 10^{18} cm^{-3}; B, barrier width; x, Al molar ratio of the quantum-well barrier; W_b, base width; B_f, analyzer thickness; x_f, Al molar ratio of the analyzer, and N_{db}, doping density in the base in 10^{18} cm^{-3}.

TSHES data for device A, whose barrier thickness is 300 Å, are shown in Fig. 6. In this figure, a narrow hot-electron injection peak is observed at $V_e = -9$ V at $T = 4.2$ K, and represents the electrons conducting through E_1 via the direct tunneling (DT) process (22). At higher temperatures, the energy of hot-electron injection rises continuously with temperature, indicating that the thermal electrons conducting through higher-energy states dominate the transport process. However, the continuous energy shift is inconsistent with the discrete-level structure of the present QWIP, and can be explained only by another transport process. The new process is known as thermally assisted tunneling (TAT) (25). In this process, the in-plane energy E_{in} is first transferred to the parallel component through large-angle scattering before tunneling occurs. With a larger parallel component, $E = E_1 + E_{in}$, the electron is then able to tunnel through the barrier with a much higher transmission coefficient. Consequently, the tunneling process depends on the total energy of an electron rather than on its parallel component, and it is possible that the TSHES detects electrons with E between E_1 and E_2. Since the average E_{in} increases continuously with temperature, so does the energy of the hot-electron injection. It should be emphasized that for a given E, the TAT process is always much smaller than a direct tunneling process because of the higher-order processes involved, and can be observed only when the direct tunneling processes are

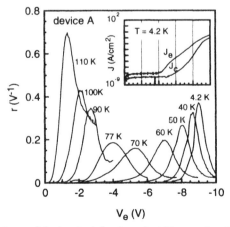

FIG. 6.   The parameter r of device A, defined as $d\alpha/dV_e$, as a function of V_e at different temperatures. The insert shots the current densities of the emitter and collector. At 4.2 K, a finite current level at a small V_e is maintained, indicating direct tunneling from E_1 is still appreciable.

suppressed for certain reasons. In the present case, DT is forbidden at any E other than E_1 because of energy quantization. At T higher than 90 K, thermal excitation to the extended states increases the parallel component E directly, thus making the direct thermionic emission (DTE) process dominant. Due to the increase of the 3-dimensional density of states with energy, the injection peak continues to shift to higher energies at higher temperatures. In this wide-barrier QWIP, miniband structure is not apparent so that the energy shift at high T is also smooth.

When the barrier thickness increases to 500 Å, the DT process from E_1 is greatly suppressed, upon which another tunneling mechanism becomes evident. The TSHES results for device B having this barrier thickness are shown in Fig. 7. Below $T = 60$ K, the hot-electron injection is observed to have a number of closely spaced peaks. These injection peaks are believed to be related to the definite impurity levels within the barriers. Electron transport in this case is through tunneling among impurities, in which process the energy components are randomized. This conduction process will be referred as electron-impurity tunneling (IT). The magnitude of the IT current depends on the concentration of impurities in the barriers, usually from the diffusion of dopants, and is relatively independent of the barrier thickness. With a large barrier thickness, IT is expected to be

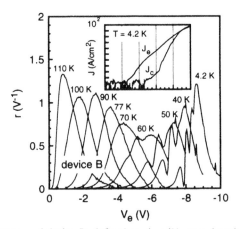

FIG. 7. The parameter r of device B, defined as $d\alpha/dV_c$, as a function of V_c at different temperatures. The insert shows the current densities of the emitter and collector. At 4.2 K, the current level at a small V_c is extremely low indicating direct tunneling from E_1 is not effective.

significant. Since the signature of an impurity-enhanced process is still clearly observable around 60 K, a fraction of TAT current at this barrier thickness also conducts through impurities. This process will be referred as thermally assisted impurity tunneling (TAIT). With a larger in-plane thermal energy, the higher-energy impurity levels can be accessed for tunneling at a lower V_e, and hence increases the TAT current at low biases. The existence of IT is consistent with the physical-barrier thickness being larger than the deduced thickness, using the standard expression for direct tunneling (22) for a thickness larger than 300 Å (26).

From the TSHES of device B, besides the tunneling from the E_1 state, two additional peaks are observed. To confirm the common nature of these impurity levels, another device (device C) is prepared with similar structural parameters as listed in Table I. In this case, V_e is fixed and the analyzer potential height is varied through an applied V_c. The 4.2-K result is shown in Fig. 8. Clearly, three distinct injection peaks are observed at each V_e, consistent with the characteristics of device B. A closer examination of Fig. 8 also reveals that the peak in the middle may consist of an unresolved doublet. The origin of these impurity levels has yet to be identified. Evidently, TSHES provides a useful diagnostic technique to investigate impurity levels in semiconductors as well.

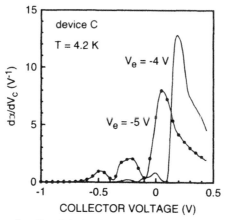

FIG. 8.   The parameter $d\alpha/dV_c$ of device C as a function of V_c for two fixed values of V_e. The observed peaks are attributed to tunneling through impurity states residing in the MQW barriers.

B. OPTICALLY STIMULATED HOT-ELECTRON SPECTROSCOPY

For detector applications, one also would like to know the transport mechanism and the trajectory of the photogenerated electrons. For this purpose, an optical-excitation measurement is performed; the technique is referred to as optically stimulated hot-electron spectroscopy (OSHES) (4). In this technique, infrared radiation of known energy $\hbar \omega$ is incident onto the device to initiate intersubband transitions while the sample is maintained at a fixed temperature. The emitter and the collector photocurrents are monitored using AC lock-in techniques to separate the photocurrent from the dark current. In this case, a hot-electron distribution peak is expected with energy $\hbar \omega$ above the E_1 subband. Although the present technique can be used only to probe a limited range of energy within the absorption peak, the ability of putting monoenergetic electrons in definite excited states by a monochromatic light source is one of the advantages of OHSES. The observed energy distribution of the photoelectron injection is then entirely due to the dynamical and the scattering effects in the transport process without the complication of the initial thermal broadening as in TSHES.

Figure 9 shows the OSHES results for device D in curve (d), in which $r_p \equiv d\alpha_p/dV_c$ is plotted as a function of V_c, where α_p is the ratio of

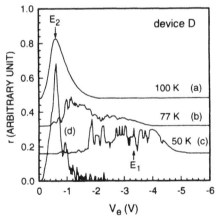

FIG. 9. The parameter r_d for dark current of device D as a function of V_c at different temperatures are shown as curves (a), (b), and (c). The parameter r_p for the photocurrent measured at 90 K is shown as curve (d). The peak position of r_p is independent of temperature. The curves have been shifted vertically for clarity.

collector photocurrent density to the emitter photocurrent density, under a
9.25-μm light source. (The spacing between E_1 and the absorption peak
energy E_2 corresponds to λ_p of 8.1 μm.) In order to compare with the
thermally excited electrons, the TSHES performed at 100, 77, and 50 K is
also shown in Fig. 9 as curves (a), (b), and (c), respectively. The large
fluctuations of r at low temperatures are reproducible structures caused by
the strong negative differential conductance (NDC) observed at these
temperatures. The NDC is pronounced in the TSHES when the barrier
thickness of the MQW is thin (22) and the analyzer barrier height is low,
such as in the present device. The TSHES data show that the dark current
at 77 K is dominated by TAT current since the hot electrons are distributed between the ground state E_1 and the resonant state E_2, and are
lower in energy than the photoelectron distribution indicated by curve (d).
The OSHES shows the energy of the photoelectron injection being very
narrow even for unbound excited states, and coincides with the direct
thermionic injection from the E_2 state.

Similarly, one can also measure photoelectron injection by changing V_c
for each V_e. Fig. 10 shows the relative positions of dark-electron injection
and photoelectron injection for device B at three different values of V_e
and at $T = 77$ K. In this figure, $\beta_d(V_c) \equiv dI_{dc}/dV_c$ and $\beta_p(V_c) \equiv dI_{pc}/dV_c$
for a fixed V_e are plotted. Since I_c is a constant under a fixed V_e, $\beta_d(V_c)$,
and $\beta_p(V_c)$ are proportional to $\rho_d(E)$ and $\rho_p(E)$, respectively, where $\rho(E)$
is the hot-electron distribution. Since a negative V_c raises the potential
barrier height of the filter, a more negative V_c represents electrons of
higher energies. The photocurrent is generated by a CO_2 laser with
photon energy equal to 133 meV modulated at 200 Hz and incident at a
45° angle. At $V_c = -1$ V, a distinctive peak in β_d is observed. This peak is
due to direct thermionic emission from the absorption peak energy E_2,
which is located at 146 meV above the ground state E_1. The value of E_2 is
deduced from the photocurrent spectrum. The rising of β_d at V_c larger
than 0.5 V is due to TAT occurring below the barrier height. At $V_g =$
-1 V, the energy of the photocurrent is actually slightly lower than that of
the dark current due to the fact that the photon energy used is less than
$E_2 - E_1$. At higher V_c (-2.5 and -3.5 V), the energy of the photoelectrons is higher than that of the dark electrons, as TAT current increases
rapidly with V_c. Note that the present sample is seven times higher in
doping than an optimized QWIP with which the DTE peak will not be
significant at 77 K.

In this section, both TSHES and OSHES show that an electron is able
to maintain its initial energy relative to the band edge when it travels
across the QWIP structure. If quasiballistic transport ever occurs in the

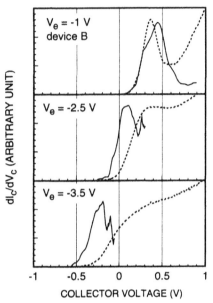

FIG. 10. The parameters dI_{cd}/dV_c (dotted curves) and dI_{cp}/dV_c (solid curves) as a function of V_c for three different values of V_e. At a higher V_c, both the dark-electron injection and photoelectron injection shift up in energy; therefore, both curves shift to a more negative V_c. The change of dark-electron distribution is due to the increasing thermally assisted tunneling current at higher V_c.

QWIP structure, more injection peaks or distribution width of the order of V_c should be observed in these spectroscopies. Comparing the OSHES and the TSHES results, the photoelectron peak is nearly coincident with the DTE peak from the same states, showing that the energy relaxation and the trajectory of the photocurrent are the same as that of the thermionic current of the same energy. One can then conclude that the transport mechanism of a given energy is independent of the source of excitation. It also implies that, if scattering does not mix high-energy and low-energy dark-electron injection, it also will not mix the photoelectrons with dark electrons of different energies.

Based on the experimental evidence shown here, a consistent physical picture can be established for the transport of the electrons in a MQW structure. Despite the presence of a large applied electric field, the electrons are in remarkable local equilibrium with the thermal and optical excitations, in the sense that the electron distribution in each quantum well is location-independent, and is given by Fermi–Dirac distribution plus a distribution created by photoexcitation. The energy of the photoexcitation distribution is determined only by the incident photon energy, and is independent on the electric field. Under an applied electric field, the momentum distribution of the electrons is slightly modified to give a conducting current. The current contribution of each subband depends on the total electron population from both excitations. No appreciable amount of electrons acquire enough energy from the applied field to create an nonequilibrium electron-energy distribution in the MQW structure. These transport processes are in striking similarity with that of the ordinary bulk band-to-band photoconductors.

An interesting question is why quasiballistic transport is not observed in the MQW structures. Here, a brief explanation is proposed. If one accepts that there is a steady state under a bias, then the current continuity requires

$$\frac{d\rho}{dt} = 0, \tag{8}$$

where ρ is the charge density in a well. The quasiballistic picture does not account for charge refilling and hence is incomplete. Furthermore, the Boltzmann equation says that

$$\frac{df}{dt} = 0, \tag{9}$$

where f is the steady-state energy distribution function. Since a quasiballistic electron is in constant acceleration under an applied field, f can never be an invariant of time. With these two steady-state conditions, the only consistent steady function f is the local equilibrium distribution function set up by the excitations. Although there is a constant exchange of electrons among different wells due to electron transport, the "scatter-in" term for each state is necessarily balanced by the "scatter-out" term so that the net population of the distribution is unaffected by the transport process, a situation analogous to the Fermi–Dirac distribution being unaffected by scattering.

IV. Infrared Hot-Electron Transistors

In studying the properties of a QWIP structure, a quantum-barrier analyzer serves as a diagnostic tool. In this section, the same quantum barrier will be used not only to observe the properties of a QWIP but actually to change them for performance improvement in infrared detection. For this purpose, the analyzer acts as an electron energy filter. A QWIP, together with an energy filter, forms a new infrared detector known as an infrared hot-electron transistor (8–10, 27–30).

A. FIGURES OF MERIT

An IHET makes numerous fundamental improvements on the properties of a QWIP. Before discussing these improvements, two widely used figures of merit should be mentioned: the detectivity D^* of a single detector and the noise equivalent temperature difference NEΔT of an imaging system. D^* is defined as

$$D^* = \frac{R\sqrt{AB}}{i_n}, \tag{10}$$

where R is the responsivity, A is the detector area, B is the measurement noise bandwidth and i_n is the noise current. NEΔT is given by

$$
\begin{aligned}
NE\Delta T &= \frac{\sqrt{AB}}{D^*(dP_B/dT_s)}, \\
&= \frac{1}{\sqrt{A}}\frac{1}{D^*\sin^2\theta}\frac{T_s}{I_0(T_s,\lambda_1,\lambda_2)}\frac{kT_s}{h\nu}\frac{1}{(2\tau_{int})^{1/2}},
\end{aligned}
\tag{11}
$$

where P_B is the optical power incident on the detector, θ is the cone angle of the cold (detector) shield, T_s is the temperature of the light source, $I_0(T_s,\lambda_1,\lambda_2)$ is the incident power per unit area between λ_1 and λ_2, $h\nu$ is the average photon energy, and τ_{int} is the integration time of a measurement. Equation (11) is a good approximation when $h\nu$ is much larger than kT_s.

D^* is a measure of single detector sensitivity because the power signal-to-noise ratio (S/N) of a measurement is proportional to D^*, and is given by

$$\left(\frac{S}{N}\right)_P = (D^*I_0)^2\frac{A}{B}. \tag{12}$$

A detector with a larger D^* will have a higher S/N. However, Eq. (12) also indicates that having a large D^* is not the only means to achieve a high S/N because it can also be obtained by decreasing B, i.e., increasing τ_{int}. Therefore, in evaluating the merits of a detector, the format of the measurement should also be taken into account. For example, in order to generate a thermal image having a resolution of $N \times N$ at a fixed frame rate, the integration time of a serial scanning system and a parallel scanning system will be shorter than that of a staring system by a factor of N^2 and N, respectively. Consequently, in order to obtain an image with the same S/N, the D^* of the detector used in a serial system needs to be N times larger than that used in a staring system. In a parallel system, it needs to be $\sqrt{N}$ larger. For focal-plane arrays (FPA) with high resolution, the differences in the D^* requirement can be substantial. If a detector can be used in the staring format, the D^* of the detector is not as critical as that used in the scanning formats. For example, assuming a QWIP staring array with a peak detection wavelength at 9 μm and an absorption width of 2 μm, a typical specific D^* of 10^{10} cm $\sqrt{\text{Hz}}$/W at 77 K will give NEΔT a reasonable value of 12 mK at a frame rate of 30 Hz, calculated from Eq. (11) with $A = 50 \times 50$ μm^2, $T_s = 300$ K, and $\theta = 28°$. Therefore, the typical D^* of a QWIP is sufficient for 10-μm thermal imaging at 77 K.

B. Dark-Current Requirement

Since the D^* of a typical QWIP is high enough to give a reasonable temporal NEΔT, a QWIP FPA is more likely limited by the fixed pattern noise due to pixel nonuniformity rather than temporal noise, provided that the full frame time can be used for signal integration, which is assumed in the previous example. However, for each readout circuit, there is a limit on the charge handling capacity; the current level of a detector under operation must be below this limit for the assumed integration time. For example, if the typical charge-handling capacity of a readout is 5×10^7 e^- per pixel, τ_{int} is 33 ms, and the detector area is 50×50 μm^2, then the maximum total current density J_{tot} must be less than 10 μA cm^{-2}. Under the normal thermal imaging condition, J_{tot} includes both the dark-current density J_d and the 300-K background photocurrent density J_p. If we further require the detection to be background limited (BLIP), i.e., $J_p > J_d$, then J_d should be much less than 10 μA cm^{-2}. On the other hand, for a QWIP with a cutoff wavelength at 10 μm and a doping density N_d of 1×10^{18} cm^{-3}, the typical J_d at 77 K is 10 mA cm^{-2}, approximately three orders of magnitude higher than this limitation. Hence, a long integration time is not feasible and the QWIP technology is less competitive at this temperature. Normally, thermal imaging using QWIP structures has been confined to $T \sim 60$ K (*31, 32*). Besides saturating a readout circuit, a large dark current also reduces the charge-injection efficiency and increases the thermal noise of the readout circuit, which will further reduce the FPA sensitivity (31). Therefore, it is critical to reduce the dark current of the detector before it is applicable at 77 K.

Due to the large dark current, a QWIP is usually not BLIP at 77 K. BLIP is important in thermal imaging because it indicates the robustness of the system against other possible noise sources. For example, if J_d is larger than J_p, a small nonuniformity in J_d will cause a large fixed-pattern noise in the FPA image, and this noise source is expected to be dominant at a long integration time. Therefore, the BLIP condition is much desired. One can quantify the degree of BLIP by defining a parameter %BLIP, which is the ratio of the background photocurrent noise to the total current noise.

C. Functions of an IHET

From the above discussion, the primary function of an IHET ought to be to reduce the current level of a QWIP without adversely affecting the detectivity. This function can be accomplished by selectively filtering the hot-electron injection from the QWIP. Since the conducting electrons maintain their original energies while traveling across the QWIP, electrons originating from different transport mechanisms will still be separated in energy when they are injected into the base. A properly designed energy filter can then be used to select only those electrons which are degenerate in energy with the photoelectrons. Consequently, not only can the total current level be reduced, but also the detectivity and the BLIP tempera-ture T_b can be increased in the process.

In principle, there are no limits on the combinations of material layers in a filter design. Nevertheless, these combinations eventually fall into two categories in terms of filter-transmission characteristics. They are either high-pass filters (HPFs) or bandpass filters (BFs). For a high-pass filter, its characteristics are described by the turn-on energy and the sharpness of the turn-on step. On the other hand, a bandpass filter is characterized by the energy of the transmission peak and its bandwidth. A desirable filter characteristic depends on a particular application. But in general, a high-pass filter is more suitable for filtering out the lower energy tunneling dark currents, such as DT, TAT, and TAIT currents. It can also be used to suppress part of the thermionic emission current with energy between the barrier height H and the cutoff energy E_c, where E_c is the energy corresponding to the cutoff wavelength λ_c. Theoretically, a well-designed

QWIP should have E_c aligned with H, thus reducing the unwanted TE current (14). However, this alignment will severely limit the choice of the absorption to a narrow band, which is a general characteristic of intersubband transitions when the resonant excited state approaches H. In practice, it also imposes a more stringent requirement on the accuracy of material growth and the limit of the applied bias under which the alignment condition remains true.

For high-temperature operation, a bandpass filter is more useful because the majority of the thermionic-emission current can be higher in energy than the photoelectrons as revealed by TSHES in Figs. 6 and 7. In this case, it is beneficial to filter out the high-energy dark current as well. Note that although the thermal dark current of a QWIP is exponentially related to its cutoff wavelength, this is not necessarily true for an IHET with a bandpass filter. By reducing the bandwidth of a filter, the amount of filtered current can be reduced arbitrarily, enabling the detector be operated at much higher temperatures without saturating a readout circuit. Of course, the detection will not be BLIP in these extreme temperatures. However, if the passband of the filter is well-aligned with the photocurrent distribution peak, the reduction of %BLIP will not be severe, as is demonstrated later. The bandwidth of a filter turns out to have little effect on the spectral width of the photoresponse due to energy broadening of the photoelectron distribution in the base.

Besides reducing dark current, another advantage of an IHET is its larger output impedance, which results in a reduced readout noise of a CMOS readout circuit (33). The larger output impedance in some cases will also magnify a signal since photovoltage is a product of photocurrent and impedance. When an increase in the impedance is larger than the photocurrent reduction at the collector, there will be a photovoltage and a photopower amplification, and hence there is a transistor action in an IHET (10).

In general, an IHET can be thought as a two-stage device, in which each stage has a unique function. The emitter stage is designed to give desirable optical properties, and the collector stage is designed to improve its electrical properties. By optimizing these two stages independently, better overall optoelectronic properties of a detector can be obtained. Besides controlling the electrical characteristics, the filter can also be used to tailor its optical properties as is discussed later. In the following paragraphs, a more detailed discussion of the characteristics of IHETs is given using several examples. In these examples, bound-to-extended-state QWIPs are adopted since they are more suitable to be integrated into an IHET.

In optimizing detector performance, there are different criteria for different detector figures of merit. If one considers only the background photocurrent and dark current g–r noise, and assumes τ_{int} to be inversely proportional to the total current density J_{tot} for a fixed charge-handling capacity, the expressions relating the figures of merit and the dark current and photocurrent densities, J_d and J_p, are given as follows

$$D^* = \frac{J_p}{I_o} \frac{1}{\sqrt{4egJ_{tot}}} \propto \frac{J_p}{\sqrt{J_d + J_p}}, \qquad (13)$$

$$NE\Delta T \propto \frac{1}{D^*\sqrt{\tau_{int}}} \propto \left(\frac{J_p}{J_d + J_p}\right)^{-1}, \qquad (14)$$

and

$$\%\text{BLIP} = \frac{\sqrt{4egJ_p}}{\sqrt{4egJ_{tot}}} \propto \left(\frac{J_p}{J_d + J_p}\right)^{1/2}. \qquad (15)$$

From these expressions, the optimizing condition for D^* is different from that for $NE\Delta T$ and %BLIP.

D. High-Detectivity IHETs

In applications such as high-speed thermal imaging, where readout charge handling is not a problem, it is advantageous to improve the D^* of a detector. When an IHET is operated as an infrared detector, there are two transfer ratios, α_d and α_p, which are defined as J_{cd}/J_{ed} and J_{cp}/J_{ep}, respectively. J_{ed} (J_{cd}) and J_{ep} (J_{cp}) are the emitter (collector) dark current density and photocurrent density, respectively. From Eq. (13), the detectivity measured at the collector D_c^* under low background condition is related to that measured at the emitter D_e^* by the expression

$$D_c^* = D_e^* \frac{\alpha_p}{\sqrt{\alpha_d}}. \qquad (16)$$

D^* will be improved if α_p^2 is larger than α_d. Therefore, it is important to maintain a large α_p in the D^* optimization.

In order to obtain a large α_p in the example of device D listed in Table I, a thin (300 Å) $In_{0.15}Ga_{0.85}As$ base layer is adopted to avoid phonon and impurity scattering (8). The presence of indium reduces the surface depletion by reducing Fermi-level pinning, and, at the same time, it increases the conduction–band discontinuity by 100 meV, which allows heavy doping ($n = 1.2 \times 10^{18}$ cm^{-3}) without causing excessive electron thermionic emission from the base to the collector. The higher doping density further reduces the surface depletion and the resistance of the base layer, and hence further decreases the thickness requirement. In addition, the reduced Γ–L scattering of the photoelectrons in the strained InGaAs base, due to the larger Γ–L valley separation, is also expected to increase α_p compared to a GaAs base. Besides adopting a thin base, the filter barrier height is chosen to be the same as that of the QWIP so as not to discriminate photoelectrons.

The QWIP performance is first characterized by measuring the properties of the emitter. The emitter dark current is shown in Fig. 11 and the responsivity is shown in Fig. 12. Together with the measured noise gain, the D_c^* can be deduced, and is plotted as a function of V_e at 77 K in Fig. 13. The maximum D_c^* is found to be 6×10^9 cm $\sqrt{\text{Hz}}$/W with a cutoff at 10.0 μm. The spectral response is shown in Fig. 14. This value of D_c^* is the same as that reported by Levine et al. (21) having a similar detector structure, except that the barrier width is 480 Å, instead of 200 Å as in this detector. This result shows that the barrier thickness does not affect the detector sensitivity at 77 K, since direct tunneling is unimportant at this temperature.

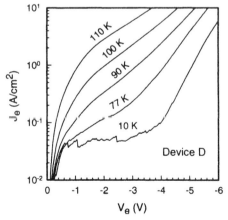

FIG. 11. The I–V characteristics of device D. At $T = 10$ K, direct tunneling dominates, with which negative differential conductance (NDC) is observed whenever there is a high field domain formation. At high temperatures, NDC is suppressed when direct thermionic emission current becomes the major transport mechanism.

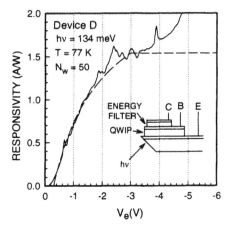

FIG. 12. The responsivity of device D as a function of V_c. The dashed curve is the theoretical fitting with contact recombination. The insert shows the device geometry and the light-coupling scheme used in this article.

In fact, TSHES performed on this QWIP (Fig. 9) indicates dark-current transport dominated by TAT near the top of the barrier at 77 K and is lower in energy than the photocurrent emitted from E_2. Therefore, by using a high-pass filter to block the TAT current but allow the photoelectrons from E_2 to pass to the collector, the D^* of the detector should increase. Figure 15 shows the filtered collector dark current and the

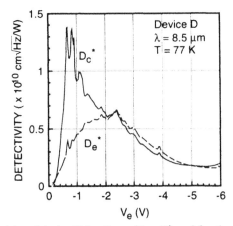

FIG. 13. The detectivity of device D for the emitter D_e^* and for the collector D_c^* at 77 K.

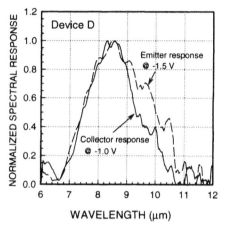

FIG. 14. The spectral response of device D for the emitter and the collector. The cutoff wavelength for the collector is shorter because the longer-wavelength photoelectrons are blocked by the filter and there is insufficient scattering in the base to mix the photoelectrons originating from a different wavelength.

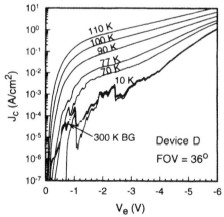

FIG. 15. The collector-current characteristics of device D at different temperatures. The figure also shows the 300-K background photocurrent measured at 10 K. The IHET is BLIP at 70–77 K depending on the applied V_e.

photocurrent at 10 K when the detector is exposed to 300-K background radiation with a field of view (FOV) = 36°. From Fig. 15, the IHET is seen to be BLIP at T = 70 to 77 K, depending o the applied V_e. In this figure, the collector voltage V_c is always kept at zero volts. On the other hand, the QWIP is not BLIP at any temperatures, due to the presence of tunneling current in this thin barrier detector. With the measured collector responsivity, the current transfer ratios at 77 K, α_d and α_p, can be determined and are plotted in Fig. 16. α_p turns on at a lower V_e than α_d because the photoelectrons have higher energies. α_p becomes constant when all photoelectrons are above the filter barrier height. At V_e = -0.85 V, α_p ($= 0.33$) is much larger than α_d ($= 9.1 \times 10^{-3}$). As a result, D_c^*, shown in Fig. 13, is a factor of three larger than D_e^* at this voltage, demonstrating the advantage of energy filtering. Because the maximum D_e^* occurs at -2 V, the net D^* enhancement is factor of two. At lower temperatures, the enhancement tends to be larger as indicated in Fig. 17 for device D because tunneling dark current can be totally suppressed when V_c is small. This result indicates that a single thick barrier filter is as efficient as the total number of thick quantum-well barriers in a QWIP in suppressing the tunneling current.

At $V_c \approx -1$ V, part of the D^* enhancement is due to the slightly shorter wavelength cutoff by 0.6 μm measured at the collector, as indicated in Fig. 14. At a small V_c, some of the photoelectrons originated from

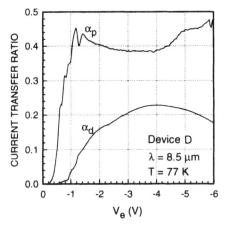

FIG. 16. The current transfer ratio of device D at 77 K. α_d is for dark current and α_p is for photocurrent. The difference in α_d and α_p is largest at -0.85 V, where TAT current is being blocked by the filter.

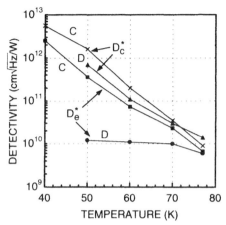

FIG. 17.   The detectivity of two devices C and D as a function of temperature. D_e^* of device D is the same as that of device C at 77 K, showing tunneling current in the thin barrier QWIP is unimportant at high temperatures. D_c^* of device C is slightly better than that reported in literature since the QWIP has been optimized such that the cutoff energy is located at the quantum-well barrier height. D_c^* of device D is higher than D_c^* of device C at all temperatures, showing that a single thick barrier filter is as effective as 50 thick quantum-well barriers in suppressing tunneling dark current. D_c^* of device C is higher than D_e^* because the TAT and TAIT current existing in the 500-Å quantum-well barriers under biases is blocked by the 2000-Å filter barrier.

long-wavelength radiation are not able to overcome the filter barrier, and hence α_p is a function of photon energy. In Fig. 18, the experimental α_p for different photon energies is shown (*27*). It is measured using a monochromatic CO_2 laser source at $V_c = -1.2$ V. α_p does not follow a step function because of energy broadening in the base. By changing the collector voltage V_c, the distribution width at half maximum (Γ) shown in Fig. 19 was found to be insensitive to V_c, and the mean value is 76 meV. However, this apparent Γ is more likely due to the broadening occurring in the thicker filter barrier (2000 Å) instead of the thinner (300 Å) base. By varying the incident photon energy at $V_c = 0$ V, the analyzer filters the photoelectrons at the front edge of the barrier, and broadening in the base can be measured. The solid curve in Fig. 18 is the fit to the experimental data assuming that the photoelectron distribution is Lorentzian and that the photoelectron injection energy increases with the incident photon energy. From the fit Γ is deduced to be 17 meV, an extremely small value considering the large applied voltage of -1.2 V. At the same time, the maximum α_p is found to be 0.45 occurring at $V_E = -1.1$ V, four times

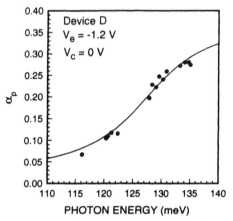

FIG. 18.   The dependence of photocurrent transfer ratio α_p as a function of incident photon energy. The solid curve shows the theoretical fitting assuming the hot-electron distribution is Lorentzian with a width of 17 meV at the front boundary of the filter.

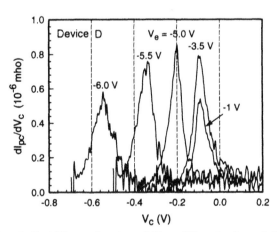

FIG. 19.   A plot of dI_{pc}/dV_c as a function of V_c at different values of V_c measured at a temperature of 10 K.

larger than a detector with a 1500-Å GaAs base (*28, 29*) because of the thinner base and the larger Γ–L valley separation for the present device.

An IHET structure is also able to improve the detectivity of a thick barrier QWIP. For device C, each quantum-well barrier is 500 Å thick, and the quantum-well parameters have been optimized, with $E_c = H$. Tunneling current is negligible at low temperatures as demonstrated in Fig. 20. Without tunneling current, the background limited temperature T_b of the QWIP is finite and is found to be 67 K. The spectral response is shown in Fig. 21. The cutoff wavelengths λ_c at 10.5 μm. On the other hand, by filtering out the TAIT current as detected in Fig. 8, T_b can be increased to 73 K as shown in Fig. 22. The cutoff wavelength of the collector is found to be the same as the emitter, which can be attributed to the larger energy broadening in the base. For the present device, the base is composed of a 300-Å $In_{0.1}Ga_{0.9}As$ layer and a 200-Å GaAs layer. The detectivity as a function of temperature is shown in Fig. 17. Although the D^* of the QWIP continues to increase at lower temperatures, the D^* of the IHET increases at a higher rate by eliminating the dominant TAIT current at these temperatures.

In summary, we have described two high-detectivity broadband infrared hot-electron transistors with an InGaAs base. At 77 K, the D_c^* for unpolarized light is 1.4×10^{10} cm $\sqrt{Hz}$ /W for device D with a λ_c of 9.5 μm, and is 0.9×10^{10} cm $\sqrt{Hz}$ /W for device C with a λ_c of 10.5 μm;

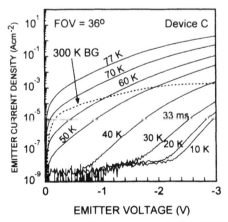

FIG. 20.   The emitter current characteristics of device C. The QWIP is BLIP at 67 K. The dashed curves is the 300-K background photocurrent density measured at T = 77 K. The dotted line is the current level that will saturate the readout if the integration time is set to be 33 ms.

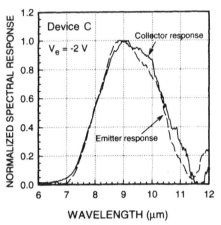

FIG. 21.   The spectral response of device C at $V_e = -2$ V. There is no significant difference between emitter and collector response due to the low filter-barrier height to accept most photoelectrons and because of larger impurity scattering.

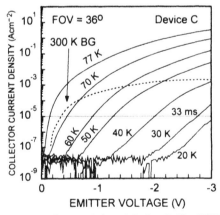

FIG. 22.   The collector current characteristics of device C. The IHET is BLIP at 73 K. The dashed curve is the 300-K background photocurrent density measured at $T = 77$ K. The dotted line is the current level that will saturate the readout if the integration time is set to be 33 ms.

both are a factor of two higher than the corresponding QWIPs. Because of the accompanying reduction of dark current, the integration time can be extended, thereby obtaining a large reduction of $NE\Delta T$ as expected from Eq. (11). For example, the $NE\Delta T$ of device D is a factor of 36 smaller than that of the corresponding QWIP at 77 K when operated at -0.85 V.

E. IHETs FOR SMALL $NE\Delta T$

In the previous examples, the goal is to preserve filtered photocurrent to achieve an improved detectivity. However, at $T = 77$ K, the dark-current levels of the collector, being equal to 0.44 mA/cm^2 for device D and 4 mA/cm^2 for device C at the operating voltages, are not low enough for full-frame signal integration. Furthermore, the detectors are not BLIP at this temperature.

In order to greatly reduce the dark current, two different filter structures were designed (9). The QWIP of the first detector (referred as device E) consists of a 6000-Å n^+ emitter contact layer, 30 periods of 50-Å GaAs well, and 500-Å $Al_{0.25}Ga_{0.75}As$ barrier, and an n^+ base contact layer. The base layer is composed of a 300-Å $In_{0.1}Ga_{0.9}As$ layer and a 200-Å GaAs layer. On the top of the base, a 2000-Å $Al_{0.29}Ga_{0.71}As$ barrier as a high-pass filter and a 1.1-μm n^+ GaAs collector are grown. The Al molar ratio of the filter barrier is graded from 0.14 to 0.29 for 300 Å on each side of the barrier to reduce the quantum mechanical reflection. The band structure is shown in Fig. 23. The graded barrier provides a sharp turn-on

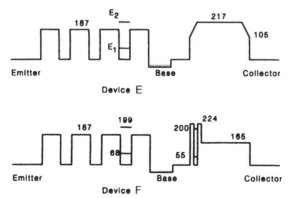

FIG. 23. The band structures of devices E and F. The numbers indicated are the energies in meV.

characteristic in the transmission coefficient at an energy $E = 217$ meV as shown in Fig. 24. The doping density N_d is 0.5×10^{18} cm^{-3} in the QWIP and 1.0×10^{18} cm^{-3} in the base. For another IHET (referred as device F), the QWIP structure is the same as for device E except that $N_d = 1.0 \times 10^{18}$ cm^{-3}. The filter consists of a 30-Å $Al_{0.3}Ga_{0.7}As$ barrier, a 65-Å GaAs well, another 30-Å $Al_{0.3}Ga_{0.7}As$ barrier, and a 2000-Å $Al_{0.22}Ga_{0.78}As$ barrier as shown in Fig. 23. The barriers form a bandpass filter with a pass band centered at $E = 200$ meV and a bandwidth of 35 meV, as shown in Fig. 24.

Figures 25 and 26 show the 77-K current-emitter voltage characteristics of devices E and F, respectively. In these measurements, V_c is kept at 0 V. J_{cd} is the dark current flowing into the emitter; it represents the dark current level of the corresponding QWIP. Note that J_{cd} of device F is about 100 times higher than that of device E at a high bias although the expected difference is a factor of 4.4 higher from the N_d difference. It has been shown using TSHES that the difference is largely due to the TAIT current through the impurity states in the AlGaAs barriers in the higher doping detector. On the other hand, the 300-K background photocurrent density measured at the emitter J_{cp} of device F is about a factor of two higher than device E, consistent with the doping level difference and indicating the material problem of a higher-density QWIP.

As usual, the background emitter photocurrent J_{cp} in Fig. 25 and Fig. 26 is measured by first exposing the detectors to room-temperature radiation

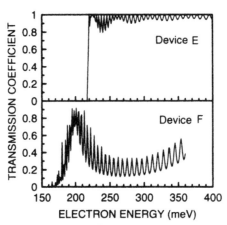

FIG. 24.   The calculated transmission coefficients for the graded high-pass filter for device E and the bandpass filter for device F.

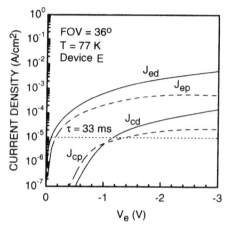

FIG. 25. The 77-K dark current density (solid curves) and background photocurrent density (dashed curves) of device E. The collector is shown to be BLIP at this temperature, and the total current level is below the limit for full-frame integration of 33 ms, shown by the dotted curve.

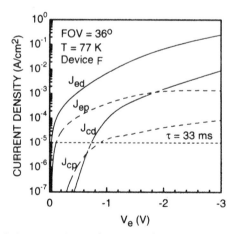

FIG. 26. The 77-K dark current density (solid curves) and background photocurrent density (dashed curves) of device F. The collector is shown to be BLIP at this temperature, and the total current level is below the limit for full-frame integration of 33 ms, shown by the dotted curve.

at detector-temperature T equal to 10 K, where the dark current is negligible, and then adjusting for a small change in the responsivity at $T = 77$ K using AC locking techniques. Infrared light is coupled to the detectors, which are without an antireflection coating, through a 45° edge with a FOV of 36°. The photocurrent level should be equal to that using a one-dimensional grating either with an antireflection coating (34) or with an AlAs cladding layer (35). In these two figures, the saturating current level is indicated as dotted lines when $\tau = 33$ ms. Obviously, J_{ed} of both QWIPs is two to three orders of magnitude higher than this limit, and since $J_{ep} < J_{ed}$ in both cases, device E is not BLIP, and device F is far from BLIP at 77 K. Note that J_{ep} itself is larger than 10 μA cm^{-2} under a small V_e, and hence it should also be reduced if a long τ is used.

In contrast, the dark-current density J_{cd} and the background photocurrent density J_{cp} after filtering are much smaller and fall within the current limit of the readout below certain voltages. For example, for device E at -0.8 V, J_{cd} is 1.68 μAcm^{-2}, 248 times lower than J_{ed} ($R_0A = 72$ and 6500 Ωcm^2 for the emitter and the collector, respectively), and J_{cd} is 0.682 μAcm^{-2}, 2180 times lower than J_{ed} for device F when $V_c = -0.55$ V. The large dark-current reduction is the result of employing a high barrier filter in device E and a narrow band pass filter in device F, so that only a very small portion of the emitter dark electrons injected into the base has energies which match the pass band of the filters and is collected at the collector.

After filtering, there is also a reduction in the background photocurrent, which is in fact necessary to avoid readout saturation for a large τ. Since J_{cp} is now larger than J_{cd} below certain voltages, both detectors are BLIP at 77 K. The improvement of the %BLIP for both detectors, defined here as $i_{np} / \sqrt{\left(i_{np}^2 + i_{nd}^2 \right)}$, is shown in Fig. 27, where i_{np} and i_{nd} are the background photocurrent noise and dark current noise, respectively. At 77 K and the stated applied voltages, both detectors can provide up to 84 %BLIP when other noise sources such as the Johnson noise and the readout noise are not considered. The smaller reduction of J_p relative to J_d is due to the fact that the average energy of the photoelectrons is higher than the dark electrons even at 77 K. Note also that the improvement of F is more substantial than E because J_d, after filtering, is unaffected by the presence of the TAIT current, which has much lower energies. This observation shows that an IHET is less affected by the doping density of the QWIP.

The detector spectral response is shown in Fig. 28 for device E at the operating voltage of -0.8 V. The larger fluctuations in the collector

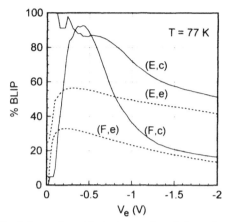

FIG. 27. The %BLIP of the emitter (dashed curves) and the collector (solid curves) for devices E and F at 77 K. Both devices provide up to 84 %BLIP at this temperature for a 36° field of view without antireflection coating.

photocurrent are due to the smaller signal level at the collector and are therefore more susceptible to system noise. From Fig. 28, the spectral responses of the emitter and the collector are shown to be the same. This can be explained by the interdiffusion of electrons that originated from different wavelengths due to impurity scattering in the base. Because of

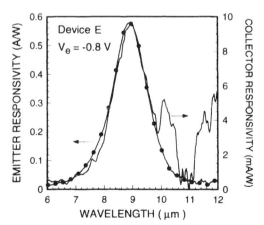

FIG. 28. The spectral response of device E at the operating temperature of 10 K and $V_e = -0.8$ V. The spectral response is not affected by energy filtering because of scattering in the last quantum-well barrier and in the base.

the large reduction in the total current level, there is a reduction on the calculated D_c^* by a factor of three for both detectors, and D_c^* at $\lambda = 9.0$ μm is 1.1×10^{10} and 0.5×10^{10} cm $\sqrt{\text{Hz}}$/W, respectively, for devices E and F at 77 K with a 9.8-μm cutoff. Note that if J_d and J_p of the present QWIP were suppressed uniformly at the readout, the D^* would have been reduced by a factor of 16 for device E and 47 for device F respectively, demonstrating the advantage of energy filtering.

Although there is a reduction of D^* in the present case, the NEΔT of the IHET detector arrays is actually improved because the integration time is inversely proportional to J_{tot} and can be made much longer than for a QWIP array. The resultant NEΔT based on the measured spectral response $\Delta\lambda = 1.7$ μm, the specific D^* and the 33-ms integration time is 14 and 26 mK for devices E and F, respectively, which is 3.2 and 8.6 times smaller than that of the corresponding QWIP arrays.

Besides improving the %BLIP and the NEΔT of an array, an IHET is also more flexible in different detection conditions since the current level of an IHET can be adjusted continuously by changing V_c. For example, by applying a positive V_c, the barrier height of an energy filter can be lowered to increase the total current collection, as shown in Fig. 29 for device E. A factor of 10 in increasing the current level can be accomplished by applying a small V_c of 0.1 V. Since this value of V_c lowers the filter barrier height by only 15 meV at a distance 300 Å from the base, the large

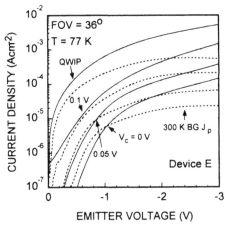

FIG. 29. The 77-K dark-current density (solid curves) and background photocurrent density (dashed curves) of device E for different V_c. At higher V_c, the D_c^* increases while the %BLIP decreases. An optimum condition may be reached for a particular application by adjusting V_c.

increase in the collector current indicates that the photoelectron energy distribution peak is located just below the filter barrier height. The increase of J_p, however, is less than the increase of J_d because of the collection of the TAT electrons, which will reduce the %BLIP of the array and may lead to a larger fixed pattern noise. On the other hand, the increase in J_{tot} also increases the D^* and tends to reduce the temporal noise. For example, at $V_c = 0.1$ V, D_c^* is 2×10^{10} cm $\sqrt{\text{Hz}}$ /W. Therefore, depending on the pixel uniformity, the readout noise, and the preferred integration time, V_c can be adjusted to obtain an optimum current level for the best performance of a particular FPA.

In summary, two IHETs have been designed and demonstrated which are background limited and whose current level is compatible with the readout circuit at 77 K. Therefore, the detectors are suitable for thermal imaging at this temperature.

F. IHET FOR HIGH-TEMPERATURE OPERATION

From TSHES measurements, it was found that, at moderate temperatures, the dark current of a QWIP is dominated by a TAT current, due to which the dark current is lower in energy than that of the photocurrent. At these operating temperatures, a single, graded high-pass filter is adequate in suppressing the TAT current and achieving a larger detectivity and a smaller NEΔT. However, in certain applications, high-temperature operation is an overriding criterion, in which case the peak energy E_p of the dark-current distribution can be higher than that of the photocurrent. As shown in Section III, when DTE current dominates, E_p continues to shift to higher energies at increasing temperatures due to the increase of the three-dimensional density of states at higher energies. Therefore, for high temperature operation, it is important to eliminate the high-energy DTE current as well as the low-energy TAT current. As shown in Fig. 30, although the HPF in device E offers the same performance as the BF in device F at 77 K, the current filtering capability differs significantly at higher temperatures. Below 77 K, the TAT current peak is just below the pass band of device F at $V_c = -0.8$ V. In this case, the BF is acting as a HPF, resulting in the same α_d as device E. On the other hand, when T is around 100 K, most of the hot electrons are high-energy electrons. The HPF of device E becomes ineffective as indicated by a large α_d, in contrast with the relatively small α_d offered by the BF of device F.

In this section, a bandpass filter is designed for device G with a bandwidth of 10 meV. It is used to selectively collect photoelectrons from

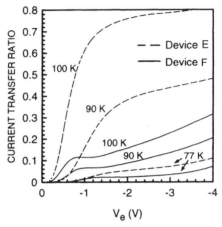

FIG. 30. The dark-current transfer ratio α_d for device E (dashed curves) and device F (solid curves) at temperatures higher than 77 K. α_d of device E rapidly increases to near unity at 100 K, because most dark currents consist of high-energy thermionic currents, which are not affected by the high-pass filter. On the other hand, α_d of device F shows bandpass characteristics and the filter remains effective at high temperatures. The α_d characteristics are also consistent with a peaked injection distribution at high T.

a QWIP with peak wavelength λ_p at 13.5 μm while suppressing the dark electrons of other energies (30). The bandwidth of the filter is narrow compared to both the dark current distribution width of 40 to 70 meV and the photoelectron distribution width of 31 meV. The nominal QWIP consists of 30 periods of a 66-Å GaAs doped well ($n = 0.5 \times 10^{18}$ cm^{-3}) and a 500-Å undoped $Al_{0.16}Ga_{0.84}As$ barrier. The entire structure is sandwiched between a 1-μm doped GaAs layer at the bottom as the emitter contact, and a 500-Å doped $In_{0.08}Ga_{0.92}As$ on the top as the base contact. The filter is placed next to the base, followed by a doped GaAs layer as the collector contact. The filter consists of a 45-Å $Al_{0.22}Ga_{0.78}As$ barrier, a 82-Å GaAs well, and a 40-Å $Al_{0.22}Ga_{0.78}As$ barrier, and is followed by a wide 2000-Å $Al_{0.12}Ga_{0.88}As$ barrier to complete the filter structure. The doping level of the contacts is $n = 1 \times 10^{18}$ cm^{-3}. The structure is grown by molecular-beam epitaxy on a semi-insulating substrate, and the nominal band structure is shown in Fig. 31. The intersubband transition between bound state E_1 and continuum state E_2 is originally designed to yield an absorption peak at $\lambda_p = 15$ μm.

The transmission characteristic of the double-barrier filter is modeled using the transfer matrix technique and is plotted in Fig. 32. The filter's

Device G

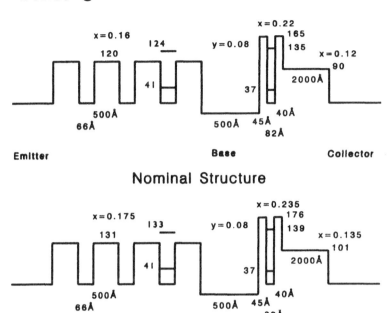

Nominal Structure

Actual Structure

FIG. 31. The band structure of device G for the nominal device parameters and for the actual parameters where the Al molar ratio differs by 0.015. The energies indicated are in meV.

bandpass transmission peak is designed with a pass-band E_2' at 135 meV (relative to GaAs conduction band edge) and with a bandwidth of 10 meV. At higher energies, the filter transmission shows a bandgap followed by a wider resonance at $E_3' = 280$ meV. The wide AlGaAs portion of the filter is designed to block the transmission through the ground state E_1' ($= 37$ meV) of the filter.

Figures 33a and 33b show the emitter and collector current–voltage characteristics, respectively. The solid curves represent the dark current, whereas the dashed curves are the 300-K background window photocurrent (using a 45° edge coupling) measured at 10 K. Notice the occurrence of the plateaus at $T \sim 50$ K for the collector dark-current density J_{cd} and the photocurrent density J_{cp}, which are the characteristics of a bandpass filter (28). After filtering, J_{cd}' at 77 K is reduced by 3 orders of magnitude

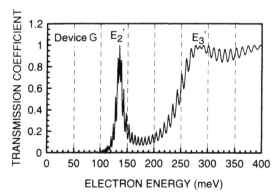

FIG. 32. The calculated transmission coefficient of the bandpass filter in device G. The first resonant state E_1' at 37 meV is not shown. The state E_2' is used for current filtering in the present detector design.

compared with J_{cd} at the emitter voltage V_c of -0.5 V, and the background limited temperature is increased from 52 to 63 K at the same voltage.

To fully understand the collector current characteristics, the dark-current transfer ratio α_d, along with the 300-K background photocurrent transfer ratio α_p, is shown in Fig. 34. α_d shows a peaked structure at a given T, and this peak shifts to a lower V_c at a higher T. The relationship between α and the hot-electron distribution $\rho(V_c, E)$ at the front boundary of the filter is given in Ref. (4).

$$\alpha = f \int_0^\infty \rho(V_c, E) T(E)\, dE, \qquad (17)$$

where f is the collection efficiency of the filter, and $T(E)$ is the transmission coefficient of the filter. It was found that the width of the injected dark electron distribution depends on the thermal distribution of the dark electrons in the QWIP at a given T plus the scattering broadening in the base, and is larger than 40 meV for $T > 40$ K. Therefore, $T(E)$ of the present filter, which has a width of 10 meV, acts effectively as a delta function, and α_d becomes directly proportional to the dark-electron distribution $\rho_d(V_c, E)$ for E less than E_3'. This situation is in contrast to the single-barrier high pass filter case where it is the $d\alpha/dV_c$ which is proportional to ρ. The characteristics of α_d shown in Fig. 34 thus directly indicate the peaked nature of the dark-current injection. As T increases, the current injection shifts to a higher energy, and hence the peak appears

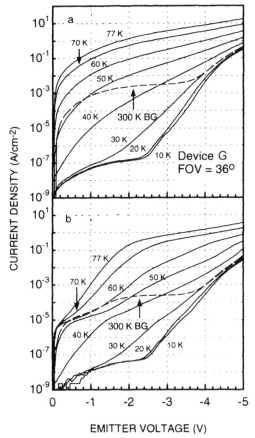

FIG. 33. (a) The emitter and (b) collector current–voltage characteristics. The dark current (solid curves) and the 300-K background photocurrent (dashed curves) characteristics of device G. The QWIP is BLIP at 52 K and the IHET is BLIP at 63 K. The IHET can be operated at 77 K with an integration time of 8.3 ms.

at a lower V_c. In fact, the dark current injection at T larger than 60 K is predominantly above the E_2' passband for $V_c < -1$ V, so that there is a large dark-current reduction above this temperature. For $V_c > -1$ V, TAT current becomes dominant at $T \sim 50$ K. Even in this case, the narrow bandpass filter is seen to provide effective filtering for these lower-energy dark electrons, leading to a very low α_d. Because there is no TAT current component in the photocurrent, α_p becomes much larger than α_d in this voltage regime.

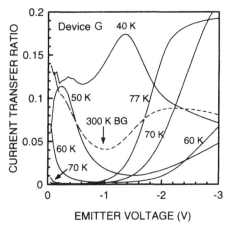

FIG. 34. The dark-current transfer ratio α_d (solid curves) and background photocurrent transfer ratio α_p (dashed curve) of device G. The figure shows that the photoelectron energy is between the 50- and 60-K dark-current energy.

The peak of α_p is observed at $V_c \sim -0.3$ V, which indicates that the photocurrent peak lies just below the filter passband in the absence of an applied V_c and is degenerate in energy with the thermionic dark current peak at $T \sim 52$ K. These results indicate that for the long-wavelength detectors, thermionic emission current may overlap or even surpass the photocurrent in energy at liquid nitrogen temperatures. By using the present bandpass filter, one can suppress this high-energy current component as well as the TAT current and achieve a higher %BLIP. For example, at $T = 77$ K and $V_c = -0.5$ V, the %BLIP for the emitter is 4.5% whereas that of the collector is 45%, a factor of ten increase. The large reduction of the dark current also allows a longer integration time at these temperatures.

The spectral photoresponse of the detector shows additional novel features of a bandpass filter. The emitter response was first measured at $T = 10$ K using a 1900-K ceramic element source and a grating monochromator. The results are shown in Fig. 35. The observed emitter peak at 13.5 μm shown in Fig. 35 is shorter than expected, indicating that the Al molar ratio x of the barriers in the actual sample is larger by 0.015 than the nominal value. This reduces the energy difference between E_2' and E_2 from the expected value of 16 meV to 6 meV, which is due to E_2 increasing faster with x than E_2'. The revised band structure with the actual Al molar ratio is shown in Fig. 31. As a result, the photoelectron

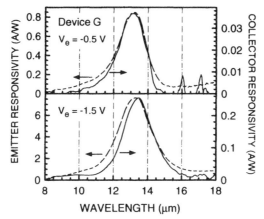

FIG. 35. The spectral response of the emitter (dashed curves) and the collector (solid curves) of device G at two values of V_c. When the photoelectron energy is aligned with the filter pass band at -0.5 V, there is no significant difference between the emitter and the collector spectral response. At the higher V_c, the higher-energy photoelectrons are rejected by the filter and there is a cutoff at the short wavelengths.

peak is aligned with the filter passband at a small V_c of -0.3 V as indicated in Fig. 34. This energy alignment is confirmed by the observation that the collector spectral response shown in Fig. 35a coincides with that of the emitter at $V_c = -0.5$ V. Note that the filter collects the full spectrum of photoelectrons from the emitter despite the fact that the pass-band width ($\Gamma_f = 10$ meV) is less than the absorption width ($\Gamma_a = 13.6$ meV). This is due to the scattering effect in the base, which causes a broadening Γ_s of ~ 17 meV for each injected electron as determined in Fig. 18. After passing through the base, photoelectrons that originated from different wavelengths diffuse into the nearby energies, so that the central portion ($\Gamma_c = \Gamma_s - \Gamma_a/2 = 10$ meV) of the photoelectron distribution (width $= \Gamma_s + \Gamma_a = 31$ meV) becomes populated with photoelectrons of different wavelengths. Therefore, the bandpass filter does not modify the spectral response when energy alignment is achieved.

At a higher V_c, the photoelectron injection shifts to a higher energy because of the depletion voltage drop V_d in the base. When the energy misalignment is more than $\Gamma_c/2$, the filter begins to receive fewer photoelectrons from the shorter wavelengths, and leads to a longer cut-on wavelength as shown in Fig. 35b, at $V_c = -1.5$ V.

Due to the large dark-current suppression at 77 K, D^* measured at the collector is improved by a factor of 1.75 compared with the emitter, and is equal to 7.0×10^9 cm $\sqrt{\text{Hz}}$/W at -0.5 V. Combined with the increased integration time because of the lower dark current, NEΔT can be improved by a factor of 55 at this temperature, and the value for the collector is 38 mK using an integration time of 8.3 ms with f/1.5 optics. When the detector is operated at its background-limited temperature of 62 K. NEΔT can be improved to 8.3 mK because of its larger $D^* = 3.2 \times 10^{10}$ cm $\sqrt{\text{Hz}}$/W at this temperature and at $V_e = -1.4$ V.

In summary, a narrow bandpass filter can be used to probe the dark-electron energy distribution and the photoelectron energy distribution directly without taking the voltage derivative. Both distributions are found to be composed of peaked structures. The energy peaks of these two distributions do not usually coincide with each other except at a particular temperature. Therefore, by aligning the filter precisely with the photoelectron peak, one obtains a high degree of dark-current rejection while maintaining a sufficient amount of photoelectron collection, and hence improves the detectivity.

These results also show that the bandwidth of the filter can be much narrower than the absorption width without affecting the collector spectral response because of the interdiffusion of photoelectrons in the base. Therefore, the present filter scheme can also be applied to detectors with wide absorption widths if the scattering width in the base is larger than half of the absorption width.

Since the filter bandwidth can be made arbitrarily small by structural design, the current level of an IHET and its impedance at a particular temperature becomes independent of its cutoff wavelength. A long-wavelength detector with a narrow bandpass filter can have a low current and a high impedance suitable for signal readout. Thus, high-temperature operation is feasible when a sensitivity requirement is not the primary concern. For the present detector without a filter, the upper operating temperature can be only 50 K for a CMOS readout circuit because of its high current level. With the narrow bandpass filter, the detector can be operated at 77 K with a 45 %BLIP level and a NEΔT of 38 mK even though the cutoff wavelength is as long as 14.0 μm.

G. Long-Wavelength IHET

One of the assets of QWIP technology is its wavelength tunability as a function of quantum-well parameters. In this section, a long-wavelength IHET with $\lambda_c = 18$ μm is described (11, 36). Long-wavelength detectors are of interest for a variety of applications including the detection of cold objects in space and for observing the earth's atmosphere. For these applications, an IHET offers several significant advantages. First, the background photon flux is usually very low in this wavelength regime. Operation of a detector requires a very low dark current for BLIP. However, in this wavelength regime, the barrier height of a QWIP is very small, which leads to a significant tunneling current even with thick barriers. Hence, the dark current of the QWIP cannot be reduced arbitrarily by lowering the temperature. A high-pass filter can be used in this case to totally eliminate the tunneling current for low background applications. Second, for thermal imaging, the thermally activated dark current of a QWIP will saturate a typical FPA readout circuit at a rather low temperature. An IHET is able to significantly increase the operating temperature. Third, an IHET reduces $1/f$ of a QWIP, which is the dominant noise source for long-wavelength QWIPs. In the following paragraphs, we describe an IHET, labeled as device H, whose dark current is two-to-three orders of magnitude lower than that of a QWIP. This IHET is capable of background-limited thermal imaging at $T = 55$ K under terrestrial background.

The QWIP consists of a 1-μm n^+ emitter contact layer, and 30 periods of 60-Å GaAs well and 500-Å $Al_{0.15}Ga_{0.85}As$ barrier. On the top of the QWIP is a 500-Å $In_{0.08}Ga_{0.92}As$ base layer, a 2000-Å $Al_{0.17}Ga_{0.83}As$ barrier, which acts as a high-pass filter, and a 1.1-μm n^+ GaAs collector. The Al molar ratio of the filter barrier is graded from 0.07 to 0.17 for 300 Å on each side of the barrier to reduce the quantum mechanical reflection of the injected electron. The doping density N_d is 0.8×10^{18} cm^{-3} in the quantum wells and 1.0×10^{18} cm^{-3} in the base. The collector and emitter areas are 2.25×10^{-4} cm^2 and 7.92×10^{-4} cm^2, respectively. The band structure of device H is given in Fig. 36. The bound state in the well is located at $E_1 = 42$ meV, while the quantum-well barrier height H and the filter barrier height H_f are 112 and 126 meV, respectively.

The measured emitter dark-current density J_{dc} and the collector dark-current density J_{dc} are shown in Figs. 37a and b as solid curves. The 300-K background photocurrents using a 45° coupling angle and a FOV of 36° are also shown. Although the QWIP is BLIP at 38 K, it can be operated only above 35 K because of the large dark current. In comparison, after energy filtering, the detector is BLIP at 55 K without saturating the readout. The improvement in T_b is due to the large difference in the current transfer ratios α_d and α_p shown in Fig. 38. At $T = 50$ K and $V_e = -0.2$ V, the dark current can be suppressed by almost four orders of magnitude,

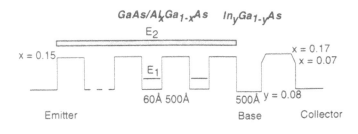

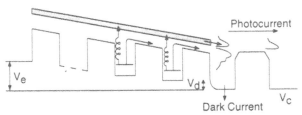

FIG. 36. The band structure of device H.

whereas the background photocurrent is suppressed by only two orders of magnitude.

In order to understand the origin of improvement in greater detail, the current densities are assumed to have the form

$$J(T) = J_0 \exp\left(-\frac{E_a}{kT}\right), \tag{18}$$

and the activation energies for the emitter E_{ac} and the collector E_{ac} are extracted and plotted in Fig. 39. The experimental value of E_{ac} at low biases is found to be 52 meV, which is in excellent agreement with the expected value of $H - E_F = 52.8$ meV. The activation energy of TAT current (25) described by

$$J_{TAT}(T, V_p) = \left\{ \frac{m^*}{\pi \hbar^2 L} \int_0^\infty \frac{D(V_p, E_1 + E_{in})}{1 + e^{(E_{in} - E_F)/kT}} dE_{in} \right\} ev \tag{19}$$

is expected to decrease with bias due to the increase of the transmission coefficient of the barrier, in qualitative agreement with the experiment. In Eq. (19), V_p is the voltage drop per period, L is the period length, D is the transmission coefficient, E_{in} is the in-plane energy, v is a fitting parameter which has the same dimension as the velocity. In order to obtain a more quantitative comparison, the theoretical activation energy $E_{ac,t}$ is ex-

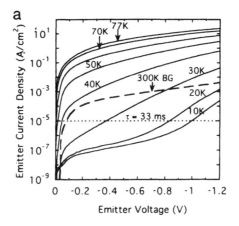

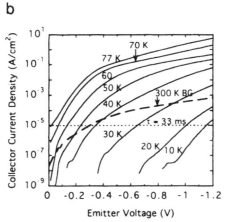

FIG. 37. The measured (a) emitter dark-current density J_{dc} and (b) collector dark-current density J_{dc}. The dark current (solid curves) and the 300-K background photocurrent (dashed curves) at $T = 10$ K of device H. The QWIP is BLIP at 38 K and the IHET is BLIP at 55 K. The dotted line is the current level that will saturate the readout if the integration time is set to be 33 ms.

tracted from the calculated J_{dc} using Eq. (19) and is plotted in Fig. 39 as a dashed curve. From Fig. 39, it is apparent that although the trend agrees with the experiment, the magnitude of the decrease in $E_{ac,t}$ is insufficient to explain the observation. Quantitatively, if we express $E_{ac,t}(V_p)$ as $E^0_{ac,t} - bV_p$, where V_p is the voltage drop per period, then the average b is equal to 0.35, and the observed b is equal to 0.9 for $|V_e| \leq 0.9$ V and 0.6 for $|V_e| > 0.9$ V.

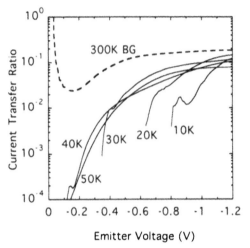

FIG. 38. The dark-current transfer ratio α_d (solid curves) and background-photocurrent transfer ratio α_p (dashed curve) of device H.

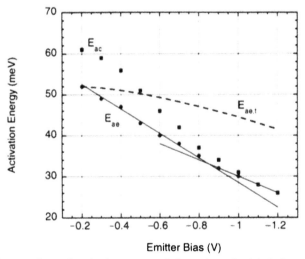

FIG. 39. The experimental activation energy of the emitter E_{ae} (circles) and the collector E_{ac} (squares) as a function of V_c. The dashed curve is the theoretical thermally assisted tunneling activation energy. The solid lines are guides to the eye.

The fact that the decrease of the observed E_{ae} is almost equal to V_p is consistent with the TAIT model described in Section III. As a result of asymmetrical dopant ion diffusion into the "top-side" barrier (37–41), the barrier assumes the potential profile illustrated in Fig. 40, where the diffused dopant ions cause the barrier potential to be pulled down by an amount Φ. Since the thin layer located at the point A is highly conductive because of the presence of the impurity states, the effective barrier height determining E_{ae} is actually located at point B in Fig. 40. When a bias of V_p is applied on a quantum-well unit, nearly all the potential drop is across the barrier so that the potential at B is pulled down by an amount approximately equal to eV_p, which explains the magnitude of decrease in E_{ae} when V_c is less than -0.9 V. E_{ae} changes slopes at $V_c = -0.9$ V, beyond which the slope tends to agree with the TAT model. Therefore, we interpret $V_p = 30$ mV as the flat-band condition illustrated in Fig. 40b and hence Φ is deduced to be 30 mV. From Poisson's equation, the diffusion length is deduced to be 68 Å, consistent with the reported value of 35 Å for a much lower doped sample, since the diffusion length is known to increase with N_d (37, 38).

As for the collector activation energy, it is 9 meV higher than E_{ae} at a low bias, indicating that the collector accepts higher-energy electrons injected from the emitter. As V_c increases, E_{ac} merges gradually with E_{ae}, indicating that the depletion voltage in the base increases consistently with V_c at a constant rate for this detector. At $V_c = -0.5$ V, E_{ac} is 52 meV, the same as that of E_{ae} at low biases. Therefore, from the observed activation energies, one can conclude that up to $V_c = -0.5$ V, the collector accepts only the thermionic emission current, which occupies higher energies in

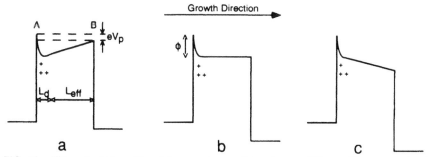

FIG. 40.   The potential profile of the quantum-well barriers at different V_c as a result of dopant diffusion.

the injected hot-electron distribution, and rejects the lower-energy TAIT current. This explains the performance enhancement.

The increase in the detector detectivity is shown in Fig. 41. In this figure, the emitter detectivity is the maximum detectivity as a function of V_e at each temperature, and the collector detectivity is the value at $V_e = -0.2$ V. There is a factor-of-ten improvement in D^* at each temperature because α_p is, in general, a factor of 100 larger than α_d. Since NEΔT is proportional to α_p/α_d, it will be greatly improved in FPA applications. The energy filtering shows no significant effect on the spectral response. The absorbance of this device as well as the emitter spectral response is shown in Fig. 42.

Another advantage of an IHET is the elimination of $1/f$ noise. Among different conduction mechanisms in a QWIP, only thermally assisted impurity tunneling TAIT current potentially causes $1/f$ noise. The $1/f$ dependence of TAIT noise is due to the large time constant of the charge exchange process between an impurity in the barrier and a quantum well. However, there will be no $1/f$ noise if the impurity levels are always empty. It is therefore easy to understand why $1/f$ noise is more acute in long-wavelength QWIPs. In these detectors, the quantum-well barrier height is lower, which brings the impurity levels in the barriers closer to the Fermi level of the well. The proximity of the levels makes the

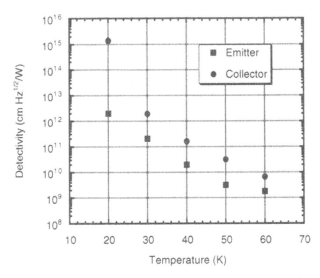

FIG. 41. The detectivity of the emitter (squares) and the collector (circles) of device H at different temperatures.

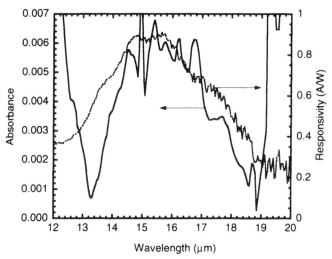

FIG. 42.   The absorbance and the emitter spectral response of device H at $V_c = -0.5$ V.

occupation number of an impurity closer to 0.5, where the impurity causes most noise (*42*). Figure 43 shows the noise spectrum of device H. The large $1/f$ noise in the emitter makes the QWIP unsuitable for long-time integration. In contrast, the noise spectrum of the IHET is free of $1/f$ noise because of the suppression of TAIT current component.

In summary, the TAIT current was demonstrated to be much more important in a long-wavelength QWIP because of the proximity of the impurity levels in the barriers and the Fermi level. The TAIT current is activated at a very low temperature and leads to a large dark current as indicated in Fig. 37. Even at $T = 10$ K, the tunneling current of a QWIP is still significant, meaning that the QWIP cannot be made BLIP by lowering the temperature in low-background detection. By placing a blocking filter next to the QWIP, it is possible to selectively filter out the TAIT current and the direct tunneling current for low-background applications.

V. Multicolor QWIP and IHET

The structural dependence of the absorption wavelength and its relatively narrow lineshape provide an opportunity to detect more than one specific wavelength simultaneously. At present, there are several QWIP structures that can provide certain features of multicolor detection. The

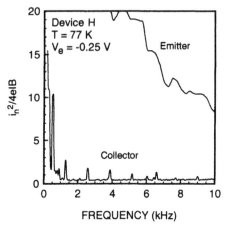

FIG. 43.   The noise spectrum of the emitter and the collector of device H at 77 K. The $1/f$ noise carried by TAIT current in the emitter is blocked by the collector filter, and is absent in the collector noise spectrum.

simplest approach is to stack several QWIP structures together (*43, 44*), each of which covers a specific range of wavelengths. Other structures such as QWIPs with asymmetrically graded barriers (*45*) or QWIPs with two occupied subbands (*46*) have also been investigated. Since multicolor quantum-well infrared detection is still in its infancy, most research efforts are directed toward the basic understanding of a particular structure rather than finding an optimum detector structure for an application. In the following paragraphs, an approach using strongly coupled asymmetric quantum wells is described (*47–49*). In these structures, two quantum wells with different well widths are joined together through a thin barrier. Since the resultant wave functions are of mixed parity, the dipole selection rule is relaxed and optical transitions to all excited states are allowed. Using this approach, it should be possible to expand the range of spectral response to cover both 3- to 5- and 8- to 12-μm atmospheric windows simultaneously. With suitable device parameters, the peak detection wavelengths can also be tuned by applying a bias. In addition, the responsivity of these structures at certain wavelengths is increased through photoexcited coherent tunneling (*48*).

Besides their usefulness in multicolor infrared detection, the coupled structures are also useful in demonstrating an analytical technique, which is known as infrared photoelectron tunneling spectroscopy (IPET) (*47*). In this technique, the photocurrent tunneling out of a quantum-well structure is monitored as a function of incident photon energy. From the spectral response, the quantized levels of the structure can be deduced. This technique provides more direct information compared with other conventional methods, such as photoluminescence, because the band structures of the holes are not involved. It is also more reliable than optical-absorption measurements since the presence of other transmission-reducing processes such as plasmon absorption in the contact layers, optical–phonon absorption in the substrate layer, and the Fabry–Perot oscillations between the epilayers will not produce a photocurrent.

Other than multicolor QWIPs, we also demonstrate wavelength selection using a bandpass filter in an IHET structure. By changing the pass-band energy of the filter through the applied collector voltage, the filter is able to intercept photoelectrons originating from different wavelengths, and hence the detection wavelength is voltage tunable. In the previous section, we had already discussed using bandpass filters in discriminating the dark current of specific energies.

A. MULTICOLOR QWIP

The first coupled quantum-well structure (labeled device I) contains 50 periods of 65-Å GaAs quantum well (doped at $n = 1.0 \times 10^{18}$ cm^{-3}), 40-Å undoped $Al_{0.25}Ga_{0.75}As$, 14-Å undoped GaAs, and 150-Å undoped $Al_{0.25}Ga_{0.75}As$. The MQWs are sandwiched between the top (0.5 μm) and the bottom (1 μm) GaAs contact layers in which $n = 1 \times 10^{18}$ cm^{-3}. The band diagram of the structure is shown in Fig. 44. If the wells are isolated, there will be two bound states ($e_1 = 50$ meV, $e_2 = 181$ meV) in the thicker well and one bound state ($e'_1 = 163$ meV) in the thinner well. In the actual structure, e_2 and e'_1 are coupled, thus shifting the global energy levels to the values $E_1 = 50$ meV, $E_2 = 154$ meV, and $E_3 = 188$ meV. The resulting wave functions of these bound states, together with that for the first continuum state (ψ_4) are depicted in Fig. 45. Based on these wave functions, the oscillator strength (f_n) between the ground state and the excited state, n, can be calculated. The results for zero bias are $f_2 = 0.46$, $f_3 = 0.45$, and $f_4 = 0.13$. The approximate equality between f_2 and f_3 is confirmed in an infrared absorption measurement.

Under an applied bias, both the energy levels and the oscillator strengths change as e'_1 shifts relative to e_1 and e_2, leading to a change in the detection wavelengths. For example, under a forward bias defined in Fig. 44, e'_1 is downshifted relative to e_1, making the spacing between E_1 and E_2 (which is originated from e'_1) smaller and shifting the correspond-

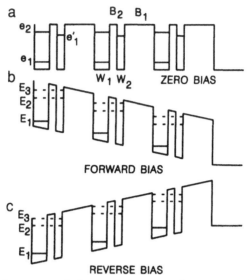

FIG. 44. The band structure of the coupled asymmetric quantum-well structure under (a) zero bias, (b) forward bias, and (c) reverse bias. The states e_1, e_2, and e'_1 are the uncoupled states. The states E_1, E_2, and E_3 are the global states. W_1 and W_2 are the thickness of the wells, and B_1 and B_2 are the thickness of the barriers.

ing transition, denoted by $2f$, to a longer wavelength. On the other hand, the transition energy between E_1 and E_3, denoted by $3f$, is relatively unchanged as E_3 is originated from e_2 of the same well. Under a reverse bias, e'_1 is uplifted by the bias. When e'_1 is well above e_2, E_3 is originated mainly from e'_1, making the E_1 to E_3 transition, denoted by $3r$, bias dependent. The transition $2r$, on the other hand, is unchanged with bias as E_2 is originated from e_2. From this discussion, the $2r$ transition should be degenerate with the $3f$ transition at large biases because both final states are the decoupled e_2 state.

The experimental photocurrent spectra at two biasing conditions are shown in Fig. 46. The dashed curve shows the spectrum under reverse bias with the potential drop per period $V_r = 63.5$ meV. The solid curve shows the spectrum under forward bias with $V_f = 58.4$ mV. Under reverse bias, three peaks in the spectrum are observed and can be assigned to the $2r$, $3r$, and $4r$ transitions. The inverted triangles in the bottom of the graph indicate the predicted transition wavelengths assuming a linear potential drop within a coupled quantum-well unit. In order to more precisely fit the observed peaks, the potential drop between the centers of the wells V_w is

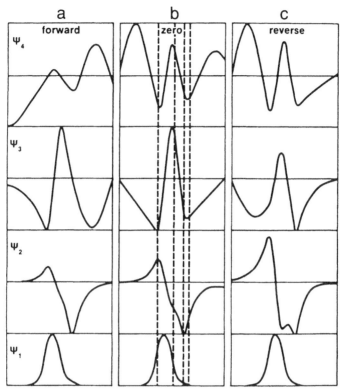

FIG. 45. The shape of the wave functions in the coupled quantum-well structure under (a) forward bias, (b) zero bias, and (c) reverse bias. Based on these wave functions, the oscillator strength can be obtained. The dashed lines indicate the locations of the bandgap discontinuity.

adjusted from 0.3 V_n (linear drop) to 0.45 V_n, and the result is indicated by the arrows. In general, the photocurrent should increase rapidly with decreasing wavelengths due to the exponentially increasing transmission coefficient p of the barrier. However, in this case, the magnitude of the $2r$ peak is quite substantial in comparison with the shorter wavelength peaks because the smaller p_2 is compensated by the larger oscillator strength. Under this bias, $f_2 = 0.76$, $f_3 = 0.11$, and $f_4 = 0.14$.

Under forward bias, only one peak with a right shoulder is found. It can be assigned to the $4f$ and $3f$ transitions. The $2f$ peak is unobservable because of the small p_2 and f_2 under this bias ($f_2 = 0.32$, $f_3 = 0.44$, and $f_4 = 0.10$). As discussed above, the $3f$ transition is nearly degenerate with

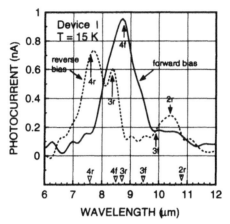

FIG. 46. The IPET spectra of device I under forward and reverse biases. The inverted triangles indicate the transition wavelengths assuming linear potential drop within each quantum-well unit. The arrows indicate the fitted transition wavelengths assuming a different V_w.

the $2r$ transition. The arrows indicate the fitted peak positions using the same nonlinear potential drop as the reverse bias case.

B. PHOTOEXCITED COHERENT TUNNELING

Next, we describe a QWIP structure (labeled as device J) that shows additional novel features. The device consists of 50 periods of 72-Å GaAs ($n = 1 \times 10^{18}$ cm^{-3}), 39-Å undoped Al$_{0.31}$Ga$_{0.69}$As, 20-Å undoped GaAs, and 154-Å undoped Al$_{0.31}$Ga$_{0.69}$As. At zero bias, $e_1 = 47$ meV, $e_2 = 183$ meV and $e_1' = 174$ meV. The calculated f_2 ($= 0.27$) is smaller than f_3 ($= 0.75$), consistent with only one pronounced peak in the optical absorption measurement. Even though f_2 is small at zero bias, it becomes larger ($f_2 = 0.74$, $f_3 = 0.26$) under a reverse bias of $V_r = 123$ mV. As a result, the $2r$ peak and the $3r$ peak are comparable under this bias as shown in Fig. 47. Another striking feature in this spectrum is the well resolved $4r$ peak at $\lambda = 5.6$ μm. It indicates that, although this level is significantly (40 meV) higher than the barrier height, the level is still relatively well defined, confirming the existence of the Stark ladder even for the extended continuum states as found in the tunneling characteristics (2, 48). The arrows in Fig. 47 indicate the fitted transition wavelengths

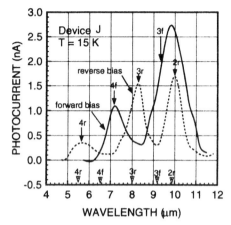

FIG. 47. The IPET spectra of device J under forward and reverse biases. The inverted triangles indicate the transition wavelengths assuming linear potential drop within each quantum-well unit. The arrows indicate the fitted transition wavelengths assuming a different V_w.

assuming V_w to be 0.45 V_p and 0.25 V_p for forward bias and reverse bias, respectively. The effect of electron screening is due to the distortion of the ground-state wave function Ψ_1. Since Ψ_1 is primarily located in the thicker well for the present higher barrier device, the larger V_w under forward bias can be attributed to the lack of electron screening in the thinner well. From Fig. 47, it is shown that with a coupled asymmetric quantum-well structure, the spectral response can be greatly expanded. For the present QWIP, it extends from 5.2 to 10.6 μm. Since the energy of the E_4 level increases with the decreasing barrier width, an optimized structure with a reduced barrier width should be able to detect specifically 3 to 5 μm and 8 to 12 μm simultaneously.

For a forward bias per period V_f of 109 mV, the large f_3 ($f_2 = 0.19$, $f_3 = 0.85$) contributes to a large photocurrent I_p near the expected $3f$ peak. The fact that the photocurrent peak does not coincide with the calculated absorption peak is a noteworthy feature of the I_p spectrum. Since I_p is proportional to both the oscillator strength and the tunneling probability at a particular energy, I_p is peaked at the maximum absorption only if p is a monotonic function of energy. For the present device, it is designed such that p peaks near $\lambda = 10$ μm, using the coherent tunneling mechanism; the resultant I_p then peaks toward the energy where p is a maximum, and the combined factors make this peak dominant over all other peaks. The detailed mechanism is described as follows.

When an electric field is applied to this QWIP structure, it is broken up into a high field domain and a low field domain as evident in the $I-V$ characteristics shown in Fig. 48. At low temperatures, electron transporting out of E_1 is via sequential tunneling in both domains (22). However, if the electrons from E_1 are photoexcited to an excited state with energy close to the uncoupled e_1' state, the conduction mechanism can be quite different. In the high field domain under forward bias, where E_1 is approximately aligned with E_2 of the adjacent period, the tunneling lifetime width of the state e_1' in the thinner well can be large (> 50 meV) compared with impurity broadening (≈ 7.4 meV) because of the large $T_1(e_1')$ and $T_2(e_1')$ at high field, where T_1 and T_2 are the transmission coefficients of the thicker barrier B_1 and thinner barrier B_2 in Fig. 44, respectively. In this case, coherent tunneling of the photoexcited electron from the thicker well W_1 through the state e_1' and out of the barriers is possible. In addition, T_1, originally smaller than T_2 at zero bias, increases more rapidly under an applied field, finally becoming greater than T_2 at large biases. Therefore, in a certain range of applied bias, $T_1 \simeq T_2$ and a large coherent enhancement is expected (50). This is in contrast to the

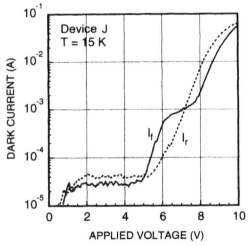

FIG. 48.   The dark-current characteristics of device J under forward bias (solid curve) and reverse bias (dashed curve) at $T = 15$ K. The area of the detector is 3×10^{-4} cm^2.

case of reverse bias, in which T_1 is always less than T_2. Using a WKB approximation, the global coherent transmission coefficient T_G is given by

$$
T_G = \frac{4}{\left(\dfrac{T_1}{T_2} + \dfrac{T_2}{T_1} + 2\right)\cos^2\theta + \left(\dfrac{T_1 T_2}{16} + \dfrac{16}{T_1 T_2} + 2\right)\sin^2\theta}, \tag{20}
$$

where $\theta = (2m^*)^{1/2} W_2 [E^{1/2} - (E_1')^{1/2}]/\hbar$ is the phase angle measured relative to the angle at the resonant energy. T_G for forward (T_G^f) and reverse (T_G^r) bias at $E = 123$ meV above E_1 can be calculated as a function of V_p. Their ratio starts from unity at small bias and rapidly increases to $\simeq 100$ at $V_p = 100$ mV, and then falls back to $\simeq 8$ at large bias. In contrast, if sequential tunneling were the dominant mechanism, then $T_{seq} = T_1 T_2$ (48), the maximum forward-to-reverse transmission coefficient ratio would only be 10.

Figure 49 shows the experimental responsivity in both biases for $\lambda = 10.3\ \mu$m. Since the high field domain formation quantizes V_f and V_r at

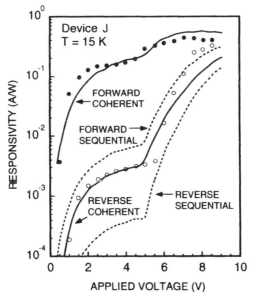

FIG. 49.   The experimental responsivity of device J under forward bias (filled circles) and reverse bias (open circles). The solid curves are the theoretical fittings assuming coherent tunneling. The dashed curves are the theory assuming sequential tunneling.

93 mV per period for a bias smaller than 5 V, due to the T_G difference, the theoretical responsivity under a forward bias R_f should be always larger than that under a reverse bias R_r by a constant factor of 100 below 5 V. This large difference is indeed observed experimentally, indicating that photoexcited coherent tunneling is operative in double-barrier structures. At higher biases, the difference between the two responsivities becomes smaller, as expected. The solid curves in Fig. 49 are the fits assuming coherent tunneling, and the dashed curves are the fits assuming sequential tunneling. Obviously, the data support the theory of coherent tunneling. Note that the present difference of R_f and R_r cannot be explained by the I-V asymmetry usually observed in thick barrier QWIPs. Due to dopant diffusion as discussed in Section III, the band bending in a thick barrier is asymmetrical, which leads to a higher dark current and a higher photocurrent when the bottom contact is biased negatively. In the present case, the difference in the dark-current characteristics between the two polarities is insignificant as indicated in Fig. 48. In fact, for biases below 5 V or above 7 V, the reverse dark current is higher than the forward dark current, in contrast to the fact that the reverse photocurrent is always much lower than the forward photocurrent. A simple explanation for the more symmetrical I-V characteristics observed in this thin barrier QWIP is the more uniform dopant distribution in the barrier as the barrier thickness approaches the dopant diffusion length.

With the coherent tunneling mechanism, the transmission coefficient can be made close to one even though the excited state is below the barrier height. Apart from the lifetime consideration, this means that the coupled well structure may have the same responsivity as the extended-state detector without the additional nonideal current, an effect very similar to bandpass filtering. Therefore, the present structure is potentially more sensitive than an extended-state detector at a specific wavelength. By adding more thin barriers to each period, the e'_1 state forms a miniband, which can conduct photoelectrons preferentially (51). Such a structure may also have the same advantages over the extended detectors.

C. MULTICOLOR QWIP WITH THICK BARRIERS

For thin barrier QWIPs such as devices I and J, several wavelengths are detected at the same time, which may not be suitable for all applications. Furthermore, the quantization of voltage drop in the high field domain limits the wavelength tunability of the structure. The finite tunneling dark current will also reduce the detectivity of the detector at low temperatures without an IHET structure. In these respects, coupled well structures with thick barriers provide complementary alternative to voltage-tunable multi-color detection. It also turns out that such a structure detects only one color at each bias.

One example of this type of detector is given by device K. Its structure is the same as that of device J except that the thickness of the barriers dividing the coupled quantum-well units is extended from 154 to 500 Å, and the number of units in this device is 30. Figure 50 shows the normalized photoresponses at three different biases. The observed wavelengths for $2r$, $3f$, and $3r$ transitions are very close to that of device J as expected because the thicker barriers should not change the energy of the bound states. On the other hand, the transitions to the extended states become much less prominent in the present device. As the barrier thickness increases, the quantized levels above the barriers become denser and the oscillator strength spreads among these levels. The resultant transition to each individual extended state becomes much weaker. The corresponding photocurrent from these states then forms a broad tail immediately above the barrier height. This feature is most pronounced when the applied bias is small and is in the reverse direction as indicated in Fig. 50. At $V_r = 200$ mV, although the $2r$ transition is stronger ($f_2 = 0.74$, $f_3 = 0.26$), the corresponding photocurrent is not observed because of the small

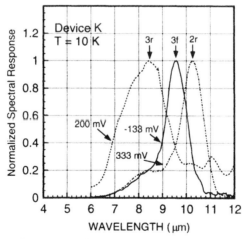

FIG. 50. The spectral response of device K under different biasing conditions. The peaks are identified to be $2r$, $3f$, and $3r$ transitions, with peak responsivity of 0.09, 0.03, and 0.0022 A/W, respectively.

p_2. The major contribution to the photocurrent in this case is from the $3r$ transition. However, because of the relatively small f_3 and the confined nature of the state, the extended-state transitions become significant, which leads to a relatively broad detection peak under this bias. At a large bias V_r of 333 mV, the effective barrier height is much lower, making p_2 and p_3 close to unity. In this case, the photocurrent is dependent only on the oscillator strength. With a further increased f_2 at this bias, the $2r$ transition is in fact dominant. From the observed ratio of photocurrent between $2r$ and $3r$ transitions, f_2 is deduced to be a factor of 5 larger than f_3 at this bias. A similar situation occurs under forward bias; the large $f_3 = 0.85$ at $V_f = -133$ mV makes the extended-state transitions relatively insignificant, resulting in a narrow lineshape for the $3f$ transition peak.

From the above discussion, device K can serve as a three-color detector within the 8- to 12-μm range. Different from devices with thin barriers, it detects only one color at a time and the color can be changed by changing the bias. This unique feature is a result of adopting thick quantum-well barriers, which broaden the extended-state transitions and make the tunneling probability of bound states highly dependent on bias. With the reduction in tunneling dark current, the device is BLIP at 60 K for forward bias up to -4 V, and reverse bias up to 10 V. At this temperature, the D^* is about 10^{10} cm $\sqrt{\text{Hz}}$ /W.

D. MULTICOLOR IHET

In this section, we discuss the function of an IHET in tailoring the optical properties of a QWIP. By adopting different filter structures or by applying a collector voltage V_c, the pass-band energy of a filter can be changed and photoelectrons of different energies can be collected. As a result, the spectral width and the cutoff wavelength of the detector can be controlled by the filter design or by changing V_c. In the case where the spectral response of the associated QWIP has more than one peak, a bandpass filter can also be used to intercept a particular photoelectron peak by adjusting V_c and can achieve voltage-tunable multicolor infrared detection. Because only a specific range of electron energy is accepted, the bandpass filter also increases the background limited temperature and the detectivity of the QWIP by filtering away part of the dark current as well as the unwanted photocurrent.

The example is device L (12), which consists of 30 periods of GaAs/ AlGaAs quantum wells with the Al molar ratio x in each barrier varied in three steps. The barrier widths of each step are 160.8, 166.7, and 172.9 Å with x equal to 0.28, 0.305, and 0.33, respectively. The wells are 50-Å thick and the doping is 1.2×10^{18} cm^{-3}. The base layer consists of a 300-Å $In_{0.15}Ga_{0.85}As$ layer and a 200-Å GaAs layer. The filter is a bandpass filter which contains a 40-Å ($x = 0.3$) barrier, a 50-Å GaAs well, another 40-Å ($x = 0.3$) barrier, and a 2000-Å $Al_{0.25}Ga_{0.75}As$ barrier. The quantum wells are clad between a top (1.1 μm) GaAs layer as the collector contact and a bottom (6000 Å) GaAs layer as the emitter contact. All of the contact layers are doped with $n^+ = 1.0 \times 10^{18}$ cm^{-3}. The energy-band structure of the device is shown in Fig. 51.

Because of the lack of parity symmetry in these step barriers, optical transitions to all of the excited states are allowed. Using the transfer-matrix method and accounting for band nonparabolicity, four bound states are found in the well. The calculated ground state E_1 is equal to 73 meV.

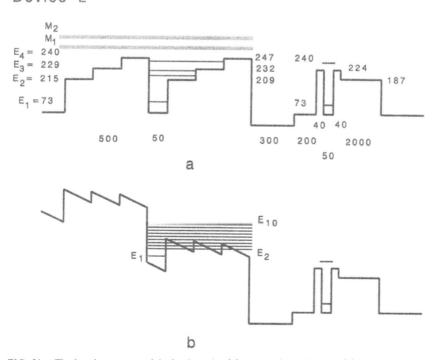

FIG. 51. The band structure of device L under (a) zero-emitter bias and (b) forward-emitter bias. The numerical values given are either energy values in meV or layer thickness in Å.

The three excited bound states are $E_2 = 215$ meV, $E_3 = 229$ meV, and $E_4 = 240$ meV. Neglecting many-body effects, the corresponding transition wavelengths λ are 8.7, 7.9, and 7.4 μm for electron transitions from the ground state to these three excited states. Besides the bound states, the first and the second minibands, M_1 and M_2, are located at 264 and 292 meV, respectively, providing $\lambda = 6.6$- and 5.7-μm detection. For the bandpass filter, the first and second pass bands are at 73 and 240 meV, respectively, if band nonparabolicity is taken into account. However, the first pass band is blocked by the 2000-Å $Al_{0.25}Ga_{0.75}As$ barrier, so that the hot electrons from the QWIP can pass only the second band. At a finite V_c, the pass-band energy can be changed as shown in Fig. 52. Using this property, the filter can be used to select different colors from the QWIP or control the detection width.

The absorption peaks of individual transitions to the excited bound states are not resolved in the Fourier Transform Infrared Spectroscopy

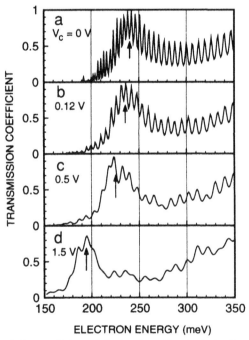

FIG. 52. The transmission coefficient of the bandgap filter in device L at different values of V_c.

(FTIR) measurement, but appear as a broad peak centered at 8.0 μm with a 2-μm linewidth, consistent with the calculated transition energies. To characterize the detector spectral response, a 45° angle was polished on the substrate to couple infrared radiation from a 1000-K glow-bar monochromator source. The power from the monochromator was calibrated using a HgCdTe detector, and the photocurrent was monitored by a current amplifier. Emitter spectral response at $T = 10$ K and $V_e = -0.5$ V is shown in Fig. 53a. Two photocurrent peaks are observed at 6.5 and 5.5 μm. These two peak wavelengths are much shorter than those expected from the bound-to-bound state transitions, but approximately match the bound to M_1 and M_2 transitions. The slightly shorter experimental value can be attributed to the depolarization effect (52, 53). Even though the oscillator strengths of these two minibands are very small (0.056 and 0.028), which result in a very small measured responsivity, the fact that the photoexcited electrons from these two states are free electrons enables them to dominate the photoresponse at a small V_c.

When V_e increases, due to the stepped nature of the barriers, the effective barrier height rapidly decreases with increasing V_c. The bound states E_2 to E_4 are sequentially exposed above the barrier and become extended, thus shifting the spectral response to longer wavelengths. On the

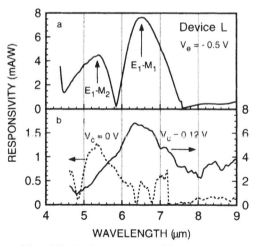

FIG. 53.   The responsivity of device L measured at (a) the emitter and (b) the collector at $V_c = -0.5$ V. Under this bias, the emitter detects $\lambda = 5.4$ and 6.5 μm simultaneously. At the collector, the response consists of only a single peak at each V_c. By changing V_c from 0 to 0.12 V, the detection peak shifts from 5.3 to 6.5 μm.

other hand, the stepped barrier forms a double-triangular-well structure under the influence of a field and creates additional bound states localized mainly in the barrier. Together with the original quantum well, the unit forms a triple-quantum-well structure as depicted in Fig. 51b. Since the oscillator strengths of the new bound states are small, they do not contribute to photocurrent significantly. Figure 54 shows the normalized emitter responsivity at $V_e = -4$ V as the dashed curve. It is peaked at $\lambda = 7.6$ μm, which corresponds to the E_1 to E_8 transition, $8f$. In the global-level scheme E_6, E_7, and E_8 are evolved from the original three excited bound states at zero bias. At $V_e = -4$ V, they become extended but still retain large oscillator strengths because of the large wave amplitudes remaining inside the quantum well. Consequently, they contribute most to the photocurrent. The arrows in Fig. 54 indicate the calculated transition wavelengths to the excited levels, in general agreement with the small features in the emitter spectral response. In this figure, E_5 is the highest bound state; its photocurrent peak could have been modified by the tunneling probability and hence its absorption wavelength appears to be shorter than that predicted. In Fig. 54, the individual peaks from each

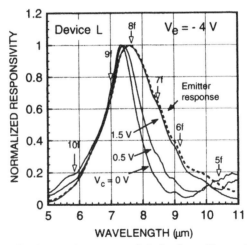

FIG. 54. The normalized spectral response of device L at $V_e = -4$ V for the emitter (dashed curve) and the collector at different V_c (solid curves). By increasing V_c from 0 to 1.5 V, the spectral width of the collector increases from 1.1 to 1.9 μm.

transition are not resolved, indicating that the level spacings are smaller than the widths of the levels in this structure. Instead, the closely spaced levels contribute to a relatively wide detection peak with a width of 2 μm. At $\lambda = 7.6$ μm, the responsivity is found to be 1.6 A/W, much larger than that at -0.5 V.

Next, the collector photocurrent spectrum was characterized at different values of V_e and V_c. Figure 53b shows the collector responsivity at $V_e = -0.5$ V for two different values of $V_c = 0$ and 0.12 V. As shown in Fig. 53a, there are two emitter photocurrent peaks at this V_e, in contrast to a single peaked response at the collector for each V_c. At $V_c = 0$ V, the filter allows mostly the M_2 photoelectrons passing to the collector. Under a positive V_c, the pass band is lowered and the filter selects mostly the M_1 photoelectrons. This experimental result demonstrates the wavelength selectivity of the bandpass filter. Nevertheless, after filtering, each detection peak is substantially broader than the individual emitter peaks. One reason for this is the relatively wide pass band of present filter ($\Gamma = 40$ meV) so that a wide range of photoelectrons is being collected. In addition, as discussed in Section IV, because of the energy relaxation in the base, the photoelectrons of different wavelengths partially overlap with each other, leading to a spectral width larger than the width of the filter pass band.

Because of energy relaxation in the base associated with hot-electron energy, there is a specific population of photoelectrons from different wavelengths, and this population is critically dependent on the relative position between the filter sampling energy and the energy of hot-electron distribution peak. As seen in Figs. 52a and b, the filter pass band is downshifted by 5 meV when V_c is increased from 0 to 0.12 V, yet the detection peak has been displaced by 29 meV. The large-wavelength shift is explained by the asymmetric energy broadening in the base. There are more electrons losing substantial energy than those gaining energy. At $V_c = 0$ V, the pass band lies just above the M_1 photoelectron peak; therefore, it receives mostly the energy relaxed M_2 photoelectrons. By lowering the pass band slightly, it is able to receive the main peak of M_1 photoelectron distribution, as well as the distribution tail of M_2 photoelectrons, thereby shifting the detection peak dramatically.

At $V_e = -4$ V, there is only one wide peak injected from the emitter. Shown in Fig. 54 are the measured collector spectral responses at different values of V_c. At $V_c = 0$, 0.5, and 1.5 V, the cutoff wavelengths λ_c were measured to be 7.8, 8.1, and 8.7 μm, respectively. Relative to the $V_c = 0$ V, the shifts of the cutoff energy are 6 and 10 meV for $V_c = 0.5$ and 1.5 V, respectively. On the other hand, the theoretical calculation shows that the pass band should be shifted by 15 and 30 meV, respectively, as shown in Figures 52c and d, almost triple the experimental values. This situation is the opposite of the lower V_c case, but is consistent with the present physical picture. Only in this case, the filter pass band is aligned with the short-wavelength photoelectrons at $V_c = 0$ V. It takes a large shift in the pass-band energy to collect most of the energy-relaxed long-wavelength electrons so that the full emitter spectral response is restored. In spite of the energy relaxation, the cutoff wavelength of a QWIP can still be changed by the filter as shown in Fig. 54.

By comparing the detector dark current and the 300-K background-window photocurrent with a 36° field of view, the IHET structure is shown to increase the background limited temperature T_b and is BLIP at a higher bias. For example, T_b at $V_c = -0.5$ V are 77 and 90 K for the emitter and the collector, respectively. At 77 K, the range of V_c at which the device is BLIP also increases from -1 to -1.7 V using the filter.

D^* for the emitter and the collector are calculated from the quantum efficiency and the optical gain obtained from the FTIR measurement. A reduction of the dark current by two orders of magnitude increases the collector detectivity. At $V_c = -0.5$ V, $V_c = 0$ V, and $T = 77$ K, the D_c^* and D_c^* are 2.5×10^{10} and 6.7×10^{10} cm $\sqrt{Hz}$/W, respectively. At $V_c = -4$ V, they are 1.9×10^{10} cm $\sqrt{Hz}$/W for the emitter and 2.2×10^{10} cm $\sqrt{Hz}$/W for the collector.

In summary, several multicolor QWIP structures have been discussed. Using strongly coupled asymmetric quantum wells, one can detect several colors simultaneously or separately, demonstrating their potential in multicolor infrared detection. Certainly, with the versatility of bandgap engineering, other structural schemes can provide similar detection capability, and they should be examined. We also describe a voltage-tunable multicolor hot-electron transistor. The bandpass filter can be used to select a detection wavelength when the QWIP has more than one detection peak, or to control the cutoff wavelength when the QWIP is a broad-band detector. In selecting the color, the filter also reduces the dark current and the unused photocurrent simultaneously, and hence increases the device detectivity. Using different QWIP parameters, the present filter design should be able to tune the detection wavelengths either within an atmospheric window or between the two atmospheric windows. Common to all multicolor detection tuned by an applied voltage, the detector current level at different biases can be significantly different, which imposes a demanding requirement on the readout design.

VI. Conclusion

In this chapter, we have summarized the optical characteristics of QWIPs. By adjusting the sample parameters, one can tune the detection wavelength from 5 to over 25 μm. The detection bandwidth can also be chosen to suit an application. With these unique optical characteristics, QWIPs should be able to widen the utility of infrared detection. The coverage of the 3- to 5-μm atmospheric window can also be extended by introducing InGaAs material in the quantum wells, thus making the technology applicable to the entire midinfrared range.

We have introduced three spectroscopies to study the energy structures and the conduction mechanisms of quantum-well structures. They are the thermally and optically stimulated hot-electron spectroscopies and infrared hot-electron tunneling spectroscopy. The same techniques should also be useful in studying other material properties such as hot-electron transport, defects, level structure of p-type quantum wells, and band-edge alignment of new materials.

We described some of the improvements made by transforming QWIPs into IHETs. The main function of an IHET is to reduce the dark current as well as the background photocurrent of a QWIP so that it can be operated at higher temperatures. If quantum efficiency can be increased to unity through a better light-coupling scheme, an IHET with a 10-μm cutoff is estimated to be background limited at 88 K with a D* of 2 × 10^{11} cm $\sqrt{\text{Hz}}$ /W, a competitive value even compared with HgCdTe detectors. Of course, the ultimate performance of QWIPs and IHETs need not be limited by these values because QWIP technology is based on well-defined physical principles; further scientific and technological advances will certainly improve these values. Other advantages of QWIP technology include high material uniformity, good compatibility with other III–V manufacturing technologies, high radiation hardness, customized detector properties and multicolor detection capability. These advantages make the technology a major contender in high-resolution infrared imaging.

There are several important topics being omitted in this chapter. Since QWIPs do not absorb light at normal incidence, coupling light into the detectors is not a straightforward issue. Although there are several coupling schemes using diffraction gratings, research still needs to be done to improve the quantum efficiency with regard to reliability and simplicity. We also did not discuss the noise properties of QWIPs and IHETs. Because of their unconventional structures, the noise properties are by no means the same as the classical noise that we have assumed in this chapter. However, the difference does not affect our conclusion significantly. In this chapter, we also pay little attention to the base of an IHET. In fact, the properties of the base have a larger effect on the detector characteristics than the collector. Since the carriers maintain local equilibrium throughout the QWIP structure, the electron transfer ratios are critically dependent on the energy and momentum relaxation in the base. Therefore, base structure optimization is an important part of the detector research. It turns out that the most important energy-relaxation mechanisms are the longitudinal optical phonon emission and plasmon emission. While phonon emission is fixed by nature, one can actually reduce plasmon emission by partitioning the base into a GaAs layer and an InGaAs layer. Due to the smaller band gap of InGaAs, the doped electrons in the GaAs layer transfer to the InGaAs layer, thus eliminating plasmon emission in the GaAs layer. The increased electron concentration in the InGaAs layer will not increase plasmon emission proportionally because the emission rate is proportional to $\sqrt{n_e}$, rather then n_e, where n_e is the electron density. In a highly doped base, this reduction is significant. The ordering of these two material layers relative to the QWIP is also important because of the difference in the satellite valleys. A detailed discussion of the base properties, however, is not the focus of this article.

References

1. B. F. Levine, K. K. Choi, C. G. Bethea, J. Walker and R. J. Malik, *Appl. Phys. Lett.* **50**, 1092 (1987).
2. K. K. Choi, B. F. Levine, C. G. Bethea, J. Walker and R. J. Malik, *Appl. Phys. Lett.* **50**, 1814 (1987).
3. L. C. West and S. J. Eglash, *Appl. Phys. Lett.* **46**, 1156 (1985).
4. K. K. Choi, M. Dutta, P. G. Newman, L. Calderon, W. Chang and G. J. Iafrate, *Phys. Rev.* B **42**, 9166 (1990).
5. M. Heiblum, D. C. Thomas, C. M. Knoedler and M. I. Nathan, *Appl. Phys. Lett.* **47**, 1105 (1985).
6. A. F. J. Levi, T. R. Hayes, P. M. Platzman and W. Wiegmann, *Phys. Rev. Lett.* **55**, 2071 (1985).
7. K. K. Choi, M. Dutta, P. G. Newman, M.-L. Saunders and G. J. Iafrate, *Appl. Phys. Lett.* **57**, 1348 (1990).
8. K. K. Choi, L. Fotiadis, M. Taysing-Lara, W. Chang and G. J. Iafrate, *Appl. Phys. Lett.* **59**, 3303 (1991).
9. K. K. Choi, M. Z. Tidrow, M. Taysing-Lara, W. H. Chang, C. H. Kuan, C. W. Farley and F. Chang, *Appl. Phys. Lett.* **63**, 908 (1993).
10. K. K. Choi, M. Taysing-Lara, L. Fotiadis, W. Chang and G. J. Iafrate, *Appl. Phys. Lett.* **59**, 1614 (1991).

11. C. Y. Lee, M. Z. Tidrow, K. K. Choi, W. H. Chang, L. F. Eastman, F. J. Towner and J. S. Ahearn, *Appl. Phys. Lett.* **65**, 443 (1994).
12. M. Z. Tidrow, K. K. Choi, C. W. Farley and F. Chang, *Appl. Phys. Lett.* **65**, 2996 (1994).
13. C. H. Kuan, K. K. Choi, W. H. Chang, C. W. Farley and F. Chang, *Appl. Phys. Lett.* **64**, 238 (1994).
14. K. K. Choi, *J. Appl. Phys.* **73**, 5230 (1993).
15. M. Helm, F. M. Peeters, F. DeRosa, E. Colas, J. P. Harbison and L. T. Florez, *Phys. Rev. B* **43**, 13,983 (1991).
16. M. Helm, W. Hilber, T. Fromherz, F. M. Peeters, K. Alavi and R. N. Pathak, *Phys. Rev. B* **48**, 1601 (1993).
17. K. K. Choi, *in* "Semiconductor Interfaces, Microstructures and Devices: Properties and Applications" (Z. C. Feng, ed.), p. 21, Institute of Physics Publishing, Philadelphia, 1993.
18. K. K. Choi, M. Taysing-Lara, P. G. Newman, W. Chang and G. J. Iafrate, *Appl. Phys. Lett.* **61**, 1781 (1992).
19. B. F. Levine, C. G. Bethea, K. K. Choi, J. Walker and R. J. Malik, *J. Appl. PHys.* **64**, 1591 (1988).
20. B. F. Levine, G. Hasnain, C. G. Bethea and Naresh Chand, *Appl. Phys. Lett.* **54**, 2704 (1989).
21. B. F. Levine, C. G. Bethea, G. Hasnain, V. O. Shen, E. Pelve, R. R. Abott and S. J. Hsieh, *Appl. Phys. Lett.* **56**, 851 (1990).
22. K. K. Choi, B. F. Levine, R. J. Malik, J. Walker and C. G. Bethea, *Phys. Rev. B* **35**, 4172 (1987).
23. B. F. Levine, *J. Appl. Phys.* **74**, R1 (1993).
24. K. K. Choi, L. Fotiadis, P. G. Newman and G. J. Iafrate, *Appl. Phys. Lett.* **57**, 76 (1990).
25. E. Pelve, F. Beltram, C. G. Bethea, B. F. Levine, V. O. Shen, S. J. Hsieh and R. R. Abbott, *J. Appl. Phys.* **66**, 5656 (1989).
26. G. M. Williams, R. E. DeWames, C. W. Farley and R. J. Anderson, *Appl. Phys. Lett.* **60**, 1324 (1992).
27. K. K. Choi, L. Fotiadis, M. Taysing-Lara, W. Chang and G. J. Iafrate, *Appl. Phys. Lett.* **60**, 592 (1992).
28. K. K. Choi, M. Dutta, R. P. Moerkirk, R. Kuan and G. J. Iafrate, *Appl. Phys. Lett.* **58**, 1533 (1991).
29. K. K. Choi, M. Z. Tidrow, M. Taysing-Lara and W. H. Chang, *in* "Quantum Well Intersubband Transition Physics and Devices" (H. C. Liu, B. F. Levine and J. Y. Andersson, eds.), p. 151, Kluwer Academic Publishers, Holland, 1994.
30. C. Y. Lee, K. K. Choi, R. P. Leavitt and L. F. Eastman, *Appl. Phys. Lett.* **66**, 90 (1995).
31. L. J. Kozlowski, *in* "Quantum Well Intersubband Transition Physics and Devices" (H. C. Liu, B. F. Levine and J. Y. Andersson, eds.), p. 43, Kluwer Academic Publishers, Holland, 1994.
32. W. A. Beck, J. W. Little, A. C. Goldberg and T. S. Faska, *in* "Quantum Well Intersubband Transition Physics and Devices" (H. C. Liu, B. F. Levine and J. Y. Andersson, eds.), p. 55, Kluwer Academic Publishers, Holland, 1994.
33. J. L. Vampola, *in* "The Infrared and Electro-Optical Systems Handbook," (J. S. Accetta and D. L. Shumaker, eds.), *Proc. SPIE* **3**, 285 (1993).
34. G. Hasnain, B. F. Levine, C. G. Bethea, R. A. Logan, J. Walker and R. J. Malik, *Appl. Phys. Lett.* **54**, 2515 (1989).
35. J. Y. Andersson and L. Lundqvist, *J. Appl. Phys.* **71**, 3600 (1992).
36. C. Y. Lee, M. Z. Tidrow, K. K. Choi, W. H. Chang and L. F. Eastman, *J. Appl. Phys.* **75**, 4731 (1994).

37. A. A. Reeder, J.-M. Mercy and B. D. McCombe, *IEEE J. of Quantum Electronics* **24**, 1690 (1988).

38. Z. R. Wasilewski, H. C. Liu and M. Buchanan, *J. Vac. Sci. Technol.* B **12**, 1273 (1994).

39. D. G. Deppe, N. Holonyak, W. E. Plano, V. M. Robbins, J. M. Dallesasse, K. C. Hsieh and J. E. Baker, *J. Appl. Phys.* **64**, 1838 (1988).

40. Ph. Jansen, M. Meuris, M. Van Rossum and G. Borghs, *J. Appl. Phys.* **68**, 3766 (1990).

41. J. E. Cunningham, M. Williams, T. Chiu, W. Jan and F. Storz, *J. Vac. Sci. Technol.* B **10**, 866 (1992).

42. P. A. Folkes, *J. of Appl. Phys.* **68**, 6279 (1990).

43. A. Köck, E. Gornik, G. Abstrieter, G. Böhm, M. Walther and G. Weimann, *Appl. Phys. Lett.* **60**, 2011 (1992).

44. I. Gravé, A. Shakouri, N. Kuse and A. Yariv, *Appl. Phys. Lett.* **60**, 2062 (1992).

45. B. F. Levine, C. G. Bethea, B. O. Shen and R. J. Malik, *Appl. Phys. Lett.* **57**, 383 (1990).

46. Y. H. Wang, Sheng S. Li and Pin Ho, *Appl. Phys. Lett.* **62**, 93 (1993).

47. K. K. Choi, B. F. Levine, C. G. Bethea, J. Walker and R. J. Malik, *Phys. Rev.* B **39**, 8029 (1989).

48. K. K. Choi, B. F. Levine, C. G. Bethea, J. Walker and R. J. Malik, *Phys. Rev. Lett.* **59**, 2459 (1987).

49. M. Z. Tidrow, K. K. Choi, C. Y. Lee, W. H. Chang, F. J. Towner and J. S. Ahearn, *Appl. Phys. Lett.* **64**, 1268 (1994).

50. B. Ricco and M. Ya. Azbel, *Phys. Rev.* B **29**, 1970 (1984).

51. L. S. Yu and S. S. Li, *Appl. Phys. Lett.* **59**, 1332 (1991).

52. T. Ando, A. B. Fowler and F. Stern, *Rev. Mod. Phys.* **54**, 437 (1982).

53. M. O. Manasreh, F. Szmulowicz, T. Vaughan, K. R. Evans, C. E. Stutz and D. W. Fisher, *Phys. Rev.* B **43**, 9996 (1991).

Quantum-Well Structures for Photovoltaic Energy Conversion

JENNY NELSON

Blackett Laboratory,
Imperial College of Science, Technology and Medicine,
London SW7 2BZ, United Kingdom

I. Introduction

The direct conversion of sunlight into electricity by the photovoltaic effect is one of the most important energy resources for the future. Photoconversion is readily achieved by charged solid-state (or solid-liquid) interfaces such as the semiconductor $p-n$ junction. Commercial silicon solar cells deliver power with a typical conversion efficiency of 16%, although efficiencies of up to 25%[1] have been achieved in more costly III–V semiconductor single-junction cells (1).

Higher efficiencies are desirable for the implied reductions in material requirements and cost. However, the conversion efficiency available with a single band gap photoconverter in a standard solar spectrum is limited to about 31% (2). This limit is imposed by the mismarriage of a broad-band solar spectrum with a single band gap photoconverter. The cell essentially reduces the energy of all photogenerated carriers to the energy of its band gap.

More efficient use of the solar spectrum requires more than one band gap. In a tandem arrangement, different solar spectral ranges are preferentially absorbed by two or more junctions of different band gaps which are electrically connected. In practice, performance has been disappointing, with the best practical efficiencies for tandem cells far below the theoretical limits for these multi-gap systems.

The quantum-well solar cell (QWSC) was initially proposed as an alternative multi-band gap photoconverter without the need for lossy electrical connections. Quantum wells included within the space charge region of a $p-n$ junction cell would act like the lower band gap cell, trapping longer wavelengths, while the original (host) cell would act like the higher gap cell of the tandem. When compared to a homogenous device without quantum wells, the QWSC would offer improved efficiencies if the extra photocurrent from the quantum wells outweighed any loss to the voltage.

Since the pioneering studies of Barnham and co-workers (3, 4) the idea

[1] In unconcentrated sunlight.

has stimulated interest from solar cell researchers worldwide. QWSC structures have been studied in the AlGaAs/GaAs, GaAs/InGaAs, InP/InGaAs, InGaP/GaAs and InP/InAsP materials systems.

Interest has been stimulated by two main questions:

1. Whether the QWSC can, *in principle*, offer power conversion efficiencies greater than the theoretical maximum for a single band gap device.
2. What *practical* advantages quantum wells can offer to real photovoltaic devices irrespective of the answer to (1). These may include extended spectral response to improve the efficiency of wide band gap solar cells; tunability of the cell's band gap to match the optimum for the solar spectrum; improved transport of photogenerated carriers through nonactive regions of the device.

Although it has been established experimentally that quantum wells can enhance the efficiency of photovoltaic conversion by wide-gap $p-i-n$ solar cells, the same result has not yet been achieved in solar cells of closer-to-optimum composition. QWSC performance is, however, evidently limited by material quality and extrapolations to ideal material are difficult. Efforts to improve material quality continue. To date the most efficient QWSCs to be produced are an AlGaAs/GaAs 50-quantum-well device which has been calibrated 14% efficient in standard test conditions (5). Some as yet uncalibrated GaAs/InGaAs 15-quantum-well devices should be more efficient (6).

The layout of this paper is as follows. Section II covers the principles of photovoltaic conversion in semiconducting diodes. Section III explains how quantum wells affect the spectral response and photocurrent of the cell and presents results of measurement and modelling. Section IV discusses QWSC dark-current and voltage behavior, presenting results in various materials systems. In Section V, different approaches to the calculation of the limiting efficiency of the QWSC are reviewed. Section VI deals with future studies of the QWSC and discusses other novel ways of using quantum wells for photoconversion.

II. Principles of Photovoltaics

A. THE $p-n$ JUNCTION SOLAR CELL

Efficient photovoltaic conversion requires strong optical absorption over the solar spectrum, a means of separating photogenerated carriers and lossless transport toward external electrical contacts.

In the standard model of a solar cell, the semiconductor $p-n$ junction (Fig. 1), the doping gradient between emitter and base creates a depleted or *space charge* region (SCR). The permanent electric field across this region provides the means of separating carriers. Illumination thus produces a photocurrent, J_p, which opposes the dark current, J_d, of the junction in forward bias. This means that, up to some threshold *open-circuit voltage* V_{oc} where J_d equals J_p, the junction will power an electrical load and so convert light into electricity. Maximum power is delivered at the operating point V_m, J_m (Fig. 2). The *short-circuit current* J_{sc} is the value of J_p at zero bias (7).

B. PHOTOCURRENT

J_{sc} depends on the form of the incident spectrum and the spectral response of the solar cell. In space, the solar spectrum resembles the spectrum of a black body at 5800 K. The Earth's atmosphere attenuates the spectrum at certain wavelengths to produce a characteristic terrestrial spectrum. At temperate latitudes, the standard for irradiation is the air

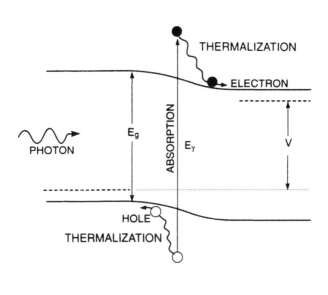

FIG. 1.   Energy-band diagram of a $p-n$ junction solar cell under illumination.

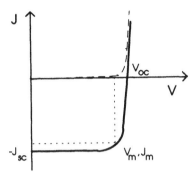

FIG. 2. Current–voltage characteristic of a solar cell under illumination, showing the short-circuit current, J_{SC}, open-circuit voltage, V_{oc}, and the maximum power point V_m, J_m. The dashed line represents the dark current $J_d(V)$.

mass 1.5 (AM1.5) solar spectrum with a power density of about 850 Wm^{-2} for direct, or 1000 Mw^{-2} for total (direct and diffuse), irradiation (2, 8). Figure 3 illustrates the terrestrial air mass 1.5 solar spectrum, the extraterrestrial (air mass 0) solar spectrum, and a 5800-K black-body radiation spectrum. For calculations of theoretical efficiency it is sometimes convenient to use the black-body approximation for solar flux, b_S. For sunlight incident on a flat plate

$$b_S(E, T_S) \, dE = \frac{2\pi}{h^3 c^2} \left(\frac{E^2}{e^{E/kT_S} - 1} \right), \tag{1}$$

with $T_S = 5800$ K.

Spectral response $SR(E)$ is defined as the number of carrier pairs collected at the contacts per incident photon of energy E. The number depends on the optical and transport properties of the material and the device geometry. Since photons are not absorbed below the semiconductor band gap E_g, this fixes a low-energy cutoff to the spectral response. At high energies, response is limited by recombination processes. Short-wavelength photons are absorbed near the surface, where the probability of carriers recombining through impurities either at the surface or within the undepleted emitter layer are high. At intermediate wavelengths, photons are absorbed in or near the space charge region and spectral response is limited mainly by the optical depth of the cell, i.e., by its reflectivity and thickness (Fig. 4).

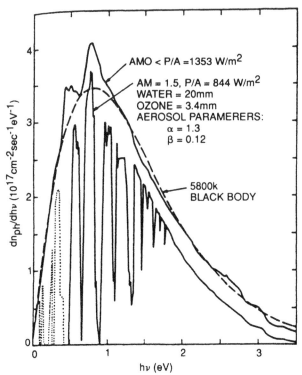

FIG. 3. The terrestrial air mass 1.5 solar spectrum, the extraterrestrial (air mass 0) solar spectrum, and a 5800-K black-body radiation spectrum.

Integrating over photon energies, the photocurrent is given by

$$J_{SC} = q \int SR(E) b_S(E, T_S) \, dE \qquad (2)$$

where q is the electronic charge. High photocurrents thus require a high spectral response which is well matched to the solar spectrum.

C. DARK CURRENT

At zero bias, in the dark, the diode is at equilibrium. Diffusion of carriers across the junction is balanced by drift currents in the other direction, and thermal generation is balanced by recombination.

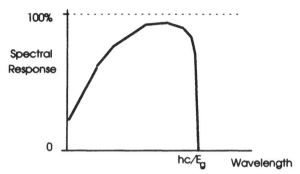

FIG. 4. Spectral response as a function of wavelength for a good solar cell of band gap E_g.

At forward bias V carriers are injected into the SCR, increasing the electron and hole populations, n and p. Both radiative and nonradiative recombination rates increase with the excess electron–hole product, $(np-n_i^2)$, where n_i is the intrinsic carrier density, and produce a net forward, or *dark*, current, $J_d(V)$. It is usual to express the nonradiative component of J_d as the sum of two parts: an injection current, $J_{inj}(V)$ which is sometimes called the diffusion current, and a recombination–generation current $J_{rg}(V)$. J_{inj} comes from the undepleted emitter and base regions where minority carriers dominate the transport, and is given by the Shockley form

$$J_{inj}(V) = J_o(e^{qV/kT} - 1). \tag{3}$$

J_{rg} is due to recombinations through impurities in the SCR and varies roughly like $e^{qV/2kT}$, as shown in Section IV. The magnitude of the recombination current is determined by the density of recombination centres and carrier lifetimes in the material, and by the band gap which determines n_i. The component of J_d due to *radiative* recombination, $J_{rad}(V)$, is masked by nonradiative processes in real cells.

In practice electrical losses at the contacts produce an effective series resistance R_S and a shunt resistance R_p so that, finally, the dark current is given by

$$J_d(V) = J_{inj}(V - JR_S A)$$
$$+ J_{rg}(V - JR_S A) + J_{rad}(V - JR_S A) + \frac{V - JR_S A}{R_p A} \tag{4}$$

where A is the cell area. The different contributions to $J_d(V)$ are shown in

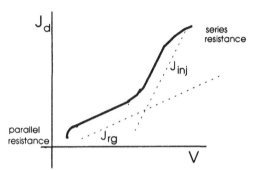

FIG. 5. Log–linear plot of the dark-current voltage characteristic of a solar cell, showing the different slopes due to recombination–generation (J_{rg}) and injection (J_{inj}) currents. Series resistance reduces $J_d(V)$ at high current levels, whereas parallel resistances are important only at very small bias.

Fig. 5. $J_d(V)$ is sometimes fitted to the form $J_{d0}e^{qV/nkT}$ where the number n is called the *ideality* factor. In direct gap III–V semiconductors the recombination–generation current dominates giving a characteristic slope of $1/2$ to the semilog plot of J_d against qV/kT, i.e., an ideality factor of 2 (9, 10).

In operation, the total current is

$$J(V) = -J_{SC} + J_d(V - JR_S A) \tag{5}$$

where the bias dependence of the photocurrent, which is usually weak, has been neglected. Power density, and hence efficiency, is maximized when

$$\frac{d}{dV}[J(V)V] = 0 \tag{6}$$

Series resistance and nonradiative recombination reduce the maximum value of the $J–V$ product and degrade power.

D. LIMITS TO EFFICIENCY

Efficient photoconversion clearly requires the maximization of absorption and the minimization of carrier recombination. Light absorption is improved in practice by various techniques. Antireflection coatings minimize reflection of light at the cell surface. Cell thickness is optimized to ensure that useful photon energies are absorbed in the active layers rather than passing through the cell. Back surface mirrors enhance absorption by trapping light in the cell. At the front surface, contact design is optimized to reduce shading losses at minimal cost to the series resistance.

Recombination losses are minimized mainly through material improvements. Good minority carrier transport reduces the injection current and a low density of recombination centers reduces nonradiative recombination in the SCR. Design features such as wide-gap window layers at the front or back surface and compositional grading are used to accelerate carriers away from lossy surfaces toward the junction.

Energy losses through the thermalization of excited carriers toward their respective band edges, through the failure to absorb light below E_g and through radiative recombination remain unavoidable and ultimately limit the power conversion efficiency.

An ideal single band gap photoconverter will thus be one which absorbs all incident light above its band gap, and which delivers one electron hole pair for every absorbed photon. Then

$$J(V) = -J_{SC} + J_{rad}(V) \tag{7}$$

and the photocurrent is the sum of absorbed solar photons

$$J_{SC} = \int_0^\infty F_\Omega a(E) b_S(E, T_S) \, dE, \tag{8}$$

where $a(E)$ is the absorptivity (i.e., the spectral response) of the cell and F_Ω is a geometrical factor which results from integrating the incident solar flux over the solid angle Ω subtended by the sun and the device surface (11, 2). The unavoidable *excess* radiative current due to spontaneous emission is the sum of emitted long-wavelength photons

$$J_{rad}(V) = \int_0^\infty F_e e(E) b(E, T_C, \Delta\phi) \, dE - \int_0^\infty F_e e(E) b(E, T_C, 0) \, dE, \tag{9}$$

where $e(E)$ is the cell emissivity and $\Delta\phi$ is the separation of the electron and hole quasi-Fermi potentials in the nonequilibrium device. Here the geometrical factor F_e is π for a one-sided flat plate emitter. The emitted flux is given by the van Roosbroeck and Shockley (12) form

$$b(E, T_C, \Delta\phi) \, dE = \frac{2\mu^2}{h^3 c^2} \left(\frac{E^2}{e^{(E - q\Delta\phi)/kT_c} - 1} \right) dE, \tag{10}$$

where μ is the refractive index of the surrounding medium.

Thermalization of carriers is implied through the cell temperature T_C. In a lossless device, carrier mobilities are infinite and the Fermi potential separation $\Delta\phi$ should be equal to the bias V across the device. The second term in Eq. (9) represents the equilibrium recombination current, J_{th}, which is equal to the total absorbed flux from the ambient medium at temperature T_C. This ensures that when no external illumination is present no current flows.

The net current is thus the difference between the two photon fluxes weighted by the probabilities of photon absorption $a(E)$ and emission $e(E)$, and integrated over the relevant geometries. Araujo and Marti (11) have shown that, for the ideal lossless device,

$$a(E) = e(E) \tag{11}$$

by the principle of detailed balance, so that

$$J_{rad}(V) = \int_0^\infty a(E)\{F_\Omega b_S(E, T_S) - F_e b(E, T_C, \Delta\phi)\} \, dE, \tag{12}$$

where the very small thermal-emission current is neglected. They find that optimum conversion is achieved for a step absorber with $a(E) = 1$ for $E > E_g$, and $a(E) = 0$ elsewhere. Conversion efficiency is then simply a function of the band gap E_g (Fig. 6).

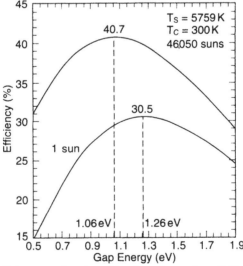

FIG. 6. Limiting efficiency as a function of band gap for a solar cell illuminated by a 5759-K black-body sun, at normal intensity (lower curve) and at the maximum concentration of 46,050 suns (upper curve). The optimum band gap shifts to lower energies under concentration. Reprinted from (13) with permission of H. S. Stephens and Associates.

At normal solar irradiation by a black-body sun at 5759 K Araujo *et al.* (*13*) predict a maximum efficiency of 30.5% at a band gap of 1.26 eV. This compares well with the limit of 31% at a band gap of 1.35 eV found by Henry (*2*) for a standard AM1.5 spectrum. Optimizing the performance of a single band gap cell is therefore a matter of choosing the right material. Gallium arsenide (GaAs) and indium phosphide (InP) have band gaps close to the optimum (1.42 and 1.35 eV, respectively, at 300 K) and are favored for high-efficiency cells. Silicon has a poorer band gap (1.1 eV) but is cheap and abundant compared to III–V materials and remains the most popular solar-cell material.

Concentration of sunlight has the effect of expanding the angular range of absorption relative to emission and improves the balance between photocurrent and radiative current so that net efficiencies are increased. For 1000 suns, a limiting efficiency of about 37% at $E_g = 1.1$ eV is predicted (*13, 2*).

The best practical efficiencies achieved to date for single-band gap cells are about 25% for a GaAs cell in standard test conditions and 27.6% under concentration (*1*). The record efficiency for silicon under standard conditions is 24% (*1*), which is close to the theoretical limit of about 27% for that band gap. Shading and series resistance account for most of the avoidable losses. Design and material improvements are not expected to improve the record greatly.

E. MULTI-BAND GAP CELLS

The limit to efficiency is due to the single band gap condition. High efficiency requires high current and high voltage. Reducing the band gap increases photocurrent but reduces the voltage by reducing the potential energy of the collected carriers.

Increasing the band gap increases the voltage but reduces current. That is the compromise represented by Eq. (12).

Better efficiencies could be obtained with more efficient use of the solar spectrum. Combining two or more cells of different band gaps into a tandem arrangement increases the amount of work done per photon. A high-band gap cell at the front of the device absorbs higher energy photons preferentially and delivers carrier pairs at a higher potential. The second, lower-gap cell receives the lower energy photons which pass through the first cell. To maximize power, the cells should be current-matched and joined in series, separated for instance by a tunnel diode (Fig. 7).

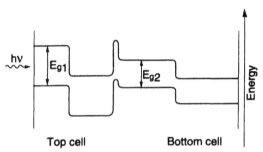

FIG. 7. Energy-band diagram of a monolithic tandem cell showing a wide band gap (E_{g1}) $p-n$ junction and a narrow band gap (E_{g2}) junction connected by a tunnel junction.

Henry (2) calculates a limiting efficiency of 50% for a two-cell tandem under concentrated light, 56% for three cells, and 72% for the impractical case of 36 cells. In practice, the best tandem-cell efficiencies are far from these limits. An indium-gallium phosphide/gallium-arsenide tandem achieved a record 29.5% in one sun, whereas in concentrated light an efficiency of 32.6% has been achieved by a gallium-antimonide/gallium-arsenide device (14, 1).

F. The Quantum-Well Solar Cell

In the quantum-well solar cell (QWSC), quantum wells of a smaller band gap material are grown within the SCR of a $p-n$ or $p-i-n$ photodiode. Normally the SCR is extended by introducing an undoped spacer layer between emitter and base. This admits more quantum wells. The structures are grown epitaxially by MBE or MOVPE and are identical to MQW structures used for other optoelectronic applications such as photodetectors (Fig. 8). In comparison with the equivalent device without wells, the *control*, the QWSC generates a higher photocurrent. If carriers could be collected at the same voltage as for the control, cell efficiency would clearly increase.

In practice, we expect quantum wells to increase carrier recombination and therefore to reduce voltage. The net effect on efficiency, then, depends on whether the benefit to current outweighs the detriment to voltage. The original calculations by Barnham and Duggan (3) showed that if voltage could be determined by the wide band gap of the host cell and photocurrent by the narrow effective band gap of the wells, then substantial efficiency enhancements would result.

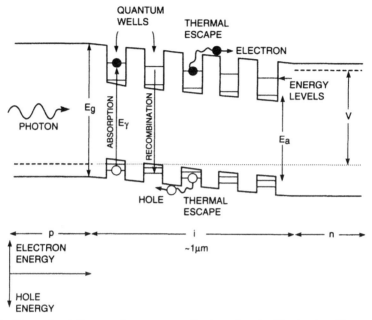

FIG. 8.   Energy-band diagram of a quantum-well solar cell under illumination. Electron-hole pairs which are generated in the quantum wells may either recombine in the well or escape to contribute to the photocurrent.

Since then, the QWSC has been studied experimentally in a variety of materials systems and investigated theoretically. It has been established that quantum wells *do* increase cell photocurrent, that they *do* degrade voltage, and that they may either increase or degrade cell efficiency. The first experimental studies of aluminum gallium arsenide ($Al_xGa_{1-x}As$)/ gallium arsenide (GaAs) test devices showed that quantum wells could substantially increase efficiency (*1*).

III. QWSC Spectral Response

The spectral response of a homogenous *p–n* or *p–i–n* device is determined by the absorption of the material and its transport properties. In Fig. 9, the spectral response for two similar *p–i–n* structures, one made from GaAs and one from aluminium gallium arsenide $Al_{0.3}Ga_{0.7}As$, are compared. For these direct gap materials, the spectral response turns on

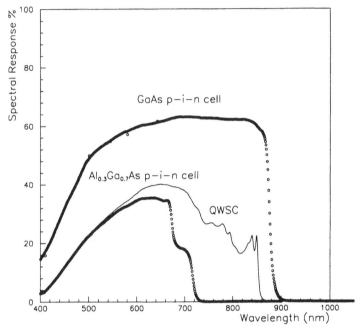

FIG. 9. Spectral response for homogenous GaAs and $Al_{0.3}Ga_{0.7}As$ p–i–n photodiodes (thick lines) and a 30-quantum-well $Al_{0.3}Ga_{0.7}As/GaAs$ p–i–n diode (thin line). Notice how the addition of wells to the $Al_{0.3}Ga_{0.7}As$ device both extends the spectral response to longer wavelengths and enhances it at energies just above the $Al_{0.3}Ga_{0.7}As$ band gap.

sharply at the band gap. The poorer response of the $Al_{0.3}Ga_{0.7}As$ device at short wavelengths is due to its poorer transport properties.

If GaAs quantum wells were added to the AlGaAs structure, we would expect the spectral response to be extended toward the GaAs absorption edge. Figure 9 shows the extra contribution made by quantum wells to the spectral response in the first experimental test (4). The samples were a pair of MOVPE grown $Al_{0.3}Ga_{0.7}As$ p–i–n test devices with 0.3-μm-thick p regions and 0.5-μm-thick undoped regions. The extended SR of the QWSC at long wavelength shows the characteristic features of quantum confinement: a blue-shifted absorption edge, sharp excitons, and a steplike spectrum. The figure also shows that the addition of quantum wells has not reduced the short-wavelength response relative to the control cell, i.e., that they have not degraded the collection efficiency.

Spectral response of the QWSC is well understood and can be fairly accurately modeled (15). At a very simple level, quantum wells enhance photocurrent by an amount which depends on the number and depth of the quantum wells. More detailed models of QWSC spectral response allow us to interpret data in terms of material and design parameters, to provide diagnostic feedback on material growth, and to optimize cell design within certain materials constraints.

The theoretical framework which follows is described in by Paxman *et al.* (*15*). Most calculations follow the same principles but are less detailed (*16, 17*).

A. CALCULATION OF SPECTRAL RESPONSE OF HOMOGENOUS SOLAR CELL

The solar cell is modeled as a $p-n$ or $p-i-n$ device with uniform layers (Fig. 10). Its spectral response is calculated by solving the semiconductor transport equations to find the photogenerated carrier densities as a function of photon energy. The solution is greatly simplified within the depletion approximation, where the SCR is assumed to be void of carriers. While the approximation is good for a highly doped $p-n$ junction, it may fail for a $p-i-n$ diode at high forward biases or for high background doping levels (*18, 19*).

The photocurrent can be expressed as the sum of three independent contributions: carriers which are generated within the depleted region and promptly collected; and carriers which are generated within each of the fieldless p and n layers and diffuse to the edges of the SCR, from where they are collected.

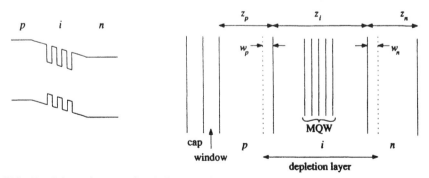

FIG. 10. Schematic energy-band diagram of a $p-i-n$ solar cell together with a layer plan showing p, i, and n regions of width z_p, z_i, and z_n and the depleted thicknesses w_p and w_n of the p and n regions. The dotted lines mark the extent of the space charge region.

For a homogenous device, the solution for the fieldless p and n regions is straightforward. For incident light of energy E and flux $b(E)$ carriers are generated at a depth z below the surface at a rate

$$G(E, z) = \alpha(E)b(E)[1 - R(E)]\exp\left(\int_0^z \alpha(E)\, dz\right) \qquad (13)$$

where $\alpha(E)$ is the absorption coefficient and $R(E)$ is the surface reflectivity. The minority carrier current and continuity equations require that

$$\frac{d^2 n}{dz^2} - \frac{(n - n_0)}{l_n^2} + \frac{G(E, z)}{D_n} = 0$$

for the electron density $n(z)$ in the p region, and

$$\frac{d^2 p}{dz^2} - \frac{(p - p_0)}{l_p^2} + \frac{G(E, z)}{D_p} = 0 \qquad (14)$$

for the hole density $p(z)$ in the n region. Here l_n, D_n and n_0 (l_p, D_p and p_0) are the minority carrier diffusion length, diffusion constant, and equilibrium concentration for the electrons (holes). Solving Eqs. (14) subject to the boundary conditions that the excess carrier concentrations vanish at the edges of the SCR, and that the recombination current at the front (back) surface is described by a surface recombination velocity $S_n(S_p)$, yields analytic solutions for $n(z)$ and $p(z)$ (9). Then the contributions to J_{SC} from the undepleted n and p layers are the diffusive currents

$$J_n = qD_n \frac{dn}{dz}$$

and

$$J_p = -qD_p \frac{dp}{dz} \qquad (15)$$

evaluated at the edges of the SCR where prompt collection is assumed.

The contribution from the SCR is simply the total net carrier generation rate in that region. For a p–i–n

$$J_{SCR} = q(1 - R)e^{-\alpha(x_p - w_p)}(1 - e^{-\alpha(w_p + x_i + w_n)}) \qquad (16)$$

where w_p and w_n are the depleted widths of the n and p layers, calculated by solving Poisson's equation, and z_p and z_i are the widths of the p and i layers (Fig. 10). The first exponential factor is due to attenuation of light in the p layer. In cases where the background doping is too high or the bias across the junction too low, the i layer may not be fully depleted and the relative contributions from fieldless and depleted regions must be adjusted. Finally the spectral response is given by

$$SR(E) = \frac{J_\mathrm{p} + J_\mathrm{n} + J_\mathrm{SCR}}{q\,b(E)} \tag{17}$$

In Fig. 11, the spectral response of the AlGaAs $p-i-n$ diode of Figure 9 is modeled. Values for l, D and $\alpha(E)$ were taken from the literature (20, 21) and S_n was fitted to the data. The very good agreement gives confidence in this rather simple treatment of the $p-i-n$ solar cell.

B. Calculation of QWSC Spectral Response

In this scheme, the introduction of quantum wells means replacing some of the absorbing material in the SCR with quantum wells. Photocurrent

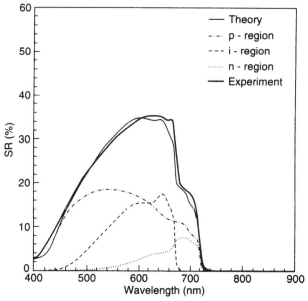

FIG. 11.   Measured and calculated spectral response for the $Al_{0.3}Ga_{0.7}As$ $p-i-n$ photodiode of Fig. 9. The broken lines show the separate contributions from the p, i, and n layers. The large p contribution shows that a lot of light is absorbed in the relatively lossy p layer before reaching the active region. The shoulder in the absorption edge indicates different Al fractions in doped and undoped layers, resulting from different growth rates.

from the quantum wells will be the sum of carriers generated in the wells, weighted by some probability η_{esc} for escape from the well.

1. Quantum-Well Absorption Spectrum. The absorption spectrum of the quantum well is calculated using a quantum mechanical model. For an unbiased quantum well, confinement quantizes the motion of electrons and holes in the growth direction, restricting their energies to a set of 2-D subbands. For AlGaAs/GaAs quantum wells, the energy spectrum is well described by a four-band Kane model in the envelope function approximation (*22, 23*). The subband-edge energies for either carrier are found by solving the resulting one-dimensional effective mass equations for each carrier type in a finite square well of width L and depth V_e (for electrons) or V_h (for holes). For quantum wells in III–V semiconductors, where the valence band degeneracy is lifted, light- and heavy-hole states must be considered separately (see Fig. 12). The joint density of states (JDoS) has a characteristic steplike form, with one step at the transition energy for each electron-hole subband pair.

In a quantum-well p–i–n device at zero bias, the quantum wells are subject to a small electric field (1–3 V/μm) and carriers are not strictly confined. The effect is to red-shift the subband-edge energies (the quantum confined Stark Effect) and broaden the excitonic peaks slightly. For all the structures modeled here, the flat-band approximation is good to within the experimental error.

The Coulombic interaction between electrons and holes creates an additional set of hydrogenic bound states below each subband edge. These excitonic states are stronger in quantum wells than in bulk material on account of the greater confinement of electron and hole. The excitonic

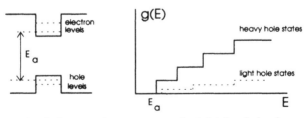

FIG. 12. Energy-band diagram of a quantum well at flat band showing confined states (subband-edge energies) for electrons and holes. Quantum confinement increases the lowest transition energy E_a over the band gap of the well material. Also shown is a schematic joint density of states function $g(E)$, showing contributions from electron-heavy-hole and electron light-hole transitions.

binding energies may be calculated by numerical methods or inferred from observation (24).

Finally, the quantum-well absorption spectrum is calculated from the JDoS by Fermi's Golden Rule. The total absorption is the sum of contributions from all electron heavy-hole and electron light-hole transitions

$$\alpha(E) = \sum_{n,m} \alpha_{e_n - hh_m}(E) + \sum_{n,m} \alpha_{e_n - lh_m}(E) \tag{18}$$

where each electron-hole subband pair contributes a step function and a set of excitons to the total absorption spectrum. For the nth electron subband and mth hole subband:

$$\alpha_{nm}(E) = \alpha_{lh/hh} M_{nm} \times \left[\sum_l r_{nml} \delta(E - E_{nm} - B_{nml}) + \theta(E - E_{nm}) \right] \tag{19}$$

where M_{nm} is the overlap integral between the electron and hole envelope functions, E_{nm} the electron-hole transition energy before Coulombic effects are included, B_{nml} and r_{nml} the binding energy and oscillator strength of the lth exciton, $\delta(x)$ is the Dirac delta function and $\theta(x)$ the Heaviside function. Only optically allowed transitions are selected, i.e., those between electron and hole subbands of like parity where $M_{nm} \neq 0$. The constants α_{lh} and α_{hh} represent the absorption coefficient on the first step edge (taking $M_{11} = 1$), which is approximately three times as strong for heavy as for light holes.

The quantum-well absorption spectrum is thus characterized by a step-like form with strong excitonic peaks, and an absorption edge E_a which is blue-shifted from the absorption edge of the quantum-well material in the bulk, E_g, by the joint electron and heavy-hole confinement energies less the exciton binding energy:

$$E_a = E_g + E_{11h} - B_{11h}. \tag{20}$$

E_a is most strongly influenced by quantum-well width and varies from the band gap of the well material, E_g, for very wide wells, to the band gap of the barrier or host material, E_b, for very narrow wells. This tunability of the absorption edge is one of the most important features of the QWSC.

At energies above E_b carriers are no longer confined in the quantum well and the simple quantum mechanical model of absorption becomes unhelpful. In this range, the absorption coefficient of the *bulk* material may be used for the quantum-well layers, as in Ref. (15), but the approximation of a sharp threshold between quasi-2-D and bulk behavior is not

good. In particular, the presence of quasiconfined states just above the top
of the well may act to enhance absorption in this region. Similarly the
simple four-band Kane model is unreliable at energies high in the quantum well and tends to underestimate absorption. Remote band effects,
valence-band nonparabolicity, and band coupling can all affect the strength
of the JDoS.

2. Carrier Escape Efficiency. The second ingredient in quantum-well spectral response is the quantum efficiency for escape from the well, η_{esc}. In
general, carriers generated in a quantum well will remain there until they
escape by tunneling or thermionic emission, or recombine. η_{esc} is the
probability of escape relative to recombination. Determination of η_{esc} for
QWSCs has received a lot of attention (*25, 17, 26, 27*). It turns out that
the question is only important for deeper quantum wells than have been
studied to date. For $Al_{0.3}Ga_{0.7}As/GaAs$, η_{esc} effectively saturates at one
(*27*).

Figure 13 compares the monochromatic photocurrent from the QWSC
of Fig. 9 at two wavelengths, one at an energy deep in the quantum well
and one at an energy above E_b. Because both currents remain constant up
to the same forward bias, we can conclude that the quantum-well pho-

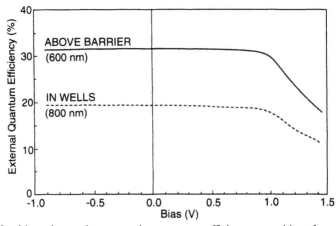

FIG. 13. Monochromatic external quantum efficiency vs. bias for a 30-well
$Al_{0.3}Ga_{0.7}As/GaAs$ QWSC at two different wavelengths, one (600 nm) corresponding to
photogeneration above the $Al_{0.3}Ga_{0.7}As$ band gap and the other (800 nm) for photogeneration in the well. Both signals remain at saturation level to the same forward bias, implying
that the probability of carriers escaping from the well is unity.

tocurrent is limited by the recombination current of the host device, and not by a reduced efficiency of carrier escape at forward bias.

This conclusion was supported by a detailed study (27) of DC photocurrent from a single quantum well (SQW) as a function of bias and temperature. The quantum efficiency for escape is defined as

$$\eta_{esc} = \frac{1/\tau_{esc}}{1/\tau_{esc} + 1/\tau_{nr} + 1/\tau_{rad}} \tag{21}$$

where τ_{esc}, τ_{nr}, and τ_{rad}, are the lifetimes for carrier escape from the quantum well and for nonradiative and radiative recombination within it. Nelson et al. (27) calculate τ_{esc} for a given carrier type, temperature, and electric field from the ratio of the escape current to the carrier density in the quantum well. That is, for electrons

$$\frac{1}{\tau_{esc}} = \frac{J_{esc}}{n} \tag{22}$$

where J_{esc} is calculated semiclassically through

$$J_{esc} = \int \frac{\hbar k_z}{m^*} g(k) f(E(k)) \, d^3k \tag{23}$$

where $g(k)$ is the quantum-well DoS function at that electric field, $f(E)$ is the Fermi–Dirac distribution function, $E(k)$ the carrier's kinetic energy, $\hbar k_z$ its momentum in the growth direction, and m^* its effective mass. The equilibrium carrier density in the quantum well is given by

$$n = \int g(k) f(E(k)) \, d^3k. \tag{24}$$

Modeling shows that escape is dominated by the fastest escaping carrier type (light holes) and is thermally activated at temperatures above 100 K. For $Al_{0.3}Ga_{0.7}As/GaAs$ quantum wells, η_{esc} saturates at unity at temperatures above 200 K where the thermionic emission time is short compared to the recombination time (Fig. 14).

Following a similar method to Nelson et al. (27), Yazawa (26) calculates τ_{esc} as a function of well depth for GaAs p–i–n cells containing indium gallium arsenide ($In_x Ga_{1-x} As$) quantum wells. Yazawa finds that η_{esc} saturates at 1 for $x < 0.15$. This agrees with his measurements of temperature dependent spectral response.

Yazawa predicts that non-unit efficiency will become important when the electron escape time becomes long compared to the recombination time, i.e., when $x > 0.15$. The compromise between increased photogener-

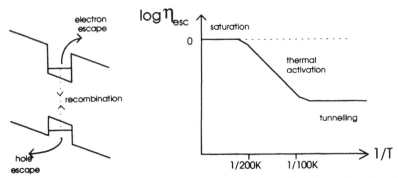

FIG. 14. Schematic of a quantum well under electric bias and an Arrhenius plot of the quantum efficiency for escape from the well η_{esc}. For $Al_{0.35}Ga_{0.65}As/GaAs$ wells, carrier escape is thermally activated above about 100 K, and saturates at unity at room temperature.

ation and increased recombination in deeper wells leads to an optimum photocurrent enhancement in $GaAs/In_xGa_{1-x}As$ quantum wells at $x \approx 0.2$. However, he assumes that escape is dominated by the *slowest* escaping carriers, in this case electrons. If instead the fastest escaping carrier were dominant, as is found by Nelson *et al.* (27) for AlGaAs/GaAs quantum wells, then η_{esc} would remain at its saturation level to much greater well depths. The latter conclusion is supported by measurements by Barnes (28), which show no discernible activation of η_{esc} for a GaAs/ $In_{0.16}Ga_{0.84}As$ SQW over the range of temperatures and biases.

This means that unit escape efficiency is likely to be a good approximation except for very deep quantum wells or low temperatures. In the case of deep wells, the favorable effect of temperature on collection efficiency may recommend QWSCs for use in concentration.

Renaud *et al.* (17) take the different approach of calculating photocurrent using a diffusion length for carriers generated in the quantum wells. This length depends on the probability of carrier escape by thermionic emission or tunneling relative to recombination. Their method should be equivalent to that outlined above where a non-unit value of η_{esc} results for deep or lossy wells.

We can conclude that in quantum wells of moderate quality ($\tau_{nr} \geq$ 1 ns), most photogenerated carriers will leave the wells by thermionic emission at room temperature. Only in deep wells where the lowest (electron *or* hole) barrier height exceeds about 200 meV will tunneling become important as recombination begins to compete with thermal escape. In such cases, the tunneling current can be maximized by optimization of the width and spacing of the quantum wells, so that resonant tunneling is achieved at the operating bias. In their original paper, Barnham and Duggan (3) proposed a multiquantum-well (MQW) system where quantum-well width increases with depth below the surface and wells are spaced so that all the $n = 1$ heavy-hole subbands line up at the operating field. This design offers the advantage of strong excitonic absorption over a range of wavelengths together with efficient collection via resonant tunneling.

Although unlikely to be advantageous for shallow-welled systems in AlGaAs/GaAs or GaAs/InGaAs, resonant tunneling structures may be useful in deep-welled systems such as $InP/In_{0.52}Ga_{0.48}As$. Mohaidat et al. (25) have studied the problem theoretically and present a method of calculating and optimizing short-circuit current, which may be useful. It is not obvious, however, how resonant tunneling would influence the recombination current.

3. QWSC Spectral Response. Following the calculation for the spectral response of a homogenous p–i–n cell, and taking $\eta_{csc} = 1$, the contribution of the quantum wells should simply be given by the photon flux absorbed in them. For N quantum wells of width L the SCR photocurrent (Eq. (16)) is modified to

$$J_{SCR} = q(1 - R)e^{-\alpha(x_p - w_p)}(1 - e^{-\alpha(w_p + x_i + w_n - NL) - N\alpha_{QW}L}) \qquad (25)$$

where α_{qw} is the absorption coefficient of the quantum well on the first electron hole subband, ($\alpha_{qw} \approx \alpha_{lh} + \alpha_{hh}$). The quantity $\alpha_{qw}L$, the absorption per level per quantum well, is approximately constant for any materials combination (15).

At energies below E_b, for few quantum wells, this reduces to

$$J_{SCR} = qN\alpha_{QW}L \qquad (26)$$

so that the spectral response should be proportional to the number of quantum wells. Experiment has confirmed this, and has placed $\alpha_{qw}L$ at around 1% (for AlGaAs/GaAs), in agreement with previous experiments and theory (15).

In Fig. 15, this model of QWSC spectral response is applied to the 30-period 87-Å MQW p–i–n cell presented in Figure 9. The same materials parameters are assumed as for the control cell in Figure 11, and the quantum-well parameters (L, V_e, V_h, m^*) are known from characterization and from the literature on AlGaAs/GaAs quantum wells.

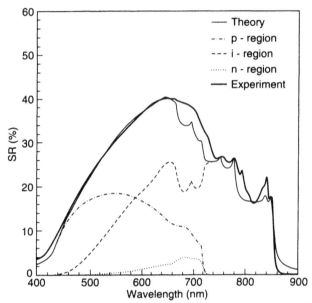

FIG. 15. Measured and calculated spectral response for the 30-well *p–i–n* diode of Fig. 9. The calculated SR shows the strong excitons and steplike form of the well density of states. Measured SR was taken at reverse bias, causing a slight distortion of the excitonic features. Notice how the contribution from the *i* region has been enhanced and extended (relative to Fig. 11) by the wells.

C. Optimization of the Spectral Response

The model provides a powerful tool for the interpretation and diagnosis of QWSC spectral response and has been applied to a range of structures and materials (*15, 29, 5*). In particular, the separation of contributions from the *p*, *i*, and *n* layers and the quantum wells helps to identify sources of loss and optimize layer thicknesses.

In the case of the samples of Fig. 9, the poor response at short wavelengths can be attributed to lossy transport in the thick *p* region and a poor surface. Because the electron diffusion length is short ($l_n \approx$ 0.1 μm) compared to the *p*-layer thickness ($x_p = 0.3$ μm), much of the *p* layer acts like a dead absorber, producing no photocurrent. Reducing the *p* thickness admits more light into the field-bearing region where collection is efficient.

In turn, a wider intrinsic layer would increase efficient light collection both by the host material at energies above E_b and by the quantum wells.

The nonnegligible n region contribution in Figure 9 shows that these structures are less than optimally thick: photon absorption in the SCR should be maximized.

The effect of quantum-well width is less obvious. Since it cannot be optimized independently of the i region thickness, it is more useful to consider absorption per unit well width. A wider well reduces the absorption edge and admits more levels, so increasing the absorption per well, but a smaller number of wells will fit in the same i region. A narrower well absorbs less light but a greater number of wells can be included. Our modeling studies indicate that the effect of greatly reducing well width is generally detrimental to the photocurrent. Increased well width can improve total absorption but at possible cost to the quantum efficiency for escape from the well, through the increased effective barrier height.

Overall response can be improved by standard techniques of solar-cell design: an antireflection coating to increase absorption, and a wide-band gap window layer to reflect carriers away from the surface and reduce surface recombination. Figure 16 shows the combined effect of some of these improvements.

1. Optimization of QWSCs in $Al_xGa_{1-x}As / GaAs$. These insights were used to design the most efficient $Al_xGa_{1-x}As/GaAs$ QWSC to date. Figure 17 shows the modeled and measured spectral response of a 50-MQW $Al_{0.33}Ga_{0.67}As/GaAs$ p-i-n cell with an $Al_{0.8}Ga_{0.2}As/GaAs$ window

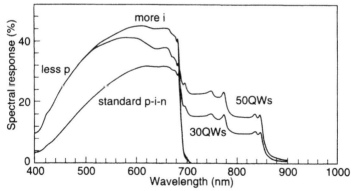

FIG. 16. Modeled SR for an $Al_{0.3}Ga_{0.7}As/GaAs$ QWSC showing the effect of various design improvements. The standard p-i-n is based on the $Al_{0.3}Ga_{0.7}As$ p-i-n of Fig. 9. The figure shows the effects of reducing the p thickness to 0.15 μm, increasing the i thickness to 0.8 μm, and introducing 30 or 50 wells.

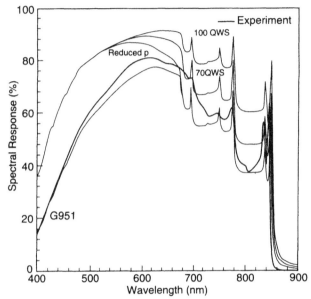

FIG. 17. Measured and calculated SR for an improved $Al_{0.3}Ga_{0.7}As/GaAs$ 50-well cell based on the design of Fig. 16 and with an antireflection coating and window layer. This sample was 14% efficient in standard air mass 1.5 conditions. The figure shows how design can be further improved by reducing the *p*-thickness, improving the surface quality and adding more wells.

layer. The *p* thickness had been reduced to 0.15 μm and the *i* thickness increased to 0.8 μm. The sample was antireflection coated and processed as a 2.5 × 2.5 mm solar cell. It was tested in standard AM1.5 conditions and found to have an efficiency of 14% (*5*).

In this structure, photocurrent losses still arise from carrier recombination in the 0.15-μm *p* layer and transmission through the 0.8-μm *i* layer. Figure 17 shows how these could be reduced with a thinner *p* layer, a thicker *i* layer, more wells and an improved surface. Such a device would be about 20% efficient if no further voltage degradation occurred (*15*).

In practice, the *p* thickness is limited by the need for good contacts and a low series resistance. The 0.05-μm *p* layer illustrated in Fig. 17 should be very heavily doped to compensate for increased R_s.

The *i* layer, however, is limited by the need for a *high* resistance! For efficient carrier collection, the whole of the *i* region of the cell should be depleted, i.e., subject to an electric field, at the cell's operating point. This means that wide intrinsic regions should have a very low level of back-ground doping N_i. The corresponding design in Fig. 16 (more i) would require $N_i \approx 10^{14}$ cm^{-3}. In practice such purities are hard to achieve in $Al_xGa_{1-x}As$.

2. Compositional Grading. As an alternative to reducing p-region thickness, short-wavelength response can be improved by reducing the *optical* thickness of the p layer with a compositional grade. Graded layers have been studied as window layers in GaAs solar cells and as emitter layers in AlGaAs/GaAs heterojunction cells (*30, 31*). The graded emitter enhances photocurrent by accelerating minority carriers away from the surface toward the depletion region, as though the diffusion length in the emitter were increased.

In a QWSC, the design does not benefit the wells except by slightly increasing the absorption of short-wavelength photons in the well layers. Its effect is mainly to improve the response of the host material. Figure 18 shows the measured spectral response of a pair of 30-period MQW $p-i-n$ samples. In one case, the 0.15-μm $Al_{0.33}Ga_{0.67}As$ p region is replaced by the same thickness of p-type $Al_xGa_{1-x}As$, where x is linearly graded from 0.67 at the front surface to 0.33 at the $p-i$ interface. The p grade enhances the spectral response by about 20% for energies above E_b, which translates to a photocurrent enhancement of 16% in an AM1.5 spectrum (*32*).

3. Alternative Host Materials. The problems of $Al_xGa_{1-x}As$ can be avoided completely by using a better material for the wide-gap regions of the cell. Gallium indium phosphide, $Ga_xIn_{1-x}P$, has longer minority carrier lifetimes than $Al_xGa_{1-x}As$, and is lattice matched to GaAs when $x = 0.52$, providing a convenient barrier material for deep, unstrained GaAs quantum wells. $Ga_{0.52}In_{0.48}P$ is already preferred as the wide-gap material in tandem calls (*14*).

The first experimental study of $Ga_{0.52}In_{0.48}P$/GaAs QWSC test structures is reported by Osborne (*33*). A set of devices was grown by MBE with a 0.15-μm p layer and a 0.95-μm i layer. The cells contained 40 GaAs quantum wells of width 80 or 120 Å. Although the short-wavelength spectral response was not, in fact, significantly greater than that of AlGaAs devices of equivalent band gap, background doping levels were lower. This allows a wider i region with more wells to be accomodated, and so improves the photocurrent enhancement.

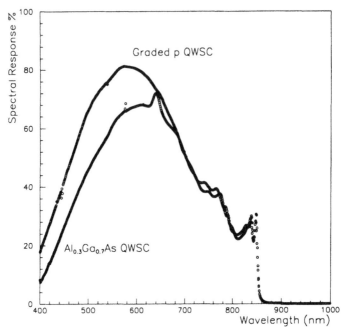

FIG. 18. Measured spectral response for an antireflection-coated 30-well $Al_{0.3}Ga_{0.7}As/GaAs$ QWSC with and without a linearly graded p region.

4. *Back Surface Reflectors.* The contribution from the quantum wells can be increased rather simply by means of a back-surface mirror. The mirror effectively doubles the optical path length through the wells. Figure 19 compares measured spectral response for a pair of AR-coated, 30-well $Al_{0.33}Ga_{0.67}/GaAs$ $p-i-n$ devices, one of which has had its substrate etched off and replaced with a layer of gold. Modeling studies show that the mirrored sample responds as though it has 60 wells, with the strongest enhancement at long wavelength.

Back-surface reflectors will be useful in almost all cases for enhancing absorption in the wells. Figure 19b shows how quantum well absorption in a $GaAs/In_{0.2}Ga_{0.8}As$ QWSC has been increased with a Bragg stack designed to reflect photon energies in the well and by a gold mirror (*28*). Renaud *et al.* (*17*) calculate that the addition of an optimally tuned Bragg reflector to an InP/InGaAs QWSC can enhance cell efficiency by about 2%.

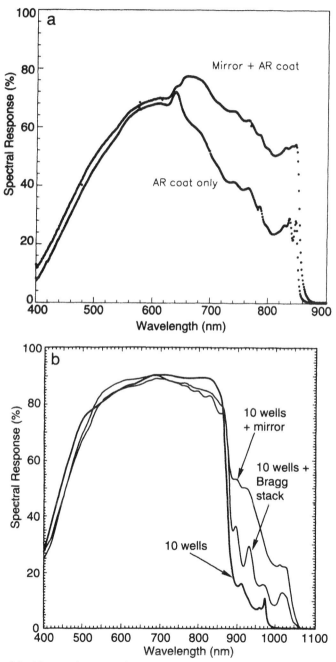

FIG. 19. (a) Measured spectral response for an antireflection-coated 30-well $Al_{0.3}Ga_{0.7}As/GaAs$ QWSC with and without a back surface mirror. (b) Measured spectral response for an antireflection-coated 10-well $GaAs/In_{0.17}Ga_{0.83}As$ QWSC showing the effects of a Bragg stack reflector and of a back-surface mirror with deeper wells.

Where the number of wells is severely limited by background doping or by strain (discussed below) more sophisticated light-trapping geometries may be designed to amplify the quantum-well response.

5. *Well Depth.* Like well width, the effect of well depth on spectral response is straightforward. A deeper well admits more subbands, thus increasing the absorption. The photocurrent should increase with well depth to the point where thermionic escape becomes as slow as recombination in the well. Yazawa shows that for $GaAs/In_xGa_{1-x}As$ QWSCs, photocurrent enhancement increases with x exactly as expected from calculation (26).

A study of $Al_xGa_{1-x}As/GaAs$ QWSCs with $0.2 < x < 0.4$ shows the quantum-well contribution increasing much as expected with x, but at a cost to the barrier contribution (32). The AlGaAs contribution decreases mainly on account of its increasing band gap but also because of worsening material quality. Recombination times and diffusion lengths fall as x approaches 35–40% where a DX center in the conduction band acts as an effective electron trap.

Table I shows how photocurrent depends on aluminum fraction for this series, using the measured spectral response integrated over a standard AM1.5 solar spectrum. Although deeper wells do not benefit J_{sc} outright, the increased host band gap E_b improves the cell voltage, and the pertinent quantity is the current–voltage product. Even though the open-circuit voltages for these samples are masked by poor material quality, the variation with well depth is encouraging. This will be discussed in Section IV.

6. *Optimizing $Al_xGa_{1-x}As$ / GaAs: Conclusions.* The most extensive studies of QWSC spectral response have been made on $Al_xGa_{1-x}As/GaAs$

TABLE I

AM1.5 PHOTOCURRENT J_{sc}, OPEN CIRCUIT VOLTAGE AND AlGaAs BAND GAP, MEASURED FOR THREE $Al_xGa_{1-x}As/GaAs$ QWSCs OF IDENTICAL STRUCTURE WITH DIFFERENT ALUMINUM FRACTION x

x	J_{sc} (A m^{-2})	V_{oc} (V)	E_b (eV)
0.20	141	0.91	1.70
0.35	136	0.97	1.89
0.46	125	1.01	2.05

$p-i-n$ diodes. This lattice-matched material system is well understood and has been a useful tool for studying the behavior of the QWSC.

For real improvements in practical solar-cell efficiencies, however, the QWSC should be compared with a homogenous cell made from a near optimum material such as GaAs or InP. Quantum wells made from InGaAs can be grown in either of these materials. More recent studies of the QWSC have focussed on the strained GaAs/In$_x$Ga$_{1-x}$As and lattice-matched InP/In$_{0.52}$Ga$_{0.48}$As materials systems.

The case of the non-optimum-gap QWSC is interesting in itself, both to study the dependence of QWSC voltage and current on band gap, and to seek a more efficient non-optimum cell, for example as the high-band gap component in a tandem.

D. SPECTRAL RESPONSE OF GaAs/In$_x$Ga$_{1-x}$As QWSCs

QWSCs in the strained GaAs/In$_x$Ga$_{1-x}$As epitaxial system have been studied by three groups: Ragay et al. (6), Yazawa et al. (26), and Barnes et al. (29). The system is interesting because GaAs with its 1.4-eV band gap currently holds the record for photovoltaic conversion efficiency in a single-band gap cell (1). The problem is that no convenient III–V semiconductor of lower band gap can be matched easily to GaAs. In$_x$Ga$_{1-x}$As quantum wells can be grown on GaAs but the mismatch in lattice constants induces a strain in the wells which in practice limits both their number and their depth.

In the strained system, the electron and hole potentials in the well system are distorted and the light hole becomes unconfined (34, 35). This reduces the absorption slightly and removes the light-hole excitonic features. Otherwise the spectral response resembles an unstrained MQW system and very clear quantum-well features are seen in spectra (24, 29).

Figure 20 shows the spectral response of a GaAs QWSC with 10 In$_{0.16}$Ga$_{0.84}$As quantum wells together with its GaAs $p-i-n$ control cell. The calculated spectral response neglects quantum well transitions entirely and so underestimates the data.

The cells are based on a GaAs $p-n$ cell which is identical to the control except that it has no i region. The efficiency of the $p-n$ cell is around 20%, which is comparable with the (practical) maximum for a single-gap cell. The great improvement in spectral response relative to Al$_x$Ga$_{1-x}$As-based cells is due to the much better transport properties of GaAs. Comparison of $p-n$ and $p-i-n$ cells in Barnes et al. (29) shows that the spectral response is not degraded by the addition of an i region but is in

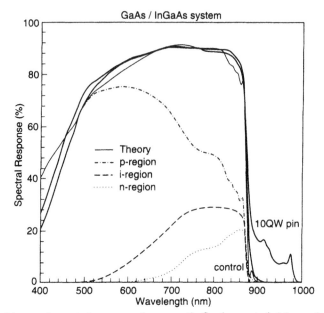

FIG. 20. Measured spectral response for an antireflection-coated 0.6-μm *i*-region GaAs *p–i–n* cell and the equivalent 10-well GaAs/In$_{0.16}$Ga$_{0.84}$As QWSC. Notice how short-wavelength response is not affected by the introduction of quantum wells and how much better the response is for GaAs than for Al$_{0.3}$Ga$_{0.7}$As.

fact enhanced at energies close to the GaAs band gap. Dark current measurements (29) show that the *i* region does not degrade the cell voltage either. The *p–i–n* structure is as good a solar cell as the *p–n*.

The quantum wells enhance the photocurrent as expected. However, if the barriers between wells become too narrow or the wells become too wide then dislocations form at the interfaces and the wells relax. The interface quality deteriorates, producing poor well spectra and greatly increased recombination. Figure 21a shows the long-wavelength spectral response for the 15-well strained sample above together with an otherwise identical 23-well sample where the wells have relaxed. Although the net photocurrent is not greatly reduced, the dark current increases by an order of magnitude, greatly reducing the cell performance (28).

The key determinant for the QWSC is the barrier width L_b. Ragay et al. (6) in a systematic study of the barrier-width dependence of MBE grown QWSCs, find that dark currents increase steadily with decreasing L_b (Fig. 21b). Barnes (28) finds that good well quality requires $L_b > 300$ Å.

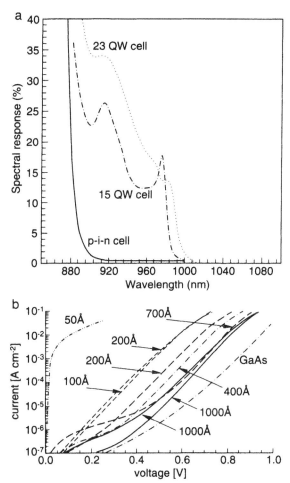

FIG. 21. (a) Measured long-wavelength spectral response for a 0.9-μm *i*-region GaAs *p–i–n* cell and GaAs/In$_{0.16}$Ga$_{0.84}$As QWSCs with 15 and 23 80-Å wells. The SR for the 23-well cell has lost its sharp excitonic features, showing that the wells have relaxed and the interfaces degraded. (b) Measured dark currents for a set of GaAs/In$_{0.2}$Ga$_{0.8}$As QWSCs of different barrier width, reprinted from Ragay *et al.* (*6*) with permission. Dark current increases steadily as barrier width between the strained wells is reduced.

Yazawa *et al.* (*26*) find consistently good well quality in a series of In$_x$Ga$_{1-x}$As wells with $0 < x < 0.15$ where $L_b = 500$ Å.

The joint constraints of wide barriers and an intrinsic layer narrow enough to bear an electric field in forward bias mean that the number of InGaAs wells is limited. The well depth is similarly limited to $x \approx 0.2$

because a higher indium fraction increases the lattice mismatch and hence the strain. Therefore, the photocurrent enhancement expected from a GaAs/InGaAs QWSC is limited.

Possible J_{sc} enhancements can be predicted from the above considerations. For Barnes' MOVPE-grown, 0.9-μm i region, GaAs p-i-n cell a J_{sc} of 284 A m^{-2} is expected in an AM1.5 spectrum. For a QWSC cell based on this control, twenty 80-Å In$_x$Ga$_{1-x}$As quantum wells may be accommodated in the 0.9-μm i region with a barrier width of 370 Å, which should be sufficient to accommodate the strain. With a back-surface mirror to double the absorption in the wells, the expected photocurrent enhancement is 11% for 20 In$_{0.17}$Ga$_{0.83}$As wells and 17% if the indium fraction is increased to 0.2.

The enhancement in J_{sc} is comparable with the voltage loss of 12% caused by adding 10 wells to a GaAs p-i-n. Whether the optimized cells achieve improved efficiency depends on how the voltage loss increases with the number of wells. Voltage is discussed in Section IV. In any case, the efficiency enhancement available by adding InGaAs wells to a GaAs host cell will not be large.[2]

E. SPECTRAL RESPONSE OF InP/In$_x$Ga$_{1-x}$As QWSCs

Indium phosphide is a favored solar-cell material on account of its near optimum band gap of 1.35 eV, and its radiation hardness, which makes it more suitable than GaAs for applications in space. In principle, InP cells should achieve efficiencies at least as high as GaAs solar cells but better material quality is needed. To date the best InP cell efficiency is 21.9% in an unconcentrated AM1.5 spectrum (*1*).

InP is interesting as the host material for a QWSC because of the availability of suitable well materials. In$_x$Ga$_{1-x}$As with $x = 0.52$ is lattice matched to InP and provides a very deep well. Wide In$_{0.52}$Ga$_{0.48}$As wells can extend the absorption range as far as 0.72 eV (1700 nm). The absence of strain means that more wells can be added than for the equivalent GaAs/In$_x$Ga$_{1-x}$As structure so that the potential current enhancements are very large. The combination of high host cell efficiency and high well photocurrent make this system the most promising for enhancements of efficiency.

---

[2]Ragay *et al.* (*6*) report an efficiency enhancement of their GaAs/In$_{0.2}$Ga$_{0.8}$As QWSCs over their GaAs control. This is because the GaAs control sample has degraded; the measured enhancement in J_{sc} cannot be explained by the contributions of the quantum wells.

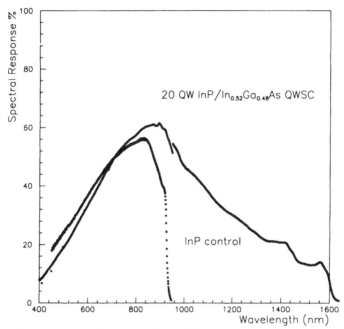

FIG. 22. Measured spectral response for a 0.56-μm *i* region InP *p–i–n* cell and the equivalent InP/In$_{0.52}$Ga$_{0.48}$As QWSC with 20 110-Å wells in the *i* regions. Notice how the wells extend the SR to much longer wavelengths, and how the quasi two-dimensional features of the well are visible only at energies near the absorption edge.

InP/In$_{0.52}$Ga$_{0.48}$As QWSCs have been studied by two groups, Freundlich *et al.* and Zachariou (*36, 37*), and were found to produce substantial photocurrent enhancements. Figure 22 shows the measured spectral response from an MBE-grown 20-well *p i n* cell with a 0.5-μm *i* region and its control. For this pair, the wells should enhance J_{SC} by 61% in AM1.5 conditions.

The spectral response reveals several problems with material quality. The poor short-wavelength response for both cells indicates poor *p*-region transport characteristics relative to GaAs. At long wavelength, although the spectral response is greatly extended, as expected, the poor excitons indicate poor quantum-well quality. Poor interfaces can degrade performance by increasing recombination rates, and may explain the much higher dark current of the well sample (see Section IV). The rather

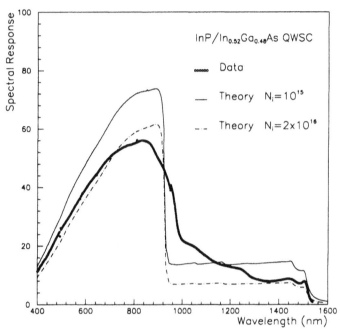

FIG. 23.   Measured and calculated SR for a 40-well InP/In$_{0.52}$Ga$_{0.48}$As QWSC with high background doping. The SR is calculated first with the design background doping level of 10^{15} cm^{-3} and then with the higher level of 2×10^{16} cm^{-3}.

indistinct InP band edge indicates that the crossover from quasi-2-D to bulk electronic behavior at E_b is blurred, and not abrupt as assumed in the spectral-response model.

Another problem is the difficulty of achieving low background doping levels. Figure 23 shows the spectral response of a device where high N_i renders the i region undepleted even at zero bias. The estimated doping level of 2×10^{16} cm^{-3} means that only two-fifths of the i region is depleted and so only two-fifths of the wells are subject to an electric field and able to contribute to the photocurrent. The modeled spectral response shows how the low response from the well is explained by high N_i.

An alternative material choice for the quantum wells is the strained InP/In$_x$As$_{1-x}$P combination, first studied by Freundlich et al. (36). The system appears to have better interface quality, while still absorbing wavelengths up to 1400 nm. Good-quality QWSCs with $x = 0.35$ offer substantial J_{SC} enhancements ($\sim 30\%$) with a smaller voltage loss than for $InP/In_xGa_{1-x}As$.

IV. QWSC Voltage Behavior

A. CURRENT GAIN VS. VOLTAGE LOSS

As explained in Sec. II, quantum wells are expected both to increase the photocurrent of a single-band gap solar cell and to reduce the voltage. Therefore, they can benefit the power conversion efficiency only when the advantage to current outweighs the detriment to voltage. For small changes this means, approximately, that

$$\frac{\Delta J_{sc}}{J_{sc}} > \frac{\Delta V_{oc}}{V_{oc}}.$$

In Table II, measured changes in J_{sc} and V_{oc} are presented for QWSCs in the range of materials combinations. The value for J_{sc} in AM1.5 light was inferred from the spectral response. In all cases J_{sc} is increased and V_{oc} reduced by the addition of quantum wells. For the wide-gap materials, efficiency is increased relative to the control, whereas for narrow-gap host materials it is reduced.

The same effect is seen in the dark-current characteristics, where the addition of wells invariably increases the dark current of the $p-i-n$ cell.

TABLE II

INCREASES IN SHORT-CIRCUIT CURRENT AND OPEN-CIRCUIT VOLTAGE LOSSES IN QUANTUM-WELL SOLAR CELLS

Host material	Quantum-well material	i-region (μm)	No. quantum wells	Quantum-well width (Å)	J_{sc} (A m^{-2})	ΔJ_{SC}	V_{oc}(V)	ΔV_{oc}
$Al_{0.20}Ga_{0.80}As$		0.48			107		0.91	
$Al_{0.20}Ga_{0.80}As$	GaAs	0.48	30	85	141	$+32\%$	0.91	—
$Al_{0.35}Ga_{0.65}As$		0.48			79		1.12	
$Al_{0.35}Ga_{0.65}As$	GaAs	0.48	30	85	136	$+72\%$	0.97	-13%
$Al_{0.46}Ga_{0.54}As$		0.48			60		1.21	
$Al_{0.46}Ga_{0.54}As$	GaAs	0.48	30	85	125	$+108\%$	1.01	-16%
$Ga_{0.52}In_{0.48}P$		0.95			69		1.17	
$Ga_{0.52}In_{0.48}P$	GaAs	0.95	40	80	111	$+61\%$	1.00	-14%
GaAs		0.60			264		0.91	
GaAs	$In_{0.16}Ga_{0.84}As$	0.60	10	80	276	$+4.5\%$	0.80	-12%
InP		0.56			162		0.69	
InP	$In_{0.52}Ga_{0.48}As$	0.56	20	60	227	$+40\%$	0.42	-39%
InP	$In_{0.52}Ga_{0.48}As$	0.56	20	110	311	$+92\%$	0.36	-48%

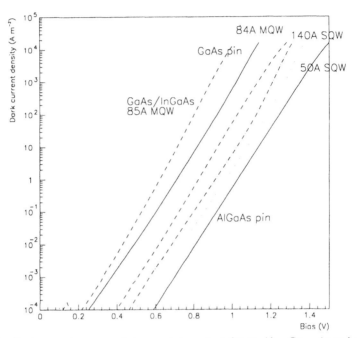

FIG. 24. Dark currents measured at room temperature for an $Al_{0.35}Ga_{0.65}As$ and a GaAs *p-i-n* cell, three $Al_{0.35}Ga_{0.6}As$/GaAs QWSC, containing 30 84-Å wells, a single 140-Å well, and a single 50-Å well, and a GaAs/$In_{0.16}Ga_{0.84}As$ QWSC containing 10 35 Å QWS.

Typical results are shown in Fig. 24 for a set of control and well cells in $Al_xGa_{1-x}As$/GaAs and GaAs/$In_xGa_{1-x}As$, all of similar *i*-region thickness (*38*).

All comparisons of J_d between a QWSC and its control reveal several features: that J_d is always enhanced by the addition of quantum wells; that the enhancement is *less* than the equivalent shift in absorption edge; and that the QWSC $J_d(V)$ rises more strongly at forward bias (i.e., it has a lower *ideality* factor). These features are discussed in Sect. IV.D. Although dark currents are highly device dependent and therefore difficult to interpret, in all cases J_d, and hence V_{oc}, were limited by nonradiative recombination, i.e., by material quality. In order to predict voltage performance in the case of better material it is necessary to isolate the effect of the wells from material properties.

In this section we discuss the interpretation and modeling of QWSC voltage characteristics. Discussion of the voltage and efficiency of an *ideal* QWSC is postponed to Section V.

B. Theory of Solar-Cell Dark Current

In Section II, the dark current of a p-n junction solar cell was given as

$$J_d(V) = J_{inj}(V - JR_s A) + J_{rg}(V - JR_s A)$$
$$+ J_{rad}(V - JR_s A) + \frac{V - JR_s A}{R_p A},$$

i.e., the sum of contributions from nonradiative recombination in the undepleted regions (J_{inj}) and the SCR (J_{rg}) and from radiative recombination (J_{rad}). The nonradiative currents are found by integrating the nonradiative recombination rate, $U(z)$, over the relevant regions of the cell. For J_{rg},

$$J_{rg} = \int_{SCR} U(z) \, dz, \tag{27}$$

and, similarly, for J_{inj}, $U(z)$ is integrated over the fieldless p and n regions. If the depletion approximation is assumed, J_d can be evaluated approximately as a function of applied bias V.

In the Shockley–Read–Hall model for a single trap state, $U(z)$ is given by

$$U(z) = \frac{np - n_i^2}{\tau_n[p(z) + p_t] + \tau_p[n(z) + n_t]} \tag{28}$$

where p_t and n_t are the equilibrium populations of trap states occupied by holes and electrons and τ_p and τ_n the respective carrier trapping times (39). The electron and hole concentrations, n and p, can be expressed in terms of their quasi-Fermi potentials ϕ_n and ϕ_p

$$n = n_i e^{q(\phi - \phi_n)/kT}$$
$$p = n_i e^{q(\phi_p - \phi)/kT} \tag{29}$$

where ϕ is the intrinsic potential and n_i the intrinsic carrier density. Then the electron hole product is given by

$$np = n_i^2 e^{q\Delta\phi/kT} \tag{30}$$

where $\Delta\phi$ is the quasi-Fermi potential splitting. Now, within the depletion approximation, the total bias applied to the junction, $(V_{bi} - V)$, is dropped across the SCR and within the SCR the quasi-Fermi level separation $\Delta\phi$ is constant and equal to the applied bias V, as shown in Figure 25. Then

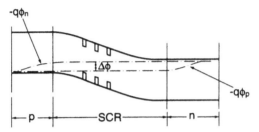

FIG. 25. Energy-band diagram of a *p–i–n* cell in the depletion approximation, showing the width of the SCR and the electron and hole quasi-Fermi levels.

evaluation of J_{rg} requires only knowledge of how $\phi(z)$ varies across the SCR.

If $\phi(z)$ is assumed to vary linearly across the SCR—an excellent assumption for *p–i–n* structures—the integral can be evaluated to yield

$$J_{rg}(V) = \frac{qn_iW}{\sqrt{\tau_n\tau_p}} \left(\frac{2\sinh(qV/2kT)}{q(V_{bi} - V)/kT} \right) \frac{\pi}{2}, \tag{31}$$

where W is the width of the SCR (*10*). The magnitude of the dark current is then simply related to the band gap (through n_i), the depletion width W, and the mean nonradiative recombination time. To a rough approximation, J_{rg} can be expressed as

$$J_{rg}(V) = \frac{qn_iW}{\tau_i}(e^{qV/2kT} - 1), \tag{32}$$

where the mean recombination time has been replaced by the single-parameter τ_i (*17*).

In the undepleted *p* and *n* regions, the majority carrier populations are assumed constant and recombination rates are dominated by the minority carrier populations; i.e.,

$$U(z) \approx n(z)/\tau_n,$$

in the *p* region, and

$$U(z) \approx p(z)/\tau_p,$$

in the *n* region. Integrating $U(z)$ yields the familiar Shockley form for the injection current

$$J_{inj}(V) = J_{d0}(e^{qV/kT} - 1),$$

where the saturation current J_{d0} varies like n_i^2 and also depends on material parameters.

Finally, the radiative current, from Eq. (9) varies with V, such that

$$J_{rad}(V) = J_{th}(e^{qV/kT} - 1), \tag{33}$$

where it is assumed that $(E_g - qV) \gg kT$. The thermal radiative current J_{th} varies with the band gap: $\sim \exp(-E_g/kT)$.

V_{oc} is then the solution to

$$J_{d0}(e^{qV_{oc}/kT} - 1) + J_{r0}(e^{qV_{oc}/2kT} - 1) + J_{th}(e^{qV_{oc}/kT} - 1) - J_{sc} = 0. \tag{34}$$

where J_{r0} is the prefactor in Eq. (32). If radiative recombination currents dominate, V_{oc} should vary, such that

$$V_{oc} = E_g + \frac{kT}{q}\ln(J_{sc}) - const., \tag{35}$$

since J_{th} varies like $\exp(-E_g/kT)$. So, in the limit of perfect material, V_{oc} should vary exactly linearly with E_g (2). The point is made by several authors with respect to QWSCs (40, 13, 16).

In practice, the recombination current in the SCR, J_{rg}, dominates in III–V materials. Then, for cells of similar structure and material parameters, we would expect

$$V_{oc} = E_g + \frac{2kT}{q}\ln(J_{sc}) - const., \tag{36}$$

since J_{r0} varies like n_i, and $n_i \sim \exp(E_g/2kT)$. However, the approximation is of limited use for materials such as $Al_xGa_{1-x}As$ where effective lifetimes vary strongly with band gap.

This model can be applied to p–i–n cells as well as to p–n cells, as long as the depletion approximation applies. In a p–i–n device, the low doping levels in the intrinsic region mean that high injection conditions are reached much sooner. Then the carriers injected into the i region distort the band profile so that electric field is lost. The effect is a reduced dependence of J_d on V at high bias. For the QWSC, the loss of electric field at high injection could prevent efficient carrier collection.

Nelson et al. (18) have tested the validity of the depletion approximation for typical $Al_{0.35}Ga_{0.65}As$ and GaAs p–i–n control cells by numerical solution of the semiconductor transport equations. In both cases, the depletion approximation is good at biases up to observed values of V_{oc} and so should be adequate to describe the behavior of such devices. However, the calculations show that high injection will become important at higher biases (1.5 V for $Al_{0.35}Ga_{0.65}As$ or 1.1 eV for GaAs) and may be pertinent to the study of better-quality materials and concentrator cells.

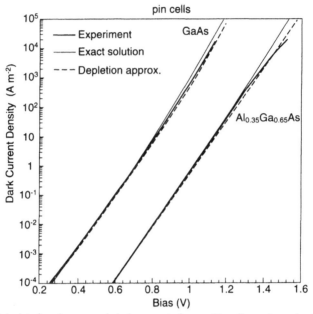

FIG. 26. Calculated and measured dark currents for an $Al_{0.35}Ga_{0.65}As$ and a GaAs $p-i-n$ cell. Exact calculations, from numerical solution of the semiconductor transport equations (full lines) are compared with the depletion approximation (dashed lines) and the data (heavy lines). Effective recombination lifetimes were taken as 1 ns for $Al_{0.35}Ga_{0.65}As$ and 7 ns for GaAs. The effects of series resistance have been ignored for clarity of presentation.

Figure 26 shows measured and calculated dark-current characteristics for an $Al_{0.35}Ga_{0.65}As$ cell with a 0.5-μm i region and a GaAs $p-i-n$ with a 0.6-μm i region. The slope of the semilog plot verifies that recombination–generation currents are dominant. Mean recombination lifetimes τ_i of 1 and 7 ns, respectively, were used to model J_{rg} from Eq. (31).

C. THEORY OF QWSC DARK CURRENT

In this framework, the addition of quantum wells is equivalent to replacing part of J_{rg} with a contribution from the wells. Calculation of the Shockley–Hall–Read recombination rates in each well requires knowledge of the electron and hole densities n_{qw} and p_{qw}, and an effective lifetime τ_{qw} for recombination.

Let us define quantum well electron and hole densities as

$$n_{qw} = f_n n_i(E_a)e^{q(\phi - \phi_n)/kT}$$

and

$$p_{qw} = f_p n_i(E_a)e^{q(\phi_p - \phi)/kT} \tag{37}$$

in analogy with the definitions for bulk material. Here $n_i(E_a)$ is the intrinsic carrier density of the bulk material of the same effective band gap E_a as the well and f_n and f_p are the factors by which the electron and hole density of states are modified by confinement.

Now, in the depletion approximation for the homogenous device, $\Delta\phi = V$ throughout the SCR. If the same condition applies when wells are present, then the electron-hole product in the well will be

$$n_{qw}p_{qw} = f_n f_p n_i(E_a)^2 e^{qV/kT} \tag{38}$$

and, in the important case when n and p are similar, the recombination rate will be increased relative to the host material by a factor

$$\frac{\tau_i}{\tau_{qw}}\sqrt{f_n f_p}\, e^{(E_b - E_a)/2kT}. \tag{39}$$

The factors f_n and f_p are close to unity and weakly dependent on field (*18*) in low-injection conditions. This means that, *if* the condition of constant quasi-Fermi levels applies, then recombination rates will be increased by a factor of $(\tau_i/\tau_{qw})\exp(E_b - E_a)/2kT$ and to a first approximation V_{oc} would be expected to vary linearly with effective band gap E_a just as for homogenous cells.

This approximation was studied in Ref. (*18*). Renaud *et al.* take a similar approach in calculations of InP/InGaAs cell efficiencies, calculating recombination currents from the carrier densities in the well for different values of τ_i. Calculations of nonradiative QWSC dark currents are not reported elsewhere, although several authors discuss V_{oc} for a QWSC in the radiative limit (see Section V).

D. ANALYSIS OF QWSC DARK CURRENT

In Ref. (*18*), a set of QWSCs are modeled within the depletion approximation as described above. Figure 27 shows calculated and measured J_d characteristics for a 140-Å SQW cell and a 30-period 84-Å MQW cell. A recombination time of 1 ns is taken for the wells in accordance with the value derived from a study of carrier escape for one of the cells (*27*). The comparison shows that the simple model overestimates recombination in

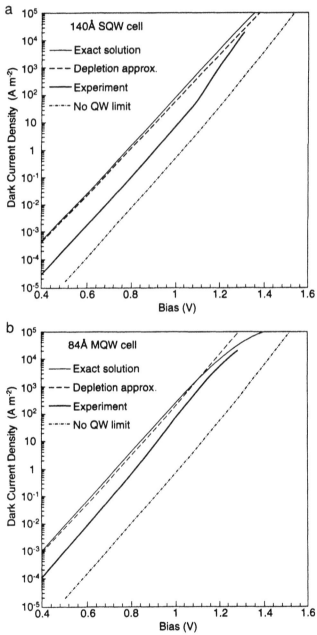

FIG. 27. Calculated and measured dark currents for two $Al_{0.35}Ga_{0.65}As/GaAs$ QWSCs (a) with a single 140-Å well in a 0.3-μm i region and (b) with 30 84-Å wells in an 0.5-μm i region. Full lines and dashed lines represent the calculations in the exact and depletion approximation when quasithermal equilibrium is assumed, and dot-dash lines represent the limit of the p-i-n cell without wells.

the well by about an order of magnitude, but that at high forward bias (low electric field) the discrepancy begins to disappear. This indicates that recombination increases more quickly with bias (with reducing field) than expected from the usual recombination–generation model.

One explanation is that the quantum well is not, in fact, in thermal equilibrium with the host cell while it is subject to an electric field. If a proportion of carriers escape irreversibly from the biased well, the net carrier population will be less than expected. As field is reduced toward flat band, escape slows down and the carrier populations tend toward their equilibrium values. Carrier escape probability depends roughly exponentially on electric field through the effective barrier height (27) and is consistent with the observed dependence of J_d on V.

E. ANALYSIS OF V_{oc}

QWSC open-circuit voltages, like dark currents, depend on both material quality and band gap. Very low V_{oc} values usually indicate a high density of recombination centres. Both Ragay et al. (6) and Haarpaintner et al. (38) show that in the strained GaAs/In$_x$Ga$_{1-x}$As system, the cell voltage degrades progressively as well separation is reduced. For highly strained systems, dislocations form near the interfaces and increase recombination, as discussed in Section III.D (28, 29). The big reduction in V_{oc} observed by Freundlich et al. (36) when unstrained In$_x$Ga$_{1-x}$As wells are added to InP may be due to poor interfaces or poor In$_{0.52}$Ga$_{0.48}$As quality.

For devices of similar material quality, V_{oc} should reflect changes in the effective band gap. In Ref. (38), studies of dependence of V_{oc} on well depth and width are reported. In Fig. 28a, V_{oc} is compared with effective band gap E_a for three Al$_{0.35}$Ga$_{0.65}$As SQW p–i–n cells with well widths of 50, 85, and 140 Å, for the Al$_{0.35}$Ga$_{0.65}$As control cell and for a GaAs p–i–n cell of similar i-region thickness. For the three cells, V_{oc} decreases linearly with E_a as expected from Eq. (36), but for all three cells V_{oc} is higher by about 0.1 V than expected from the change in effective band gap, relative to either control. In Figure 28b, three MQW cells of different well depth are presented in comparison with the three Al$_x$Ga$_{1-x}$As control cells and a GaAs p–i–n. Again, all the cells show a voltage advantage relative to the simplistic $E_a - V_{oc}$ model.

The dependence of V_{oc} on the number of wells is not clear on account of quality variations between samples, but the general trend is for a lower voltage in multiple- than in single-well devices. This is consistent with the

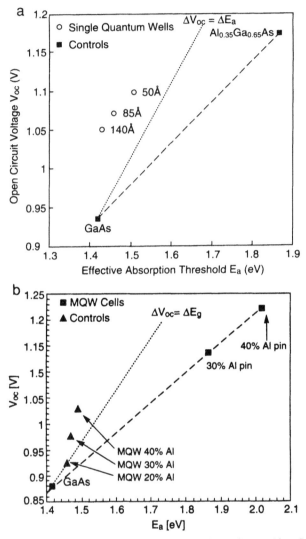

FIG. 28. (a) Open-circuit voltage versus effective band gap for an $Al_{0.35}Ga_{0.65}As$ $p-i-n$ cell, a GaAs $p-i-n$ cell and three single quantum-well $Al_{0.35}Ga_{0.65}As/GaAs$ cells of different well width. The dotted line indicates the variation expected if V_{oc} depends only on E_a. (b) As (a) for a set of $Al_xGa_{1-x}As/GaAs$ quantum-well cells of different composition and their $Al_xGa_{1-x}As$ control cells. Notice how V_{oc} for the control cells varies linearly with absorption edge, but for the QWSCs is consistently higher than expected (*38*).

idea of increased recombination through the increased total electron hole product due to the wells. Nevertheless, when compared with homogenous $p-i-n$ devices in either host or quantum-well material, QWSCs consistently display a voltage advantage.

V. Limiting Efficiency of the QWSC

Data for J_{SC} and V_{oc} of the QWSC and control cells are listed in Table II. These data alone would suggest that wells benefit the performance of wide band gap ($Al_xGa_{1-x}As$) host cells but degrade the performance of optimum band gap (GaAs and InP) cells. However, we have seen how the limitations on well number and material quality have limited the current enhancement or aggravated the voltage loss in the latter cases.

The question of the *theoretical* limit to QWSC efficiency has provoked the most interest. This limit depends on V_{oc} in the ideal case where all avoidable losses have been eliminated. In a homogenous cell with no nonradiative losses, V_{oc} is given by Eq. (35) and lies close to the band gap E_g. In their original paper, Barnham and Duggan (3) consider V_{oc} for a QWSC to be limited by the values for the host cell and for a control cell made from the well material. The upper limit requires that *no* recombination takes place in the wells. Several authors (Araujo *et al.* and Corkish and Green) have argued that this condition confounds the principle of detailed balance and is invalid (13, 41).

A. APPROACHES TO THE CALCULATION OF LIMITING EFFICIENCY

The first attempt to calculate V_{oc} and hence efficiency from the recombination rate in the well was made by Corkish (16). In this treatment, perfect material quality is assumed so that carrier mobilities are infinite and quasi-Fermi levels are constant across the device and equal to the applied bias V, as shown in Fig. 29.

Corkish considers a single well as an additional absorber of effective band gap E_a in parallel with a homogenous host cell of band gap E_b. Under illumination, in open-circuit conditions, the well produces an extra photocurrent J_{sc}^{qw} and an extra radiative dark current J_{rad}^{qw}. The photocurrent is given by

$$J_{sc}^{qw} = q \int_{E_a}^{E_b} b_s(E) a(E) \, dE \qquad (40)$$

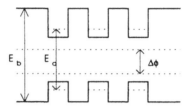

FIG. 29.   Schematic of a QWSC at flat band with constant quasi-Fermi levels.

where $b_s(E)$ is the incident solar flux density and $a(E)$ the absorptivity of the well. This is simply Eq. (2) with the absorption truncated at the host band gap. The radiative current is given by

$$J_{rad}^{qw} = \left(\frac{n_{qw} p_{qw}}{n_i^2} - 1 \right) J_{th}, \tag{41}$$

where J_{th} is the thermal radiative current as introduced in Eq. (9). When integrated over all emission angles,

$$J_{th} = \frac{8\pi\mu^2}{h^3c^2} \int_{E_a}^{E_b} \left(\frac{e(E)E^2}{e^{E/kT} - 1} \right) dE. \tag{42}$$

Corkish defines the quasi-Fermi level separation V through

$$n_{qw} p_{qw} = n_i^2 e^{qV/kT}, \tag{43}$$

where n_i is the intrinsic carrier density in the well material.

For the ideal case, Corkish assumes maximum absorptivity [$a(E) = 1$ for $E > E_a$], equates the carrier generation rate to the excess recombination rate, and solves for n_{qw} and p_{qw} in terms of V. This requires knowledge of the density of states for electron and hole and is equivalent to calculation of the confinement factors f_n, f_p in the dark-current model discussed above. Then J_{rad}^{qw} can be found for a given bias and incident spectrum.

For the composite cell, the $J(V)$ characteristic becomes

$$J(V) = J_{rad}(V) + J_{rad}^{qw}(V) - J_{sc} - J_{sc}^{qw}, \tag{44}$$

yielding an open-circuit voltage of

$$V_{oc} = \frac{kT}{q} \ln\left(1 + \frac{J_{sc} + J_{sc}^{qw}}{J_{rad}(0) + J_{rad}^{qw}(0)} \right). \tag{45}$$

Corkish calculates QWSC efficiencies in this model for a range of host band gaps E_b and well geometries. The host cell is assumed to operate in

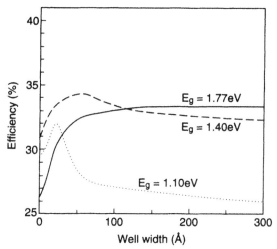

FIG. 30. Variation of QWSC efficiency with well width, according to Corkish (*16*) for three different band gaps. The results for $E_g = 1.40$ eV and above have since been shown to be overestimates.

the radiative limit. Figure 30 shows calculated efficiencies as a function of well width, i.e., of absorption edge, and E_b for an AM1.5 spectrum. The conclusion is that for host-cell band gaps at or above the optimum value, quantum wells can increase conversion efficiency above the limit for a single-gap device.

This conclusion has since been dismissed by Araujo and Marti in their detailed balance theory of the QWSC (*13, 42*). The error in Corkish's analysis is that he uses different values of absorptivity and emissivity. By detailed balance, $a(E) = e(E)$ for a system in equilibrium. Moreover, if equilibrium is upset but the quasi-Fermi level separation is still constant, then the same condition must apply (*13*). Corkish uses unit absorptivity to represent maximal light trapping but uses an emissivity which is based on the absorptivity of a single well with no trapping. That is

$$a(E) = 1 \quad \text{for } E > E_a$$

but

$$e(E) = 0.006 \quad \text{for } E_a < E < E_b.$$

Araujo and Marti show that if the detailed balance condition is restored, then the quantum well delivers a current

$$J_{\text{rad}}^{\text{qw}}(V) - J_{\text{sc}}^{\text{qw}} = \int_{E_a}^{E_b} a(E)\{F_\Omega b_s(E, T_s) - F_e b(E, T_c, \Delta\phi)\}\, dE.$$

When this is combined with the contribution from the host cell, the original $J(V)$ for an ideal photoconverter in the radiative limit—Eq. (12) —is recovered. $a(E)$ now refers to the spectral absorptivity of the composite host cell. The problem of optimising $J(V)$ becomes identical to the problem outlined in Sect. II. It appears that the maximum-efficiency QWSC should absorb all of the light above the optimum band gap for the solar spectrum E_{opt} and none below. This means that efficiency enhancements relative to the wide-gap control cell should be expected for $E_b > E_{opt}$ because the introduction of quantum wells simply represents a reduction in effective band gap toward the optimum value.

B. EXPERIMENTAL STUDY OF THE RADIATIVE LIMIT

The above conclusion has been studied experimentally for both $Al_xGa_{1-x}As/GaAs$ and $GaAs/In_xGa_{1-x}As$ QWSCs through measurements of the cells' radiative losses as a function of bias (40, 6). Although dark currents are dominated by nonradiative effects, the radiative recombination current—from Eq. (10)—is determined only by the absorptivity spectrum $a(E)$ and the quasi-Fermi level separation V. At room temperature, emission comes mainly from the absorption edge and

$$J_{rad}(V) \sim e^{-(E_g - \Delta\phi)/kT}$$

where $\Delta\phi$ is the quasi-Fermi-level separation. Now if the detailed balance approach is applied to the QWSC, then J_{rad} should vary like $e^{-(E_a - V)/kT}$ where E_a is the effective band gap of the QWSC and V the applied bias.

Ragay et al. (40) measured emission spectra for three $Al_xGa_{1-x}As/GaAs$ QWSCs of different well widths, the $Al_xGa_{1-x}As$ control, and a GaAs $p-i-n$ cell. They found that the emission peaks coincided with measured and calculated absorption edges. In Fig. 31, the bias-dependent radiative currents are compared with calculation, and seen to agree to within about 10 meV. The same measurements were carried out on a series of GaAs/InGaAs QWSCs which had very different nonradiative currents but the same absorption edge. In this case, the emission spectra for the QWSCs were coincident, and shifted from the GaAs $p-i-n$ spectrum by the difference in effective band gap.

These measurements show that radiative recombination in a QWSC is dominated by emission from the wells. They are consistent with the detailed balance argument that QWSC efficiency is limited by the same considerations as a homogenous cell. However, they do not prove the case.

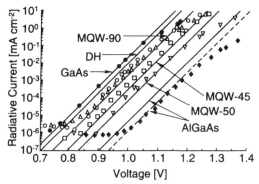

FIG. 31. Radiative recombination current vs. applied bias for a set of $Al_{0.2}Ga_{0.8}As/GaAs$ QWSCs of different well width (30, 45, and 90 Å), an $Al_{0.2}Ga_{0.8}As$ $p-i-n$ cell and a GaAs $p-i-n$ cell, all of the same i-region thickness, reprinted from Ragay et al. (40) with permission of H. S. Stephens and Associates.

C. The Question of Limiting Efficiency

These studies of QWSC behavior in the radiative limit leave several unanswered questions. First, the detailed balance argument is developed for a QWSC in open-circuit conditions: flat-band profile, constant quasi-Fermi levels and no net carrier flow. At the operating point of a QWSC there must still be an electric field across the active region in order to separate the cariers. If the wells are still biased, the carriers may be able to escape from the wells irreversibly and upset the quasiequilibrium condition. If the carrier density is reduced in the well, the radiative losses will be reduced. The theoretical analysis of QWSC dark currents reported in Sec. IV supports this notion of broken equilibrium.

Second, the emission spectra in Refs. (40) and (6) are given in arbitrary units. Absolute measurements are necessary in order to establish that the generation and recombination rates are equal and that the quasi-Fermi-level separation is indeed equal to the applied bias. A method of measuring $\Delta\phi$ from emission spectra is currently being developed (43).

Finally, the detailed balance theory implies that for cells operating in the radiative limit, open circuit voltage should increase linearly with effective band gap. In practical cells, where nonradiative processes dominate, V_{oc} should still be linear with E_a to within a factor determined by the relative nonradiative lifetimes.

From this last point we would expect the open circuit voltage of a QWSC to fall by $(E_b - E_a)$ if we replace a homogenous $p-i-n$ cell of band gap E_b with a QWSC of well band gap E_a.

Alternatively we could expect V_{oc} to increase by $(E_a - E_g)$, the sum of the electron and hole confinement energies, if a $p-i-n$ cell made from the well material were replaced by the same QWSC.

However in *every* case for the QWSCs reported here, V_{oc} is higher than expected by either argument. In Fig. 32a, V_{oc} values are compared with E_g for QWSCs in the important $InP/In_{0.53}Ga_{0.48}A$ system, demonstrating for the first time that voltage enhancements can be achieved when $E_a < E_{opt}$. Figure 32b shows the same effect for the series of cells reported in Ref. (*40*) as does Fig. 28 for a range of $Al_xGa_{1-x}As/GaAs$ cells. A higher voltage than expected by detailed balance indicates a lower recombination rate and hence a less-than-equilibrium population in the quantum wells. Although this effect may disappear, for instance, at higher illumination levels or for better-quality host materials, it indicates that the simple detailed balance theory does not adequately describe real QWSC behavior.

VI. Future Work and Other Novel Approaches

A. Practical Applications for the QWSC

It has been pointed out (*13, 17*) that if QWSCs cannot, in fact, surpass the theoretical limiting efficiency of a single-gap cell, they may allow the practical maximum efficiencies in near-optimum materials to be surpassed. In Fig. 33, the theoretical maximum efficiency as a function of band gap is compared with actual record efficiencies for various materials.

The optimum band gap for unconcentrated light (1.26 eV) is slightly smaller than that of GaAs. A cell of this band gap could be produced from the ternary alloy $In_xGa_{1-x}As$, but carrier collection from $In_xGa_{1-x}As$ cells is poorer than expected (*44*), leading to rather disappointing efficiencies. As an alternative, a $GaAs/In_xGa_{1-x}As$ QWSC could be designed with the optimum effective band gap, but with superior transport properties to the ternary alloy. Under concentration, as the optimum band gap shifts to lower energy, QWSCs could become very interesting alternatives to bulk $In_xGa_{1-x}As$ solar cells.

Renaud *et al.* (*17*) have considered the QWSC as an alternative to InP cells. InP remains the most favorable material for space applications on account of its radiation hardness even though poor material quality limits cell efficiencies to less than 22%. Renaud *et al.* calculate the performance of an $InP/In_{0.52}Ga_{0.48}As$ QWSC in comparison with an InP cell using realistic material parameters. They show that if the recombination time is long enough (> 30 μs) the addition of quantum wells enhances photocur-

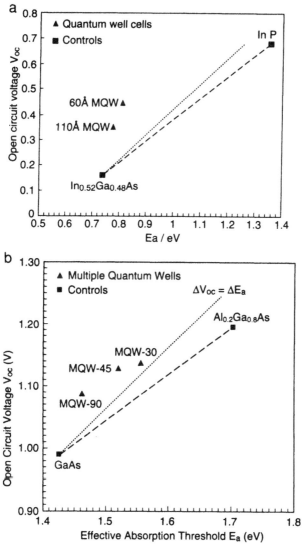

FIG. 32. (a) Open-circuit voltage vs. effective band gap for an InP $p-i-n$ cell, an $In_{0.52}Ga_{0.48}As$ $p-i-n$ cell, and two 20-well $InP/In_{0.52}Ga_{0.48}As$ QWSCs, one with 110-Å wells and one with 60-Å wells, all with a 0.56-μm i region. The dotted line indicates the variation expected if V_{oc} depends only on E_a. (b) Open-circuit voltage vs. effective band gap for the set of $Al_{0.2}Ga_{0.8}As/GaAs$ QWSCs and control cells presented in Fig. 31.

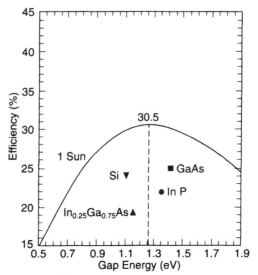

FIG. 33. Limiting solar-cell efficiency vs. band gap for unconcentrated light (as in Fig. 6), also showing the best practical efficiencies achieved in standard conditions for single-junction cells in GaAs (25%), InP (21.9%), InGaAs (19.2%), and silicon (24%). Reprinted from (*13*) with permission of H. S. Stephens and Associates.

rent to a greater degree than it increases recombination. Wells of optimum width and depth would improve the efficiency by about 2%.

B. OTHER NOVEL APPROACHES

1. MQW Structures for Improved Transport. Rather than forming part of the active region of the solar cell, MQW structures could be used to enhance transport through the nonactive regions of the device. MQW tunneling structures can exhibit superior electronic transport in the growth direction —via superlattice minibands—when compared with equivalent bulk materials. This makes MQW tunneling structures attractive as materials to connect the active region of a solar cell to the external contacts, or to connect the two components of a monolithic tandem cell together. The latter was in fact one of the first applications suggested for MQW structures in photovoltaics (*45, 46*).

Varonides (*47*) considers the use of quantum-well structures to improve transport of minority carriers through an InP based *p–i–n* cell. The emitter is replaced with a ten-period 45-Å $InP/In_{0.52}Ga_{0.48}As$ superlattice and the i layer with a linearly graded $In_xGa_{1-x}As$ layer. He calculates that, with suitable doping, open circuit voltages of 0.6 V and efficiencies of 15% can be achieved. These voltages are higher than expected for bulk $In_xGa_{1-x}As$ devices—on account of the increased Fermi-level separation in the wells—and mean that the structures could be included in tandem calls as a superior alternative to bulk $In_xGa_{1-x}As$ junctions.

2. MQW Structures in Photoelectrochemical Cells. The potential of quantum-well structures to enhance open circuit voltage is considered in the very different context of the photoelectrochemical (PEC) cell by Pelouch *et al.* (*48*) and Parsons *et al.* (*49*). The PEC cell consists of a semiconducting photoelectrode in contact with an electrolyte containing a suitable redox couple. Carriers photogenerated at the interface give up their energy to excited states in the solution. If the circuit is completed, the reaction is reversed at the second electrode and a photocurrent is produced. The voltage depends on the redox potential and ultimately on the energy with which the carriers are injected into solution. If the carriers are slow to equilibrate in the semiconductor; i.e., if they are *hot*, then they are injected into solution with a higher energy and produce a higher voltage.

Pelouch *et al.* propose using MQW or superlattice structures as photoelectrodes to enhance the efficiency of the PEC cell and have studied a variety of MQW and SQW devices. They observe slower electron cooling times in $Al_{0.48}Ga_{0.52}As/GaAs$ MQW structures than in bulk GaAs (*48*) implying a voltage advantage. This observation is also consistent with the suggestion, reported in Section IV.D, that biased wells in a QWSC are not in thermal equilibrium with the host cell.

3. MQW Structures in the Doped Regions. Ragay *et al.* (*50*) have studied the effect of growing wells within the doped layers—rather than the SCR—of a p–i–n solar cell. The introduction of wells *outside* the SCR should have a negligible effect on voltage when compared to the wider band gap control cell. This is because, in III–V solar cells, voltage is dominated by recombination in the SCR where electron hole product is greatest. The wells should increase photocurrent by extending the spectral response of the p and n layers if the quantum efficiency for escape from the (now unbiased) wells is high enough.

Ragay *et al.* (*50*) report measurements on an $Al_{0.1}Ga_{0.9}As/GaAs$ cell of this design in comparison with an $Al_{0.1}Ga_{0.9}As$ and a GaAs homogenous cell. The extra photocurrent from the wells, relative to the $Al_{0.1}Ga_{0.9}As$ control, is disappointing, and leads to a very slight enhancement of efficiency. This cell design would, therefore, seem to have no advantage since the efficiency of $Al_xGa_{1-x}As$ cells can be very substantially increased by the addition of wells to the *active* region. Nevertheless, the voltage loss caused by adding the wells was small and supports the idea that QWSC recombination currents may be minimized through thoughtful location of the quantum wells.

Acknowledgments

I am grateful to Gerardo Araujo, Geoff Duggan, Craig McFarlane, Antonio Marti, and Peter Ragay for helpful discussions; to Jenny Barnes, James Connolly, Paul Griffin, and Alexander Zachariou for communicating experimental results prior to publication; to Keith Barnham and all other members of the Imperial College Solar Cell Group for support and encouragement; to Neal Powell for help with the figures; and to the Engineering and Physical Sciences Research Council (EPSRC), the Greenpeace Trust and Tata Limited for financial support. The Imperial College samples discussed in the paper were provided by the EPSRC III–V Facility at the University of Sheffield, Philips Research Laboratories, Redhill, and the Interdisciplinary Research Centre for Semiconductor Materials. Thanks are due to Chris Button, Tom Foxon, Mark Hopkinson, Malcolm Pate, Christine Roberts, and John Roberts for growth and processing.

References

1. M. A. Green, K. Emery, K. Bucher, and D. L. King, *Progress in Photovoltaics: Research and Applications*, **3**, 51 (1995).

2. C. H. Henry, *J. Appl. Phys.* **51**, 4494 (1980).

3. K. W. J. Barnham and G. Duggan, *J. Appl. Phys.* **67**, 3490 (1990).

4. K. W. J. Barnham, B. Braun, J. Nelson, M. Paxman, C. Button, J. S. Roberts, and C. T. Foxon, *Appl. Phys. Lett.* **59**(1), 135 (1991).

5. K. W. J. Barnham, J. Barnes, G. Haarpaintner, J. Nelson, M. Paxman, C. T. Foxon, and J. S. Roberts, *MRS Bulletin XVIII*, **10**, 51 (1993).

6. F. W. Ragay, J. H. Wolter, A. Marti and G. L. Araujo, *Experimental Analysis of GaAs-InGaAs MQW Solar Cells*, "First World Energy Conference on Photovoltaic Energy Conversion," Hawaii, Dec. 1994.

7. S. M. Sze, "Physics of Semiconductor Devices," Wiley, New York, 1981.

8. F. C. Treble, ed., "Generating Electricity from the Sun," Oxford: Pergamon Press, 1991.

9. H. J. Hovel, "Semiconductor and Semimetals," Vol. 11, Academic Press, 1975.

10. C.-T. Sah, R. N. Noyce and W. Shockley, *Proc. Instn. Radio Eng.* **45**, 1228 (1957).

11. G. L. Araujo and A. Marti, *Proc. 11th EC Photovoltaic Solar Energy Conference*, 142 (1992).

12. W. Van Roosbroeck and W. Shockley, *Phys. Rev.* **94**, 1558 (1954).

13. G. L. Araujo, A. Marti, F. W. Ragay, and J. H. Wolter, *Proc. 12th EC Photovoltaic Solar Energy Conference*, **II**, 1481, H. S. Stephens and Associates (1994).

14. K. A. Bertness, S. R. Kurtz, D. J. Friedman, A. E. Kibbler, C. Kramer, and J. M. Olson, *Applied Physics Lett.* **65**, 989 (1994).

15. M. Paxman, J. Nelson, B. Braun, J. Connolly, K. W. J. Barnham, C. T. Foxon, and J. S. Roberts, *J. Appl. Phys.* **74**, 614 (1993).

16. R. Corkish, "Limits to the Efficiency of Silicon Solar Cells," Ph.D. Thesis, University of New South Wales, Australia, 1993.

17. P. Renaud, M. F. Vilela, A. Freundlich, A. Bensaoula, and N. Medelci, *Modeling p–i(Multi Quantum Well)–n Solar Cells: A contribution for a near optimum design,* "First World Energy Conference on Photovoltaic Energy Conversion, Hawaii, Dec. 1994."

18. J. Nelson, K. Barnham, J. Connolly, and G. Haarpaintner, "Quantum well solar cell dark currents," *Proc. 12th EC Photovoltaic Solar Energy Conference,* **II**, 1370 (1994).

19. M. S. Tyagi, "Introduction to semiconductor materials and devices," John Wiley, New York, 1991.

20. H. C. Hamaker, *J. Appl. Phys.* **58**, 2344 (1985).

21. D. E. Aspnes, S. M. Kelso, R. A. Logan, and R. Bhat, *J. Appl. Phys.*, **60**, 754 (1986). Also D. E. Aspnes, "Properties of GaAs," 2nd ed., Institution of Electrical Engineers, London, 1990.

22. G. Bastard, *in* "Wave Mechanics Applied to Semiconductor Heterostructures," *Editions de Physique,* pp. 63–101 (1986). Also G. Bastard and J. A. Brum, *IEEE J. Quantum Electron.* **22**, 1625 (1986).

23. G. Bastard and J. A. Brum, *IEEE J. Quantum Electron.* **22**, 1625 (1986).

24. K. J. Moore, G. Duggan, K. Woodbridge, and C. Roberts, *Phys. Rev. B,* **41**, 1090 (1990).

25. J. M. Mohaidat, K. Shum, W. B. Wang, and R. R. Alfano, *J. Appl. Phys.* **76**, 5533 (1994).

26. Y. Yazawa, T. Kitatani, J. Minemura, K. Tamura, and T. Warabisako, *Carrier Generation and Transport in InGaAs/GaAs Multiple Quantum Well Solar Cells,* "First World Energy Conference on Photovoltaic Energy Conversion, Hawaii, Dec. 1994."

27. J. Nelson, M. Paxman, K. W. J. Barnham, J. S. Roberts and C. Button, *IEEE J. Quantum Electron.* **29**, 1460 (1993).

28. J. M. Barnes, "An Experimental and Theoretical Study of GaAs/InGaAs Quantum Well Solar Cells and Carrier Escape from Quantum Wells," Ph.D. Thesis, University of London, 1994.

29. J. Barnes, T. Ali, K. W. J. Barnham, J. Nelson, E. S. M. Tsui, J. S. Roberts, M. A. Pate, and S. S. Dosanjh, *Proc. 12th EC Photovoltaic Solar Energy Conference,* **II**, 1374 (1994).

30. R. J. Schuelke, C. M. Maziar, and M. S. Lundstrom, *Solar Cells* **15**, 73 (1986).

31. G. Sassi, *J. Appl. Phys.* **54**, 5421 (1983).

32. J. Connolly, "Theory and Experiment of High Efficiency $Al_xGa_{1-x}As$ Solar Cells," Condensed Matter and Materials Physics Conference, Warwick, England, 1994. Also J. Connolly, Private Communication, 1995.

33. J. Osborne, M.Sc. Thesis, University of London, 1994.

34. J. W. Matthews and A. E. Blakeslee, *J. Cryst. Growth* **27**, 118 (1974).

35. G. Duggan, *Proc. SPIE* **1283**, 206 (1990).

36. A. Freundlich, V. Rossignol, M. F. Vilela, and P. Renaud, *InP-Based Quantum Well Solar Cells Grown by Chemical Beam Epitaxy,* "First World Energy Conference on Photovoltaic Energy Conversion, Hawaii, Dec. 1994."

37. A. Zachariou, Private Communication, (1995).

38. G. H. Haarpaintner, J. Barnes, K. W. J. Barnham, J. Connolly, S. Dosanjh, J. Nelson, C. Roberts, C. Button, G. Hill, M. A. Pate, and J. S. Roberts, *Voltage Performance of Quantum Well Solar Cells in the $Al_xGa_{1-x}As/GaAs$ and the $GaAs/In_xGa_{1-x}As$ Material*

Systems, "First World Energy Conference on Photovoltaic Energy Conversion, Hawaii, Dec. 1994."

39. R. N. Hall, *Phys. Rev.* **87**, 387 (1952). Also W. Shockley and W. T. Read, *Phys. Rev.* **87**, 835 (1952).

40. F. W. Ragay, J. H. Wolter, A. Marti and G. L. Araujo, "Experimental Analysis of the Efficiency of MQW Solar Cells," *Proc. 12th EC Photovoltaic Solar Energy Conference*, **II**, 1429, H. S. Stephens and Associates (1994).

41. R. Corkish and M. Green, *Proc. 23rd IEEE Photovoltaics Specialists Conference*, **I**, 675 (1993).

42. G. L. Araujo and A. Marti, *Electroluminescence Coupling in Multiple Quantum Well Diodes and Solar Cells*, in press.

43. E. S. M. Tsui, J. Nelson, and K. W. J. Barnham, *Determination of the quasi-Fermi level separation in single quantum well photodiodes*, manuscript in preparation.

44. J. G. Werthen, B. A. Arau, C. W. Ford, N. R. Kaminar, M. S. Kuryla, M. L. Ristow, C. R. Lewis, H. F. MacMillan, and G. F. Virshup, "Recent Advances in High Efficiency InGaAs Concentrator Cells," *Proc. 20th Photovoltaic Specialists Conference*, **I**, 640 (1988).

45. A. E. Blakeslee and K. W. Mitchell, U.S. Patent No. 4,278,474 (1981).

46. R. J. Chaffin, G. C. Osbourn, L. R. Dawson, and R. M. Biefield, *in* "Proceedings of the 18th Photovoltaic Specialists Conference," IEEE, New York, 1985, p. 776.

47. A. C. Varonides, "A New Heteroepitaxial InP/In(0.53)Ga(0.47)As (MQW) Novel $n-i-p$ Solar Cell for Space Applications," *Proc. 12th EC Photovoltaic Solar Energy Conference*, **II**, 1366, H. S. Stephens and Associates (1994).

48. W. S. Pelouch, R. J. Ellingson, P. E. Powers, C. L. Tang, D. M. Szmud, and A. J. Nozik, *Phys. Rev. B* **45**, 1450 (1992).

49. C. A. Parsons, M. W. Peterson, B. R. Thacker, J. A. Turner, and A. J. Nozik, *J. Phys. Chem.* **94**, 1381 (1990).

50. F. W. Ragay, E. W. M. Ruigrok, and J. H. Wolter, *GaAs Heterojunction Solar Cells with Increased Open Circuit Voltage*, "First World Energy Conference on Photovoltaic Energy Conversion, Hawaii, Dec. 1994."

Author Index

Gulari, E., 207(108), 235
Gunapala, S., 137(48), 139(48), 154(48),
160(48), 179(48), 234
Gunapala, S. D., 38(78, 81), 39(78), 40(78),
49(78), 73, 94(25), 111, 113–237, 128(30),
129(30), 130(30, 42), 131(30), 132(30),
133(42), 135(30), 136(30), 137(30),
138(30), 139(30), 140(30), 141(30),
142(30), 143(30), 144(30), 145(30),
146(30), 147(30), 148(30), 149(30),
150(30, 54, 55), 151(42), 153(42), 154(42),
155(59), 158(59), 159(59), 160(30),
163(70), 164(70), 165(70), 166(70),
181(55), 188(84), 189(84), 190(84),
192(89), 193(89), 194(89), 195(89),
197(97, 98), 199(97), 200(97), 201(97),
202(99, 100), 203(99), 204(99, 101),
205(101), 207(97, 99, 105), 208(112),
209(112), 210(112), 211(42), 211(112),
212(112), 213(112, 116), 214(116),
216(116), 224(141), 233, 234, 235, 236,
237
Guptill, M. T., 36(70), 73
Guth, G., 221(135), 222(135), 236
Gutierrez, D. A., 162(69), 234
Guy, D. R. P., 128(36), 233

H

Haarpaintner, G., 313(5), 325(18), 334(5),
336(5), 351(18), 353(18), 366, 367
Haarpaintner, G. H., 348(38), 355(38),
356(38), 367
Haegel, N. M., 5(19), 72
Hale, J., 68(127), 75
Hall, R. N., 348(39), 368
Haller, E. E., 5(3), 18(3), 36(69, 72), 37(69),
71, 73
Hamaker, H. C., 327(20), 367
Hamm, R., 197(97, 98), 199(97), 200(97),
201(97), 207(97), 213(116), 214(116),
216(116), 235, 236
Hamm, R. A., 150(54), 202(99, 100), 203(99),
204(99), 207(99), 234
Hara, H., 21(42), 72
Harbeke, G., 121(18), 123(18), 232
Harbison, J., 184(81), 235
Harbison, J. P., 241(15), 309
Harmon, L. D., 52(104), 60(110, 111), 74

Harris, J. S., Jr., 160(66), 234
Hartline, H. K., 62(115), 63(115), 74
Harwit, A., 160(66), 234
Hasnain, G., 106(45), 107(45), 108(45), 112,
117(6), 118(14), 121(21), 123(21), 128(14),
133(14), 137(14, 48), 139(48), 141(14),
149(14), 150(14), 154(48), 158(65),
160(14, 48), 172(14), 179(48), 193(14),
194(21), 195(21), 196(95), 209(14),
215(118), 232, 233, 234, 235, 236, 244(20,
21), 261(21), 272(34), 309
Hautman, J., 98(33), 112
Hayes, T. R., 24(6), 308
Hedin, L., 98(32), 112
Heiblum, M., 240(5), 308
Heiman, D., 155(61), 234
Heitmann, D., 96(27), 111, 155(64), 234
Helm, M., 241(15, 16), 309
Henry, C. H., 312(2), 315(2), 319(2), 321(2),
322(2), 351(2), 366
Henry, J. E., 193(91), 235
Hertle, H., 78(4), 111
Hilber, W., 241(16), 309
Hill, G., 348(38), 355(38), 356(38), 367
Hirsch, J., 54(106), 59(106), 74
Ho, P., 190(86), 191(88), 235, 290(46), 310
Hobson, W. S., 128(30), 129(30), 130(30),
131(30), 132(30), 135(30), 136(30),
137(30), 138(30), 139(30), 140(30),
141(30), 142(30), 143(30), 144(30),
145(30), 146(30), 147(30), 148(30),
149(30), 150(30), 160(30), 233
Holonyak, N., 287(39), 310
Hong, M., 163(70), 164(70), 165(70), 166(70),
234
Hong, S. C., 207(108), 235
Houghton, D. C., 81(12), 111
Hovel, H. J., 318(9), 366
Hseih, S. J., 103(39), 107(39), 112, 221(136),
237
Hsieh, K. C., 287(39), 310
Hsieh, S. J., 118(14), 121(21), 123(21), 128(14),
133(14, 44), 137(14), 141(14), 149(14),
150(14), 160(14), 172(14), 193(14),
194(21), 195(21), 209(14), 232, 233,
244(21), 249(25), 261(21), 284(25), 309
Hu, B. H., 121(19), 123(19), 232
Hu, M., 128(37), 233
Hubbard, P., 5(1), 71

Subject Index

Recent Volumes in this Serial

Maurice H. Francombe and John L. Vossen, *Physics of Thin Films*, Volume 16, 1992.

Maurice H. Francombe and John L. Vossen, *Physics of Thin Films*, Volume 17, 1993.

Maurice H. Francombe and John L. Vossen, *Physics of Thin Films, Advances in Research and Development, Plasma Sources for Thin Film Deposition and Etching*, Volume 18, 1994.

K. Vedam (guest editor), *Physics of Thin Films, Advances in Research and Development, Optical Characterization of Real Surfaces and Films*, Volume 19, 1994.

Abraham Ulman, *Thin Films, Organic Thin Films and Surfaces: Directions for the Nineties*, Volume 20, 1995.

ISBN 0-12-533021-9

9 780125 330213

90018

Printed and bound by CPI Group (UK) Ltd, Croydon, CR0 4YY

03/10/2024

01040425-0012